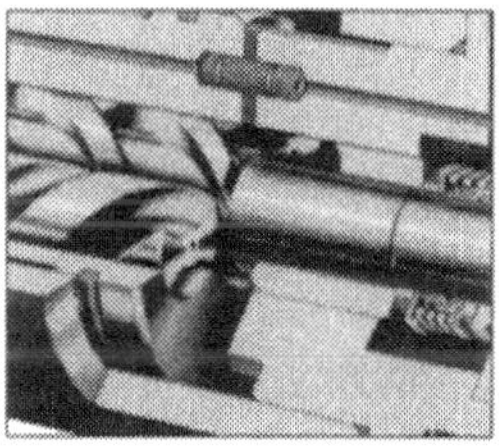

Forsthoffer's Rotating Equipment Handbooks

Forsthoffer's Rotating Equipment Handbooks

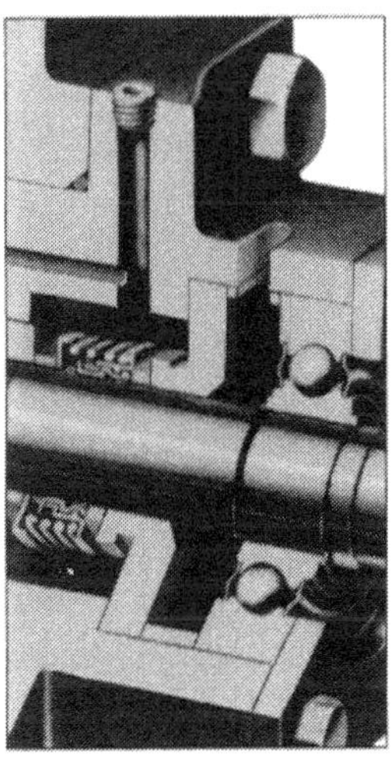

Volume 1: Principles of Rotating Equipment

W E (Bill) Forsthoffer

President, Forsthoffer Associates
Washington Crossing, PA, USA

ELSEVIER

UK Elsevier Ltd, The Boulevard, Langford Lane, Kidlington, Oxford OX5 1GB, UK
USA Elsevier Inc, 360 Park Avenue South, New York, NY 10010-1710, USA
JAPAN Elsevier Japan, Tsunashima Building Annex, 3-20-12 Yushima, Bunkyo-ku, Tokyo 113, Japan

ISBN 1-85617-467-0

British Library Cataloguing in Publication Data
A CIP Catalogue record for this book is available from the British Library

Published by
Elsevier Advanced Technology,
The Boulevard, Langford Lane, Kidlington, Oxford OX5 1GB, UK
Tel: +44(0) 1865 843000
Fax: +44(0) 1865 843971

Typeset by Land & Unwin Ltd., Northamptonshire

Transferred to digital print 2008

Printed and bound in Great Britain by CPI Antony Rowe, Eastbourne

Contents

Preface

This series has evolved from my personal experience over the last 40 years with the design, selection, testing, start-up and condition monitoring of rotating equipment. Most of the concept figures were originally written on a blackboard or whiteboard during a training session and on a spare piece of paper or I beam during a start-up or a problem solving plant visit.

My entire career has been devoted to this interesting and important field. Then and now more than ever, the cost of rotating equipment downtime can severely limit revenue and profits. A large process unit today can produce daily revenues in excess of 5 million US dollars. And yet, the operators, millwrights and engineers responsible for the safety and reliability of this equipment have not been afforded the opportunity to learn the design basis for this equipment in practical terms. I have also observed in the last ten years or so, that the number of experienced personnel in this field is diminishing rapidly.

Therefore the series objective is to present, in User friendly (easy to access), practical terms (using familiar analogies), the key facts concerning rotating equipment design basis, operation, maintenance, installation and condition monitoring to enable the reader (engineer, operator and millwright) to:

- Understand the effect of process & environmental changes on equipment operation, maintenance and reliability
- Condition monitor equipment on a component basis to optimize up-time, mean time between failure (MTBF) and mean time to repair (MTTR)
- Select, audit and test the equipment that will produce highest safety and reliability in the field for the lowest life cycle cost.

The hope is that the knowledge contained in this series will enable plant operations, maintenance and engineering personnel to easily access the material that will allow them to present their recommendations to management to solve existing costly problems and produce new projects of optimum reliability.

This volume, Principles of Rotating Equipment, is an overview of the main types of rotating machinery in industry (pumps, compressors, turbines and auxiliary systems). Each equipment type is presented with a practical emphasis on the design basis for each major component and covers such aspects as performance parameters and field monitoring, system dynamics, surge control, vibration, bearing, seal, auxiliary system design and condition monitoring of all major components and systems.

Acknowledgements

This series is a result of interactions with literally thousands of dedicated engineers, machinists, operators, vendors, contractors and students who have been an integral part of my career.

I consider myself very fortunate to have been associated with the best of the best mentors, business associates and dear friends throughout my career. Most especially, in chronological order Dick Salzmann, Bob Aimone, Merle Crane, Walt Neibel, the late Murray Rost, Mike Sweeney and Jimmy Trice. Bob, Merle, Murray and Mike have contributed specifically to the material in this series while Dick, Walt and Jimmy have tactfully kept me on track when necessary.

Special thanks to all of the global machinery vendors who have allowed me to use their material for my training manuals and now this publication.

Last but certainly not least; my career would not have been possible without the support, encouragement and assistance from my wife Doris and our children Jennifer, Brian, Eric, Michael and Dara.

A special additional note of thanks to Michael who helped assemble the material, and hopefully learned some in doing so, since he has elected to pursue a career in rotating machinery.

About the author

Bill Forsthoffer began his life-time career in rotating machinery in 1962 with De Laval Turbine Inc. as a summer trainee. After obtaining a Bachelor of Arts degree in Mathematics and Bachelor of Science degree in Mechanical Engineering, during which time he worked for De Laval part time in the Test, Compressor and Steam Turbine Departments, he joined De Laval full time in the Compressor Engineering Department in 1968. He was responsible for all phases of centrifugal compressor component and auxiliary design and also made many site visits to provide field engineering assistance for start up and problem resolution.

Bill joined Mobil Oil Corporate Engineering in 1974 and was responsible for all aspects of rotating equipment specification, technical procurement, design audits, test, field construction, commissioning, start-up and troubleshooting.

After 15 years at Mobil, Bill founded his own consulting company in 1990 and has provided rotating equipment consulting services to over 100 companies. Services include: project reliability assurance, training (over 7,000 people trained) and troubleshooting.

Bill is active in the industry as President of Forsthoffer Associates Inc., frequently writes articles for Turbo Machinery International Magazine and conducts many site specific and public workshops each year.

He can be contacted at bill@forsthofferassociates.com

DEDICATION

TO DORIS AND OUR CHILDREN
FOR THEIR TOTAL FAITH
IN MY ENDEAVORS

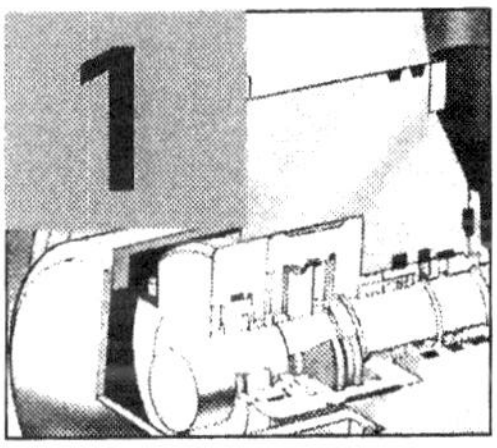

1 Rotating equipment overview

- Introduction
- Definition of rotating equipment
- Classifications of rotating equipment
- Site equipment examples
- Performance and mechanical design similarities
- The equipment 'train' or 'unit'
- Important fundamentals

Introduction

Take a minute and list all the different types and kinds of rotating equipment you can think of. Even if you have not been involved with rotating equipment for a long time, when you consider the types of equipment that you come in contact with every day, your list will be sizeable. Imagine if we pursued our objectives by looking at each individual piece of equipment. You would never remember all the aspects and the book would be long and very boring. We will not attempt this approach. Rather, this section will divide all types of rotating equipment into four major classifications. The function of each individual classification will be defined. Throughout this book we will cover many types of rotating equipment.

One good thing to remember is to always ask yourself what the function of this particular type is, what does it do? We will find that many aspects covered in this book will have the same common function.

Our approach therefore, will be to observe the similarities in both performance and mechanical aspects of various types of equipment. We will see that many of these relationships apply regardless of the type of equipment that is considered.

Now think of any rotating equipment unit that you have come in contact with and review that unit considering the different components that comprise it. You will find that every unit of rotating equipment consists of a driven machine, driver, transmission device and is supported by auxiliary equipment. That is, each unit is made up of all the classifications of rotating equipment that are described in this section. This is an important fact to remember in troubleshooting equipment. In essence then, each unit is a system.

Definition of rotating equipment

DEFINITION
Rotating equipment moves products
Products = solids, liquids, gases

Figure 1.1 Definition

Figure 1.1 presents a basic definition of rotating equipment.

As we shall see, there are different classifications of rotating equipment. Regardless of the classification, rotating equipment moves product. More properly stated:

Rotating equipment moves money!!

Stop the equipment and the source of revenue stops! This is a very important fact to remember. If you want management to approve your recommendation, you must be able to justify it economically! The form of any recommendation to management should be as shown in Figure 1.2.

A successful recommendation

Must state clearly and simply:

- The problem
- The problems' cost to the company (parts, labor and loss revenue)
- The recommended solution
- Its' cost
- Proof that the solution will work (has it worked somewhere else?)
- Savings to company by implementing proposed solution

Figure 1.2 A successful recommendation

If you proceed as shown in Figure 1.2 you will be able to obtain and maintain management support. Remember, you can learn a great deal in this book. However, if you cannot implement what you have learned the information is totally useless to the company. If you cannot obtain management support, you will never implement any action plan.

Classifications of rotating equipment

There are four (4) basic function classifications of rotating equipment. Refer to Figure 1.3 which defines the classifications of rotating equipment.

Classifications of rotating equipment

- Driven
- Drivers or prime movers (provide power)
- Transmission devices
- Auxiliary equipment

Figure 1.3 Classifications of rotating equipment

Figure 1.4 is a partial listing of some rotating equipment types grouped according to function.

Major types of rotating equipment

I. Driven equipment
- A. Compressors
  - 1. Dynamic
    - Centrifugal
    - Axial
    - Integral Gear
  - 2. Positive displacement
    - Screw
    - Rotary lobe
    - Reciprocating
    - Diaphragm
    - Liquid Ring
- B. Pumps
  - 1. Dynamic
    - Centrifugal
    - Axial
    - Slurry
    - Integral Gear
  - 2. Positive displacement
    - Plunger
    - Diaphragm
    - Gear
    - Screw
    - Progressive cavity
- C. Extruders
- D. Mixers
- E. Fans

II. Drivers – prime movers
- A. Steam turbines
- B. Gas turbines
- C. Motors
  - Induction
  - Synchronous
  - Vari-speed
- D. Engines
  - Internal combustion
  - Diesel
  - Gas

III. Transmission devices
- A. Gears
  - Helical
  - Double helical
- B. Clutches
- C. Couplings

IV. Auxiliary equipment
- A. Lube and seal systems
- B. Buffer gas systems
- C. Cooling systems

Figure 1.4 Major types of rotating equipment

Site equipment examples

Following is an example of typical site rotating equipment.

Figures 1.5, 1.6, 1.7, 1.8 show examples of each rotating equipment classification.

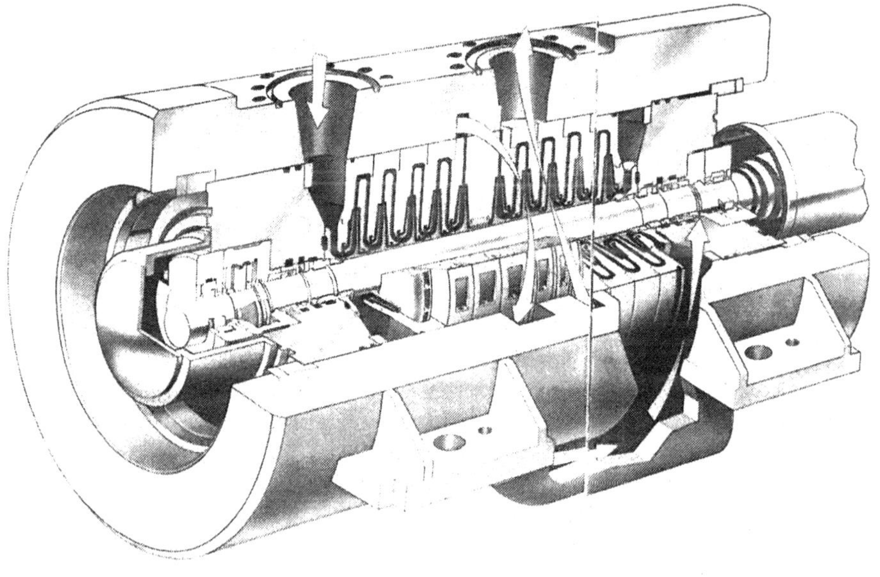

Figure 1.5 High pressure centrifugal compressor (Courtesy of Dresser Rand)

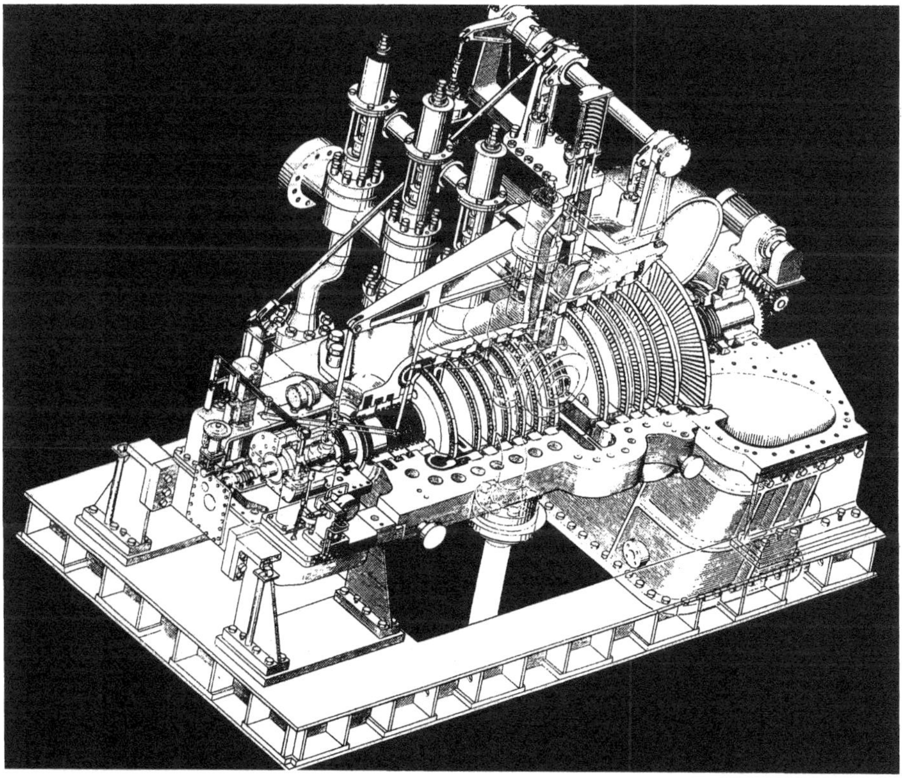

Figure 1.6 Extraction – condensing steam turbine (Courtesy of MHI)

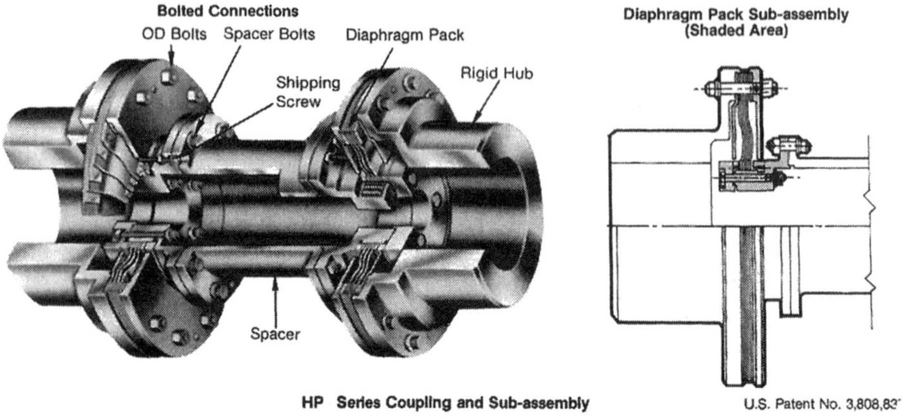

Figure 1.7 Multiple, convoluted diaphragm-spacer coupling (Courtesy of Zurn Industries)

Figure 1.8 Horizontal oil console arrangement (Courtesy of Oltechnique)

Performance and mechanical design similarities

During this book we will be examining many different types of rotating equipment. However, the task will be a lot easier if we begin our study by first focusing on the similarities of the equipment and then the specific differences.

As an example, we have chosen to first present pumps then compressors as topics. As was just discussed, both pumps and compressors are driven types of equipment and move product. Regardless of the product phase or state, their functions are identical. Refer to Figure 1.9 which compares dynamic pump and compressor performance.

Pumps and compressors

- **Both pumps and compressors** move a fluid from one energy level to another.
- **A pump** moves an incompressible fluid – a liquid. For our purposes, the volume of a liquid does not change with pressure and temperature.
- **A compressor** moves a compressible fluid – a gas. The volume of a gas changes with pressure, temperature and gas composition.
- The principles of dynamic machines apply both to pumps and compressors. However, since gases are compressible, the volume flow rate and hence the gas velocity in a passage is affected.

Figure 1.9 Pumps and compressors

The same comments can be made concerning mechanical components. Refer back to Figures 1.5 and 1.6 and ask are the functions of the casings, internal seals, shaft end seals and bearings the same? Absolutely! A bearing performs the same function whether it is in a pump, compressor, gearbox, etc.

Figure 1.10 shows how both performance and mechanical functions are similar regardless of the classification or type of equipment.

Classifications of rotating equipment

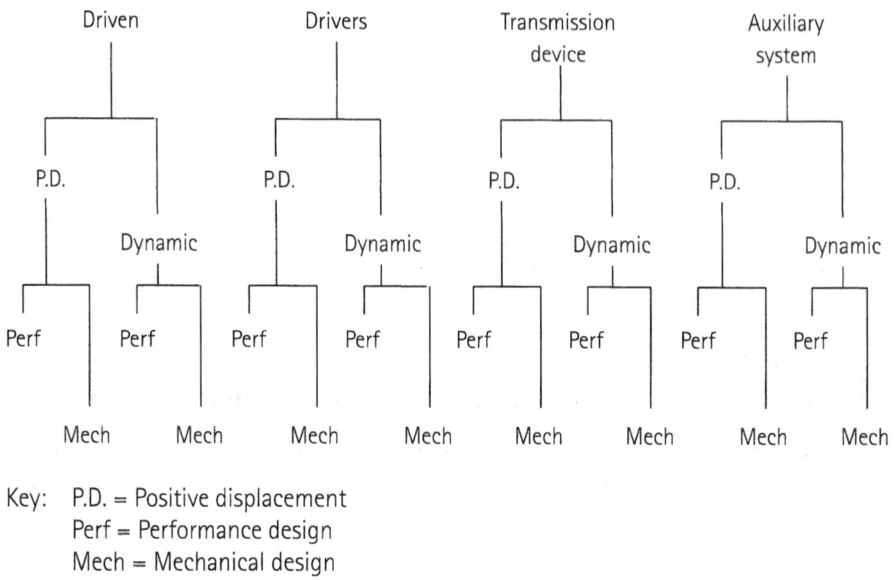

Figure 1.10 Classifications of rotating equipment

As we proceed through this book, we will discover that positive displacement or dynamic performance principles will be the same regardless of the type of equipment (pump, compressor or turbine). Also, the mechanical principles presented for bearings, seals, etc. will apply whether the component is in a pump, turbine, gear, etc.

The equipment 'train' or 'unit'

As stated, the objective is to learn the functions of equipment and major components so that they can be effectively condition-monitored to maximize site safety and reliability. Having defined the four (4) classifications of equipment, how many of these classifications are present in a pump unit or compressor train? ... All four (4)!!

Regardless of the type of unit or train, a driven, driver, transmission(s) and auxiliary system(s) must always be present. When you are asked to inspect G-301 or K-101 you are actually inspecting G-301 pump unit or K-101 compressor train. Failure to recognize this fact will severely limit your troubleshooting scope and ability.

As an example, a call from the unit shift manager may state that G-301 discharge pressure is zero – what are possible causes? A few are:

- Process change
- Pump wear
- Coupling failure
- Pump shaft failure
- Driver shaft failure
- Pump or driver shaft seizure (no oil)
- Pump seal failure
- Process valve closed
- Steam inlet valve closed (if driver is a steam turbine)

Do you get the point! ... The entire unit or train, all four machinery classifications must always be considered in rotating equipment design, revamps and troubleshooting.

Important fundamentals

Before discussing specific facts concerning all the rotating equipment on site, some important fundamentals need to be presented. The environment or surroundings for any piece of rotating equipment play an important part in determining the availability of that particular item (Refer to Figure 1.11).

This figure shows that the rotating equipment environment is the process unit in which the equipment is installed. The surroundings of the equipment will be defined early in the project. Proper design of process conditions, piping and foundations, selection of other components (drivers, transmission devices and auxiliaries) and proper specification of ambient conditions all must be considered. If any of these items are not taken into account, the end user of the equipment will be faced with a history of an unreliable process and will pay dearly in terms of lost product revenue.

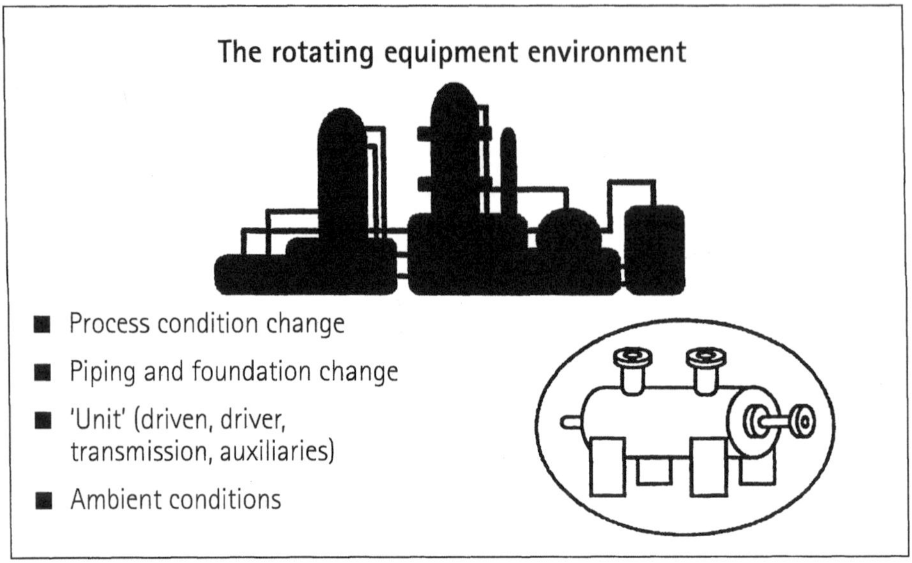

Figure 1.11 The rotating equipment environment

It is important to understand that the life span of rotating equipment is extremely long compared to the specification, design and installation phase. Refer to Figure 1.12. A typical installation will have a specification design and installation phase of only approximately 10% of the total life of the process unit. Improper specification, design or installation will significantly impact the maintenance requirements, maintenance cost and availability of a particular piece of machinery.

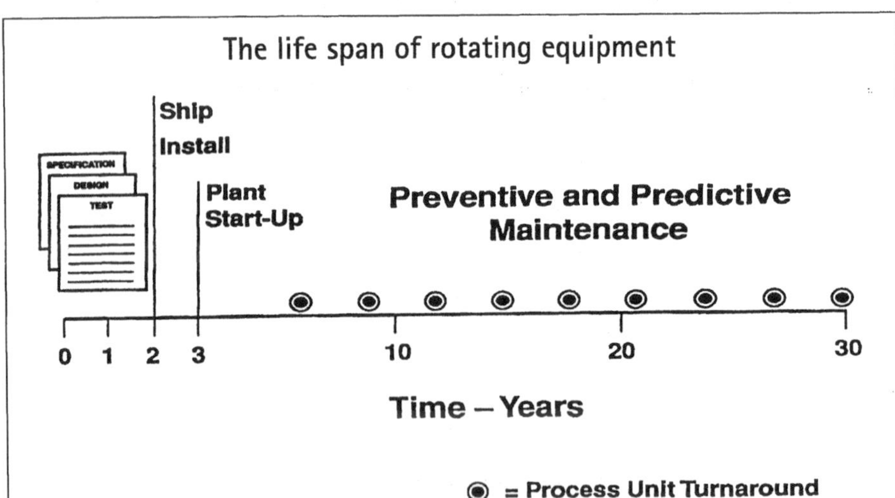

Figure 1.12 The life span of rotating equipment

The objectives of the end user are shown in Figure 1.13.

The objectives

- Maximum reliability (on stream time)
- Maximum product throughput
- Minimum operating costs

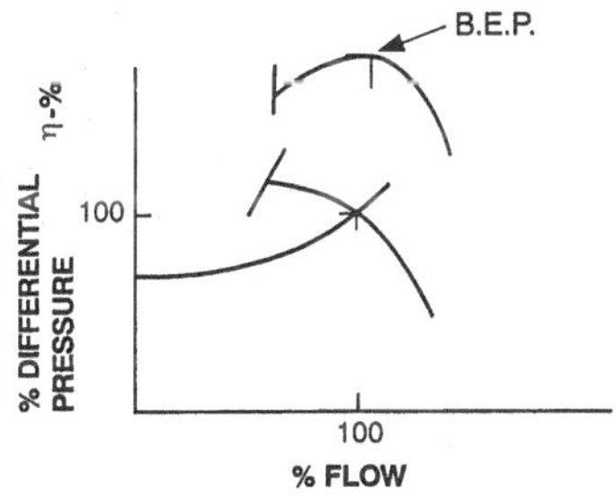

Figure 1.13 The objectives

In order to maximize the profit, a piece of machinery must have maximum reliability, maximum product throughput and minimum operating cost (maximum efficiency). In order to achieve these objectives, the end user must play a significant part in the project during the specification and design phase and not only after the installation of the equipment in the field. Effective field maintenance starts with the specification phase of a project. Inadequate specifications in terms of equipment operating conditions, mechanical design, instrumentation and the location of the instrumentations will reduce equipment availability.

The definitions of reliability and availability are shown in Figure 1.14.

Definitions: Reliability and availability

Reliability	The amount of time equipment operates in one (1) year
Reliability (%) =	$\frac{\text{Operating hours per year}}{\text{8700 hours}}$
Availability	The amount of time equipment operates in one (1) year – the planned downtime
Availability (%) =	$\frac{\text{Operating hours per year}}{\text{8700 – planned downtime (hours)}}$
Note:	The values should be greater than 95%.

Figure 1.14 Definitions: Reliability and availability

Reliability does not take into account planned down time. Availability considers planned downtime (turnarounds, etc.). Both values are stated in percentage. Typical equipment availabilities when properly specified are 97% +.

Availability directly affects the product revenue. Product revenue is the value obtained from the product produced in one day. For refineries, the loss of revenue can exceed a million U.S. Dollars a day. This will occur if a critical piece of equipment (unspared) is shut down. Remember, even though the unit may be down for a short period, the time necessary to bring the process back within specification may be significantly longer. It is very valuable to obtain the product revenue figure for the unit in which you are working. As will be shown, this value can significantly influence management in terms of decision making.

It is important to understand that upset conditions with any piece of rotating equipment can occur in a very short period of time. In addition, the requirements for reliable operation (a minimum of three (3) years continuous run) require enormous amounts of equipment revolutions all on a very thin film of lubrication oil. Failure to maintain this film is one of the major reasons for reduced equipment availability. Figure 1.15 presents a typical fact sheet concerning a turbo-compressor.

Interesting facts

- A typical turbo-compressor that runs at 6,000 RPM rotates at 100 times per second.
- It also rotates 8,640,000 times per day.
- And ... 3,153,600,000 times per year.
- All on an oil film of less than one thousandth of an inch.

Figure 1.15 Interesting facts

As an exercise you may want to determine how far this shaft would travel in one year if its diameter were four inches. There are 5,280 feet in one mile and one revolution around the earth equals 25,000 miles.

2 Compressor characteristics

(Positive displacement vs dynamic)

- Introduction
- Positive displacement compressors
- Actual volume, standard volume and mass flow
- Dynamic compressors

Introduction

In this section we will discuss the two principle compressor characteristics, positive displacement and dynamic compression. In addition, the concepts of volume flow, mass flow and standard flow will be covered. Although this chapter covers compressors, the characteristics of positive displacement and dynamic are equally applicable to pumps.

Positive displacement compression is defined as the increase of the pressure of a gas caused by operating on a fixed volume in a confined space. Types include; reciprocating, rotary liquid piston, rotary lobe and screw compressors. This concept can best be envisioned by using a simple syringe. As one moves the plunger into the syringe, the volume inside is displaced. It will be displaced regardless of the resistance in which the compressor operates, provided sufficient power is available and the design of the compressor can meet the pressure requirements. A schematic of a positive displacement reciprocating compressor is viewed and one can see that gas will not enter the cylinder until the pressure inside the cylinder is lower than the suction pressure. Conversely, gas will not exit the cylinder until the pressure inside the cylinder is greater than the discharge pressure. The valves shown in this figure are merely check valves. The suction valves act as check valves preventing the compressed gas from escaping back into the suction line. The characteristics then, of a positive displacement compressor are fixed volume, variable pressure capability (energy or head) and not self limiting. By this we mean the compressor will stall or damage itself unless a pressure or power limiting device is included in the compressor scheme. This is usually achieved by using a relief valve.

Before proceeding to the concept of dynamic compression, actual flow, mass flow and standard flow will be discussed as follows.

In the design of any compressor actual volume flow must be used. This is necessary since the design is based on an optimal gas velocity. Gas velocity is the result of a given volume flow acting in a specific area. Think of any compressor, dynamic or positive displacement, compressing a volume of one actual cubic foot per minute and let us assume that the temperature of compression remains constant. If the particular compressor in question has a compression ratio of two, one actual cubic foot per minute entering the compressor will be compressed to a discharge volume of exactly one half cubic foot per minute assuming that the gas is dry.

Standard volume is defined as one volume always referenced to the same pressure and temperature conditions. In customary units, standard pressure is defined as 14.7 pounds per square inch and standard temperature is defined as 60°F. Standard cubic feet then is the ratio of the actual pressure to the referenced standard pressure and the referenced standard temperature to the actual temperature multiplied by the actual volume. Referring back to the previous example of a compressor with a compression ratio of two and no compression temperature increase, one can see that the standard cubic feet per minute in this compressor would remain the same assuming a dry gas. This is because even though the actual volume of the gas does decrease by one half, the discharge standard volume is the ratio of the discharge pressure to the standard atmospheric pressure multiplied by the discharge volume. This will result in the same exit standard volume as the inlet.

Mass flow is the product of the actual volume flow and the density of the specific gas. The concept of mass flow and standard volume flow are the same. That is, the mass flow into the compressor example cited above will be exactly equal to the exit mass flow provided the gas is dry.

Both standard volume flow and mass flow are used to describe process capacities and are used in power calculations. After all, it would be very difficult to charge customers for produced products (gas) unless either a standard volume measure or weight measure were used.

The next concept to be covered is that of dynamic compressors. Dynamic compressors increase the pressure of a gas using rotating blades to increase the velocity of the gas. Types include axial and centrifugal compressors. As we will see later, there are basically two velocities which determine the performance of a dynamic compressor.

Both of these velocities occur at the exit of the moving blade. They are tip speed (the product of blade tip diameter and RPM) and relative velocity (velocity of the gas between two blades). The curve of a centrifugal compressor is significantly different from that of a positive displacement compressor. The centrifugal compressor is a machine of variable capacity, fixed energy or head for a given flow and is self limiting. That is, there is a maximum pressure which the compressor can produce and a maximum horsepower which it requires. Observing the curve of a typical dynamic compressor, one can see that the only place that a dynamic compressor can develop higher energy or head is at a lower flow for a given speed. We will see in the next chapter how the system combined with the compressor operating curve will determine the operating point. It should be noted that the operating curve of an axial compressor at times can approach that of a positive displacement compressor in that its volume range is small and almost approximates the positive displacement curve.

The concept of an equivalent orifice is introduced in this section. Any typical axial compressor blade row or radial compressor impeller can be reduced to a series of equivalent orifices. The inlet of the blade row or impeller being one orifice, the exit being another and the eye and hub seals being additional equivalent orifices.

It should also be noted that both the suction and discharge process system for a given point in time and flow rate can be thought of as equivalent orifices placed at the inlet and discharge of the compressor flanges respectively.

In summary, the characteristics of positive displacement and dynamic compressors are as follows: Flow fixed for positive displacement, variable for dynamic. Head, variable for positive displacement, fixed for dynamic. Limiting characteristics for dynamic compressors, not self limiting for positive displacement (requires relieving device).

Positive displacement compressors

The definition and types of positive displacement compressors is given in Figure 2.1. Figure 2.2 shows a typical process reciprocating compressor. Reciprocating compressors are generally used for the following applications:

- Low flow
- High pressure
- Low molecular weight
- Varying molecular weight

Positive displacement compressors
Increase the pressure of a gas by operating on a fixed volume in a confined space

Types include:

- Reciprocating
- Rotary-liquid piston
- Screw
- Sliding-vane
- Rotary lobe

Figure 2.1 Positive displacement compressors

Figure 2.2 Penn process compressors (Courtesy of Cooper Industries)

The internal construction of a typical reciprocating compressor cylinder is shown in Figure 2.3. As the piston moves from left to right, a constant volume of gas will be displaced. The discharge pressure developed will be the pressure at the discharge flange caused by the process pressure and system resistance. That is, the gas in the cylinder will not exit until the pressure developed in the cylinder is greater than the pressure at the compressor discharge flange.

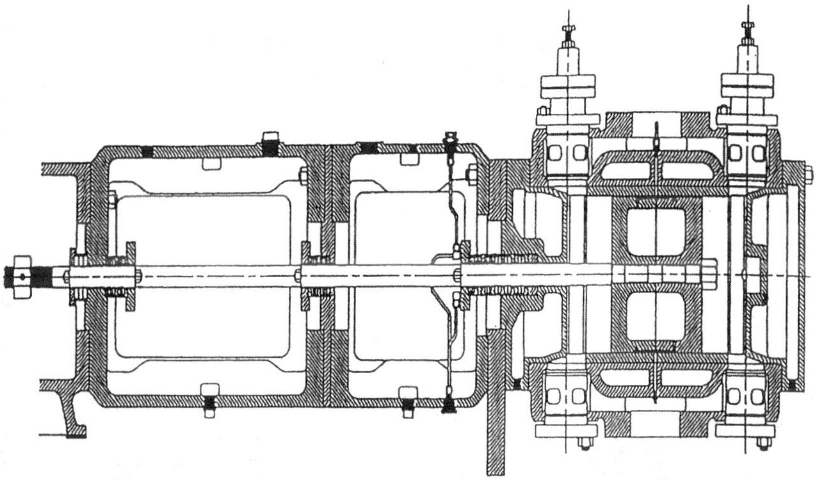

Figure 2.3 Penn process compressors (Courtesy of Cooper Industries)

Actual volume, standard volume and mass flow

Every compressor is designed on the principle of volume flow. Since compressor performance must consider gas velocity, actual flow is always used for design. Figure 2.4 shows the concept for actual volume flow. The volume flow changes directly with temperature and inversely with pressure.

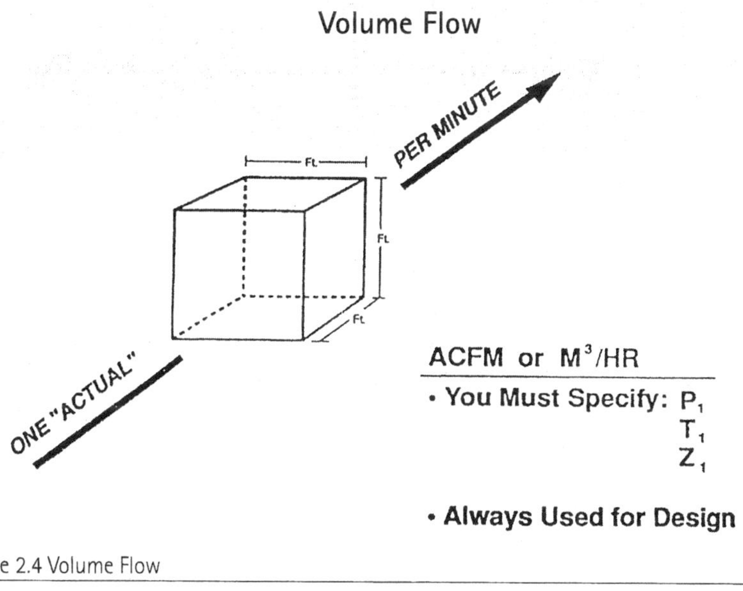

Figure 2.4 Volume Flow

Standard volume and mass flow are used to describe process conditions. If there is not any condensate formed within a compressor section, the standard and mass flow through a compressor remains unchanged. The relationships between standard volume and mass flow are shown in Figure 2.5.

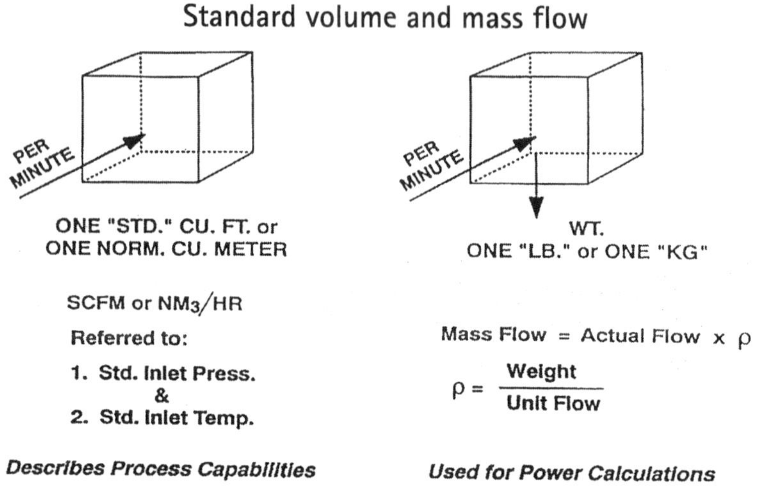

Figure 2.5 Standard volume and mass flow

Figure 2.6 depicts a simple process system.

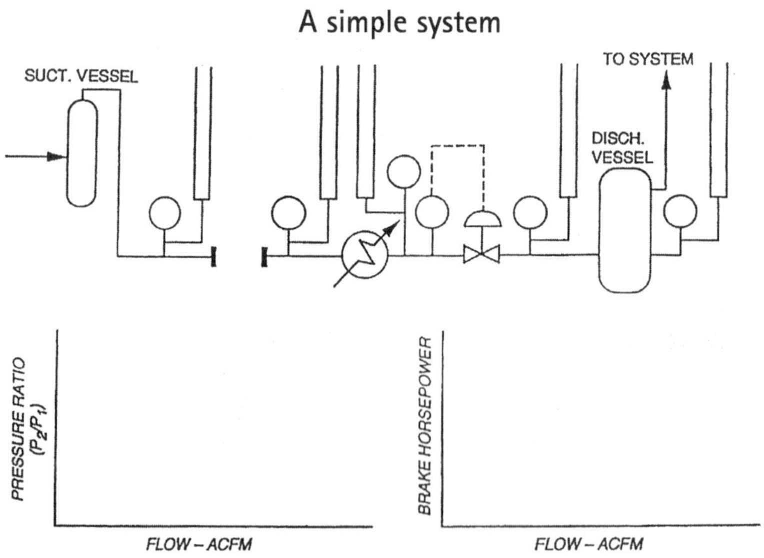

Figure 2.6 A simple system

To demonstrate the characteristics of a positive displacement compressor, insert the symbol for a positive displacement compressor into the space provided in Figure 2.6. The circles shown represent pressure gauges and the vertical tubes represent manometers. For this example pencil in the following set of initial pressures.

- P suction = 20 psig
- P downstream valve = 110 psig
- P disch. vessel = 100 psig
- P upstream valve = 135 psig
- P discharge = 140 psig

Pressures can also be indicated by coloring in the manometers to indicate the relative pressure according to height of the column in the manometer. Therefore the manometer heights will steadily decrease from the discharge to the discharge vessel.

At the conditions noted above, assume that the compressor flow is 500 ACFM and the brake horsepower is 250 BHP. The compression ratio based on the suction and discharge pressure is 4.43. Plot ACFM vs pressure ratio and ACFM vs brake horsepower on the appropriate graphs in Figure 2.6.

Now assume the control valve in the discharge process system is slightly closed. With this action assume the pressure upstream of the control valve increases to 175 psig and the discharge pressure increases to 180 psig. The horsepower also increases 300 BHP. What has happened to the ACFM in this case? Plot the new pressure ratio and brake horsepower in Figure 2.6. Notice that both pressure ratio and brake horsepower curves are essentially a vertical line.

Observing these plots, the characteristics of any positive compressor can be defined as shown in Figure 2.7.

Compressor characteristics

Positive displacement

- constant volume delivery
- variable head capacity
- not self limiting

Figure 2.7 Compressor characteristics

Note that head (energy) is proportional to pressure ratio. 'Not Self Limiting' means that the pressure ratio (head) and brake horsepower will continue to rise until the case pressure is exceeded and/or the driver maximum horsepower is exceeded. To 'limit' these parameters, a pressure relief valve is always required in a system incorporating any type of positive displacement compressor.

Dynamic compressors

The definition and types of a dynamic compressor is presented in Figure 2.8.

Dynamic compressors

Increase the pressure of a gas by using rotating blades to increase the velocity of the gas

Types include:

- Axial
- Centrifugal

Figure 2.8 Dynamic compressors

Figure 2.9 is the assembly drawing of an axial compressor. Axial compressors are used for high flow, low head applications usually above 100,000 ACFM inlet flow. The most common applications for axial compressors are F.C.C. air blowers, MTBE compressors and gas turbine air compressors.

A typical curve for an axial compressor, % volume vs % compression ratio, is shown in Figure 2.10. Note that the pressure rise, from 100% flow to surge flow, is always greater than the corresponding pressure rise for a centrifugal compressor.

An assembly drawing of a multistage horizontal split centrifugal compressor is shown in Figure 2.11.

A centrifugal compressor can also be called a radial compressor because the exit flow from the impellers is radial as opposed to axial. A typical percentage flow vs head centrifugal compressor performance curve is shown in Figure 2.12.

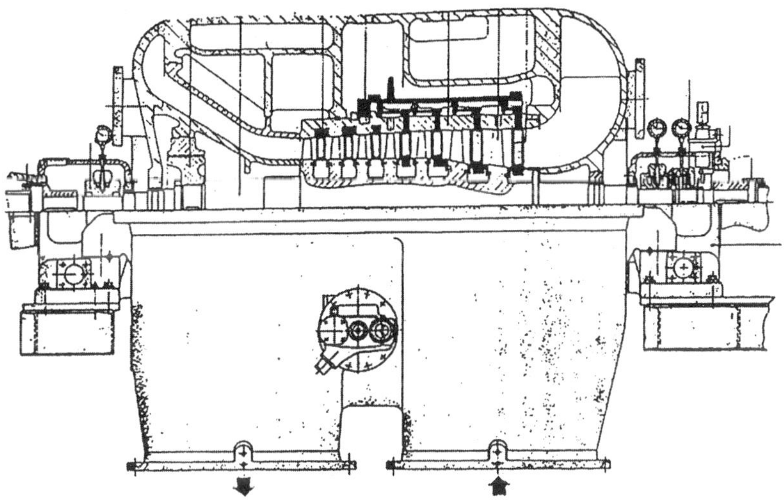

Figure 2.9 Axial compressor (Courtesy of Man Turbo)

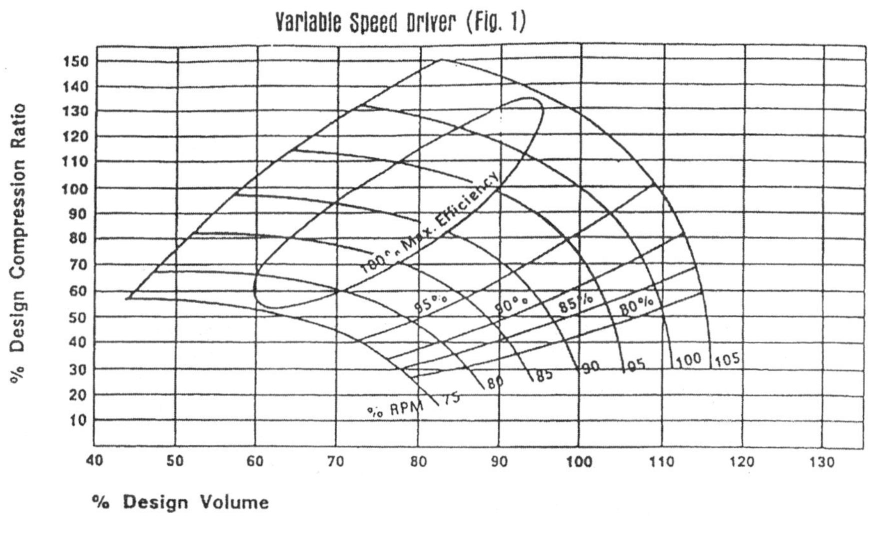

Figure 2.10 Typical Axial curve (Courtesy of Man Turbo)

As a comparison, the head (or pressure) rise shown in Figure 2.12 for the centrifugal compressor is approximately 10% as compared to 30% for the axial compressor shown in Figure 2.10.

The principle of operation of a dynamic compressor always seems more difficult to understand than that of a positive displacement compressor. As stated in the dynamic compressor definition in Figure 2.8, dynamic

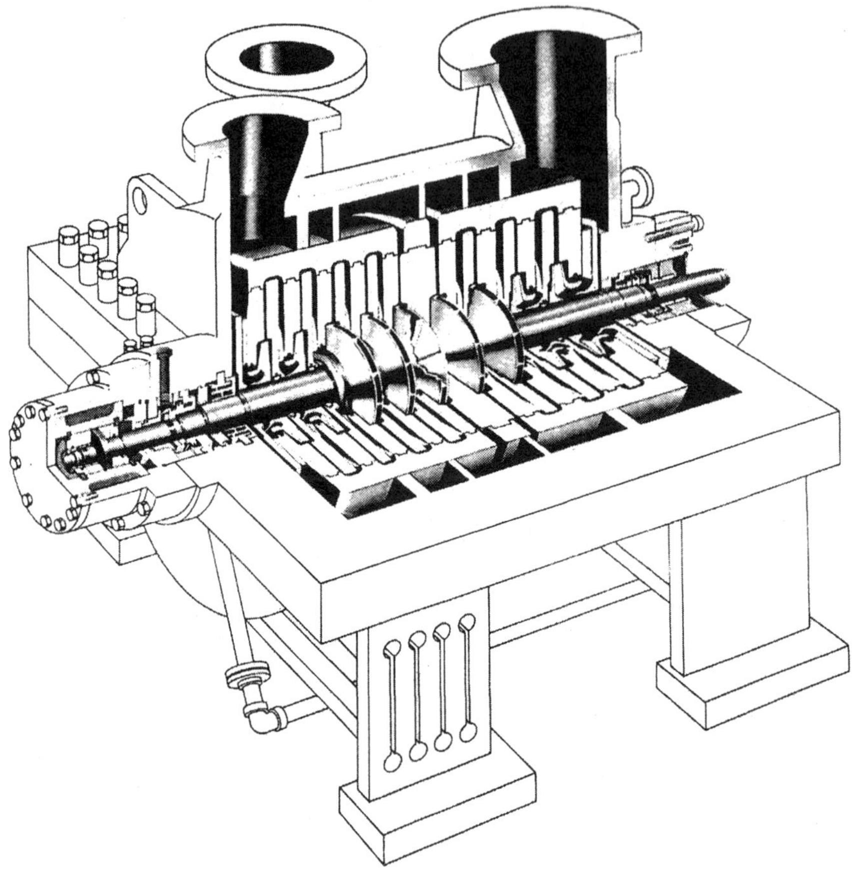

Figure 2.11 Cutaway of horizontally split fabricated compressor (Courtesy of IMO Industries Inc)

compressors increase pressure by using rotating blades to increase the velocity of the gas. Figure 2.13 shows the upper half of a centrifugal compressor impeller with the side plate removed. The velocity vectors 'U' (blade discharge tip speed) and 'V' (discharge gas – 'relative' velocity) are shown. Simply stated, the characteristic shape of any dynamic performance curve (rising head vs decreasing flow rate) is a direct result of the relative velocity through the impeller or blades. The higher the relative velocity, the less energy input from the impeller blades. The lower the relative velocity, the greater the energy input from the impeller blades. The relative gas velocity in any impeller or blade row is determined by the gas passage areas.

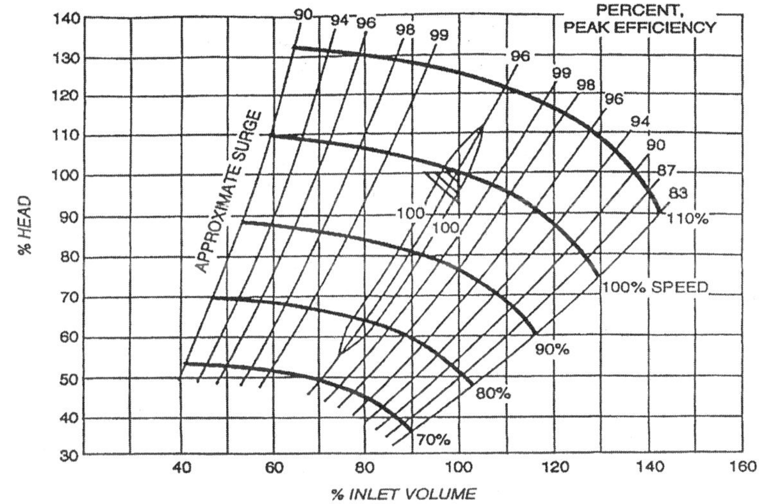

Figure 2.12 Typical multistage compressor curve

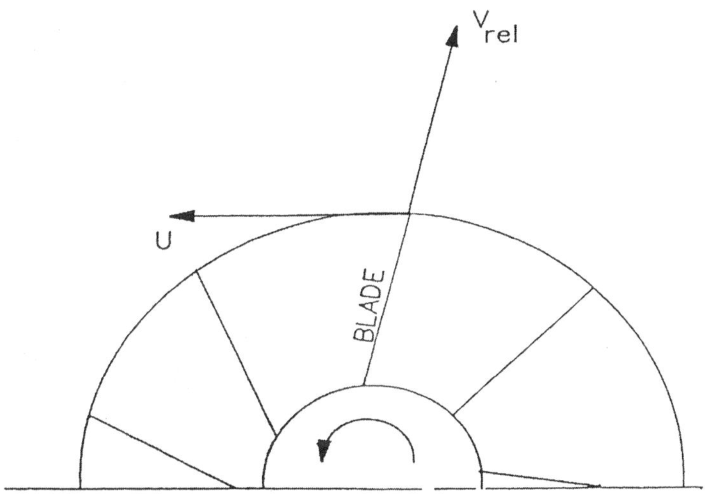

Figure 2.13 Impeller with side plate removed

Figure 2.14 shows how the various sections of any impeller (inlet and discharge) can be interpreted as 'equivalent orifices'. The designer selects the flow areas to reach the optimum gas velocities that will result in the highest impeller efficiencies based on the design flow rate.

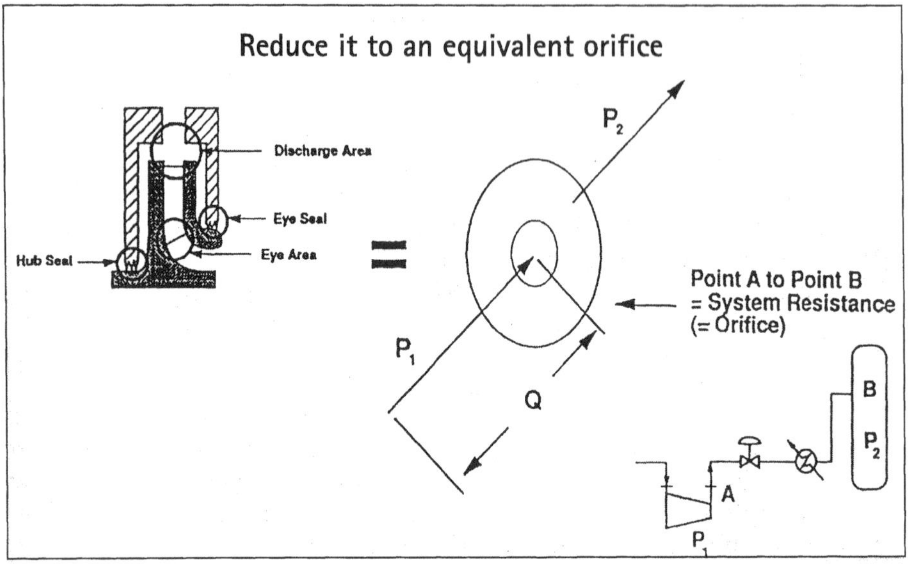

Figure 2.14 Reduce it to an equivalent orifice

Figure 2.15 presents the same simple process system used in Figure 2.6. However, for this exercise insert a dynamic compressor in the space provided. Use the same pressures as in Figure 2.6. The flow rate and horsepower will be larger than that for the positive displacement

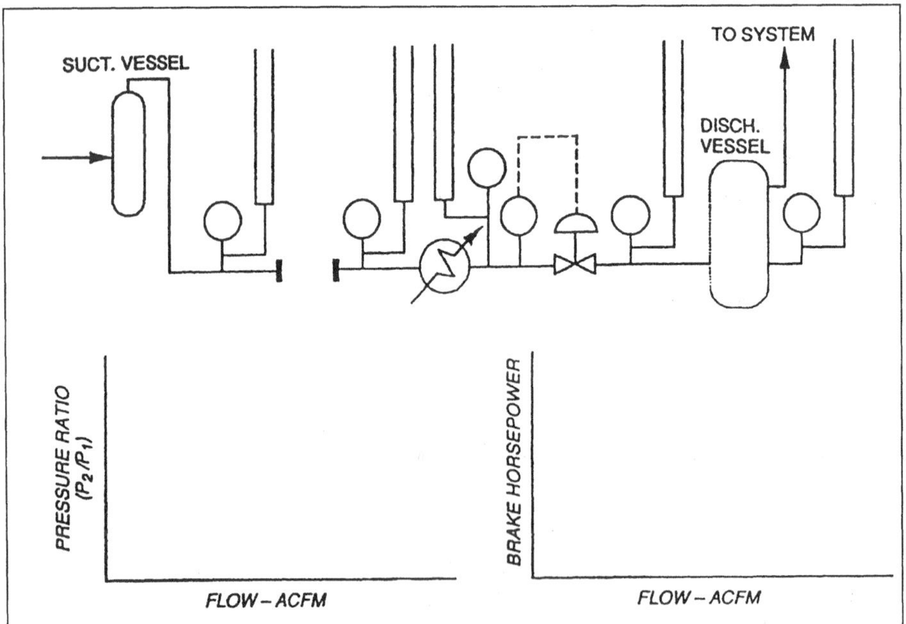

Figure 2.15 A simple system

compressor of Figure 2.6 since dynamic compressors usually handle inlet flow rates above 5000 ACFM. Assume the inlet flow is 20,000 ACFM and the horsepower is 10,000 BHP. Plot ACFM vs pressure ratio and BHP on the appropriate graphs of Figure 2.15.

Again, assume the control valve in the discharge process system is closed slightly. This action causes the compressor discharge pressure to rise to 180 psig. The horsepower however decreases to 6000 BHP. Why? The answer of course, is that the characteristic performance curve of any dynamic compressor is different from a positive displacement compressor. A positive displacement curve is essentially vertical since the actual volume flow is constant. Therefore, the brake horsepower will increase as the pressure ratio does. In the case of a dynamic compressor, increasing the pressure ratio will result in a lower compressor flow rate since the only way a dynamic compressor can develop a higher pressure ratio is at a lower gas velocity (lower flow rate). Simply stated, the longer the gas is in contact with the energy producer (blades), the higher the pressure ratio. For a dynamic compressor, this means the flow rate must decrease. Therefore, plotting the higher-pressure ratio and BHP point on Figure 2.15 will yield a higher pressure ratio, lower flow point.

The characteristics of any dynamic compressor are presented in Figure 2.16.

Compressor characteristics

Dynamic

- Variable volume delivery
- Fixed head capacity (for a certain flow)
- Self limiting

Figure 2.16 Compressor characteristics

As opposed to a positive displacement compressor, the dynamic compressor is self-limiting and does not require a relief valve to protect the compressor and driver.

Figure 2.17 summarizes the characteristics of positive displacement and dynamic compressors by presenting their typical performance curves. Note that the vendors usually do not issue positive displacement curves since volume flow is constant.

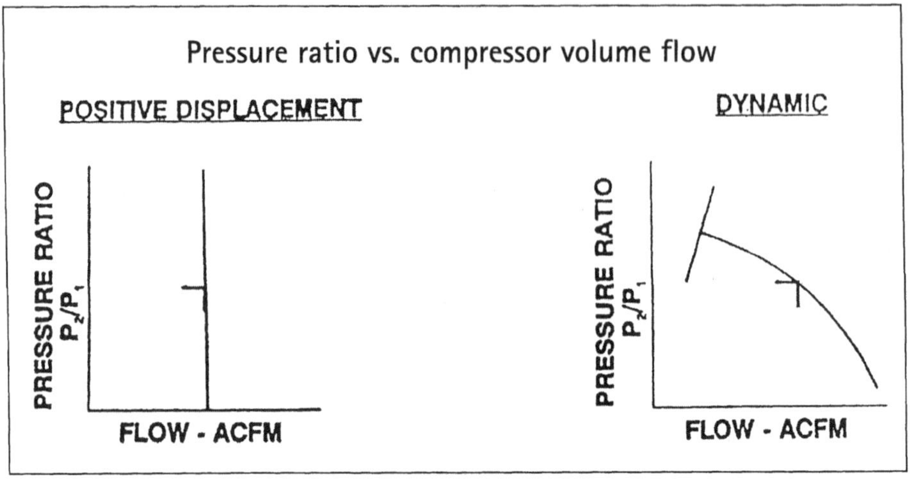

Figure 2.17 Pressure ratio vs. compressor volume flow

3 Operation of a compressor in a system

- Introduction
- The definition of a process system
- System resistance curves
- The operating point
- A positive displacement compressor in the process system
- A dynamic compressor in the process system

Introduction

In this chapter we will define a system, examine system resistance curves (head required by the process) and define the operating point of a compressor. Again, the principles discussed can be applied to a pump operating in a system.

A system is a set of connected things or parts that work together. In a typical process a set of components (vessels, exchangers, furnaces, control valves, etc. and piping) work together to produce a resistance to flow at the turbo-compressor flanges. Examining a simple system one can see that any process system relative to the compressor is comprised of two parts, the suction system and the discharge system. The objective of any compressor in a system is to remove the flow from the suction vessel at the same rate that the flow enters that vessel. In doing so, the pressure is reduced from the suction vessel to the compressor flanges.

In the discharge system the objective is to push the required amount of flow through the system resistance to achieve the final discharge system terminal pressure. This pressure may be in a vessel, may be in a ship or could be in a pipe line. The discharge pressure at the compressor flange then is the accumulation of the system resistance (pipe, valve, exchanger, etc.) from the terminal point (delivery point) back to the compressor discharge flange. An important point to remember is that the compressor differential head or energy required is the net effect of the discharge system resistance and suction system resistance. Increasing discharge system resistance with a constant suction resistance will result in higher compressor energy required. Also increasing suction system resistance with a constant discharge system resistance will result in increasing compressor energy required. Many times the concept of increasing suction system resistance is misunderstood (suction throttling).

If we review the previous chapter and examine the operation of both positive displacement and dynamic compressors in a simple system, two important facts concerning the operation of these machines in a system become evident.

A positive displacement compressor, since it increases the energy of the gas by operating on the gas in a confined space, will always increase the energy, provided sufficient power is available and the machine is designed to meet this objective. Therefore a positive displacement compressor will be relatively insensitive to gas composition changes. In addition, since the positive displacement compressor as previously defined has variable head capability, it will be relatively insensitive to system changes. As a result, a positive displacement compressor will operate at approximately the same delivered volume flow regardless of system resistance or gas composition.

On the other hand, since the characteristics of a dynamic compressor are to increase gas energy (a function of mass and velocity) by working on the gas with blades. Velocities and gas mass (density) play an important part in the make up of this compressor's characteristics. Anything that will result in a velocity and/or density change at the tip of the dynamic compressor blade will result in a different produced differential pressure and a corresponding flow change. Therefore, the dynamic compressor will be extremely sensitive to gas composition changes since these changes will produce mass and velocity changes within the compressor blades or impellers. In addition, this compressor type will also be sensitive to system resistance changes since an increased system resistance requirement will force the compressor to operate at a lower volume throughput. This is because the only way this compressor can produce higher delivered energy is at a lower velocity throughput.

Therefore one can readily see that the operating point of a dynamic compressor will be much more sensitive to both changes in system resistance and the velocities inside the impeller. Both system resistance and the dynamic compressor curve can change when plotted on flow vs pressure coordinates. As will be explained later flow vs head coordinates reflect a dynamic compressor which changes very little provided that the gas composition changes within the limits of approximately 20%. The compressor operating point can be defined as the equilibrium between the required net process system energy and the compressor produced energy. Remember, that the positive displacement compressor characteristic results in relatively constant flow regardless of system required energy, whereas the dynamic compressor characteristics results in significantly large flow changes for changes in system resistance. One additional point to understand is that the process system characteristic curve can also change.

The system resistance curve (head required by the process) is a result of the terminal pressure, that is the discharge pressure at zero flow and the system resistance (square function) as the flow changes. A recycle loop for instance is comprised of friction loss only and will have a relatively steep system resistance curve, whereas an instrument air compressor pumping into a receiver vessel with a relatively small system resistance will have a system resistance curve that will approach a horizontal line. One can see that the combination of a flat dynamic compressor curve and a flat system resistance system curve can lead to a very unstable situation. Remember that the operating point is determined by the intersection of a compressor and system resistance curves. For turbo-compressors both of these curves can change and certainly do.

The Definition of a Process System

The definition of a generic and process system are presented in Figure 3.1. The process system characteristics and the gas composition determine the head required by the process system.

System – Definition

'A system is a set of connected things or parts that work together'

In our case, a set of components (vessels), exchangers, furnaces, control valves, etc. and piping) work together to produce a resistance to flow at the turbo-compressor flanges.

Figure 3.1 System – definition

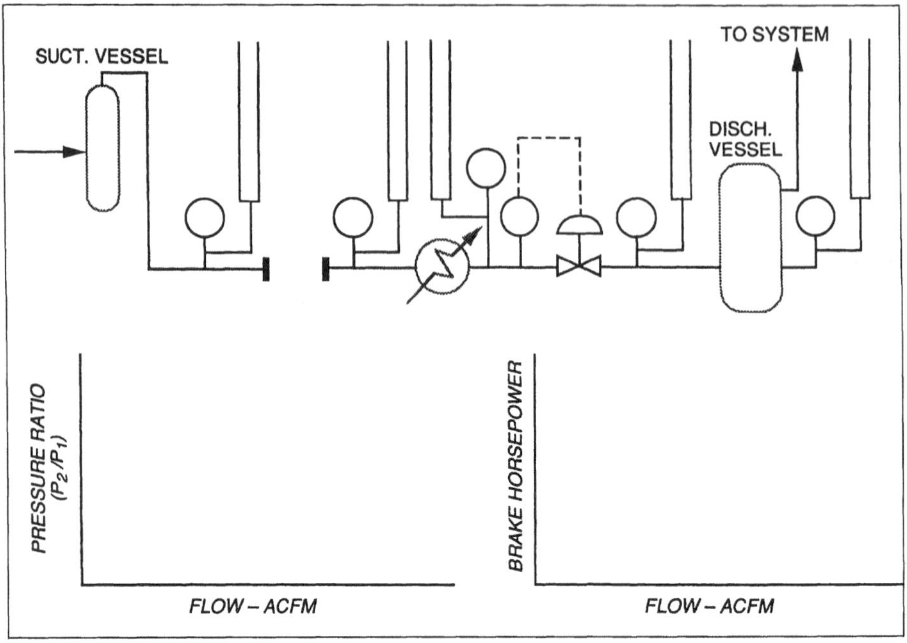

Figure 3.2 A simple system

A simple process system schematic is presented in Figure 3.2. Based on the characteristics of positive displacement and dynamic compressors presented in the previous chapter, draw both the pressure ratio vs flow and brake horsepower vs flow on the appropriate graphs.

System resistance curves

Figure 3.3 shows a system where the system resistance curve (head required by the process) intersects the compressor curve (head produced) at the design point (100% flow).

In Figure 3.4, two different system resistance curves representing two different process systems are shown. The curve starting at zero differential pressure represents a process system where only system resistance is present. When the compressor is shut down, the pressure in the system is zero. This would be the system resistance curve that would describe a recycle process. The curve that starts at approximately 50 PSI differential pressure shows that a pressure differential of approximately 50 PSI exists when the compressor is shutdown in this system. This process system would be typical of a system in which both vessel pressure and system resistance are present. One can see from this figure that the higher the percent of vessel pressure in the process system, the flatter the system resistance curve is.

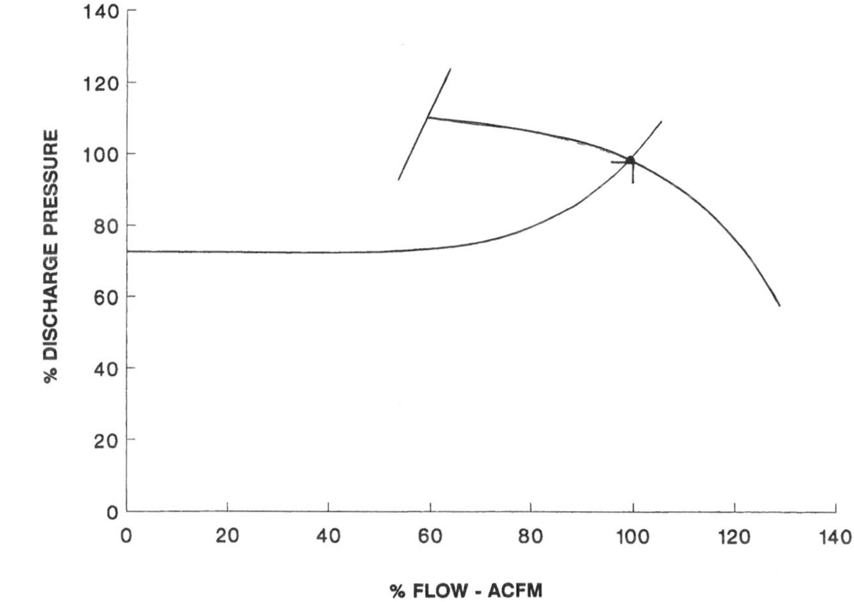

Figure 3.3 % Discharge pressure vs. % flow

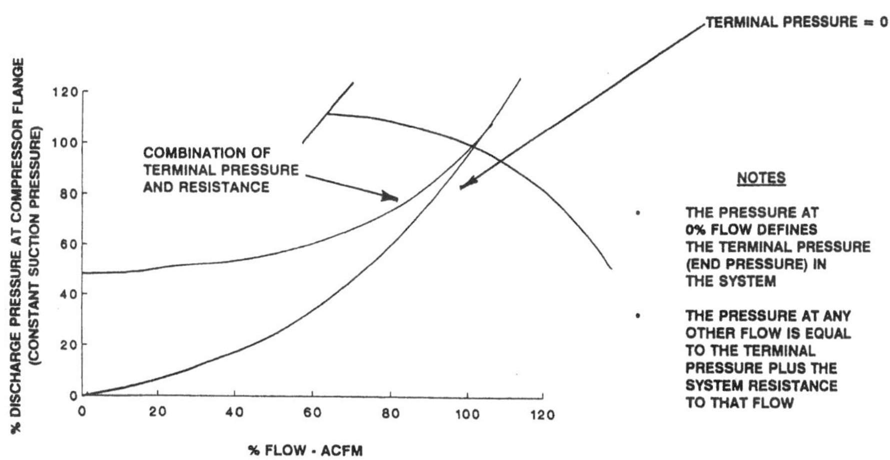

Figure 3.4 The system resistance curve for the discharge system

System resistance characteristics for the three basic types of process systems are shown in Figure 3.5. Note that the steepest curve, the recycle loop, will result in the most stable operation regardless of the compressor curve shape. The instrument air compressor process system curve on the other hand tends to be very flat due to high terminal pressure in the discharge (receiver) vessel. This system will experience instability if the compressor characteristic curve is flat.

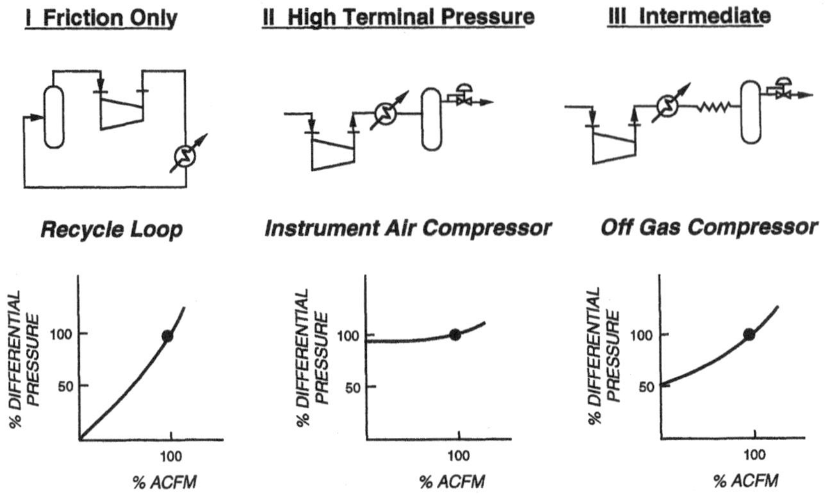

Figure 3.5 System resistance characteristics

Many times, a compressor has not been considered technically acceptable for purchase because of a relatively flat head curve (less than 5% head rise to surge). If the process system that the compressor will operate in is either Type I or Type III in Figure 3.5, this compressor should be considered acceptable based on the fact that the process system steadily rising head characteristic curve will provide a stable

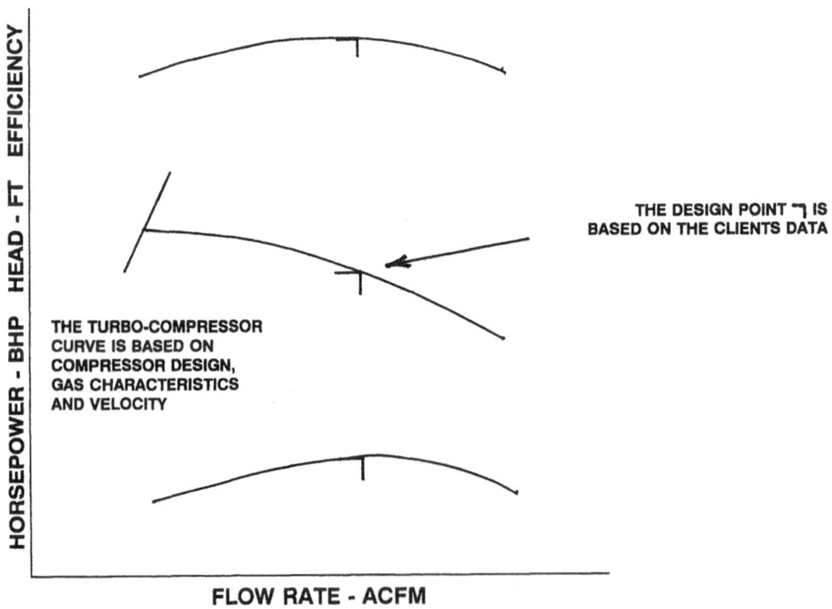

Figure 3.6 Turbo-compressor characteristics for a constant speed – the design point

operating point. Figure 3.6 shows that the compressor curve selected is based only on a single operating point specified by the client. Industry specifications only require that one operating point be guaranteed. Therefore the compressor curve shape is not guaranteed.

The operating point

The operating point is defined as the equilibrium condition that exists between the head produced by the equipment and the head required by the process system. (See Figure 3.7).

The operating point

Is determined by the intersection of the turbo-compressor and system resistance curves – both can change and do!

Figure 3.7 The operating point

If the process conditions that actually exist are properly specified and if the equipment is properly selected the operating point will occur at the best efficiency point. Figure 3.8 shows a case where the head required by the process is greater than the head produced by a centrifugal compressor at the rated point.

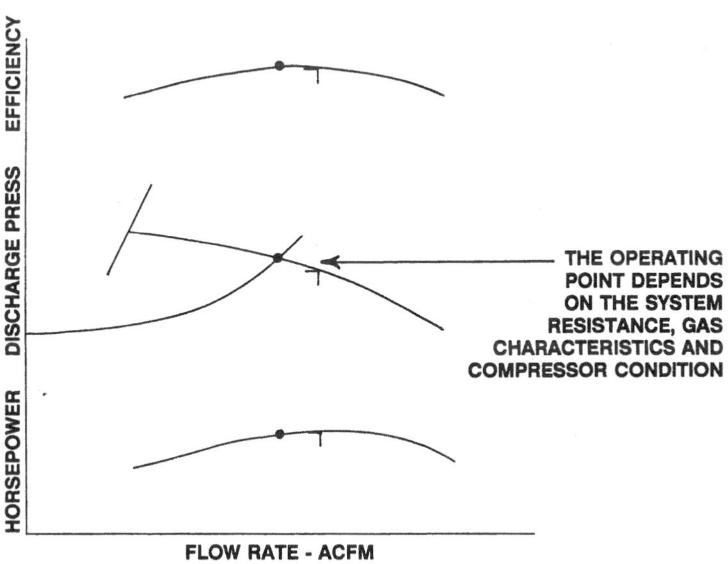

Figure 3.8 The operating point

The result is that the equilibrium point falls to the left of the best efficiency point resulting in a lower efficiency, lower flow rate and lower horsepower.

A positive displacement compressor in the process system

Referring back to Figure 3.2 of this chapter, draw any of the three typical system resistance curves on the pressure ratio vs flow graph. Notice that regardless of the change of resistance, the flow rate of a positive displacement compressor does not change. In addition, the flow rate will remain constant regardless of the density of the gas (molecular weight, inlet temperature or inlet pressure). Refer to Figure 3.9 which presents the characteristics of a positive displacement compressor.

Compressor characteristics

Positive displacement

- Constant volume delivery
- Variable head capacity
- Not self limiting
- Not flow sensitive to system resistance
- Not flow sensitive to gas density

Figure 3.9 Compressor characteristics

Based on the results drawn in Figure 3.2, a positive displacement compressor is not flow sensitive to either system resistance or gas density.

A dynamic compressor in the process system

Again, refer to Figure 3.2 and observe how the system resistance curves previously drawn affect a dynamic compressor's operating point. A dynamic compressor's flow rate is affected by system resistance change. In addition, a change in the gas density will also influence the system resistance (head required curve) and therefore change the flow rate.

Refer to Figure 3.10 which presents the characteristics of a dynamic compressor.

Compressor characteristics

Dynamic

- Variable volume delivery
- Fixed head capacity (for a certain flow)
- Self limiting
- Is flow sensitive to system resistance
- Is flow sensitive to gas density

Figure 3.10 Compressor characteristics

Based on the results of Figure 3.2, a dynamic compressor is flow sensitive to both system resistance and gas density.

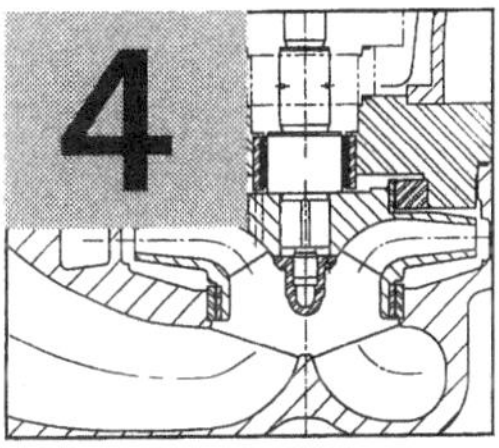

4 Pump types and applications

- Definition of pump types
- Positive displacement pumps
- Dynamic pumps
- When to use positive displacement or centrifugal pumps

Definition of pump types

A pump is defined as a device that moves a liquid by increasing the energy level of the liquid. There are many ways to accomplish this objective. At the conclusion of this chapter one will be able to identify all of the different types of pumps on site and to state the function of each specific type. Refer to Figure 4.1 and observe all of the different types of pumps which will be discussed in this chapter. Note that the pumps are divided into two distinct groups. One group pumps the liquid by means of positive displacement – the other group pumps the liquid by means of dynamic action.

Types of pumps

Positive displacement

Reciprocating

- Power
- Diaphragm
- Metering
- Direct acting

Rotary

- Screw
- Gear

Dynamic

Single stage

- Overhung
- Inline
- Integral gear
- Centrifugal
- Double flow
- Sump
- Submersible
- Magnetic drive

Multistage

- Horizontal split
- Barrel
- Canned
- Sump
- Submersible

Figure 4.1 Types of pumps

Refer to Figure 4.2 for the definition of positive displacement and dynamic action and the characteristics of each type of pump.

Positive displacement – Dynamic pump comparison

	Positive displacement	Dynamic
Definition	Increase pressure by operating on a fixed volume in a confined space	Increases pressure by using rotary blades to increase fluid velocity
Types	Screw, gear, reciprocating	Centrifugal axial
Characteristics	■ Constant volume ■ Variable differential head ■ Relatively insensitive to liquid properties ■ Relatively insensitive to system changes ■ Not self-limiting	■ Variable volume ■ Constant differential head ■ Sensitive to liquid properties ■ Sensitive to system changes ■ Self-limiting
Characteristic flow vs. differential head curves	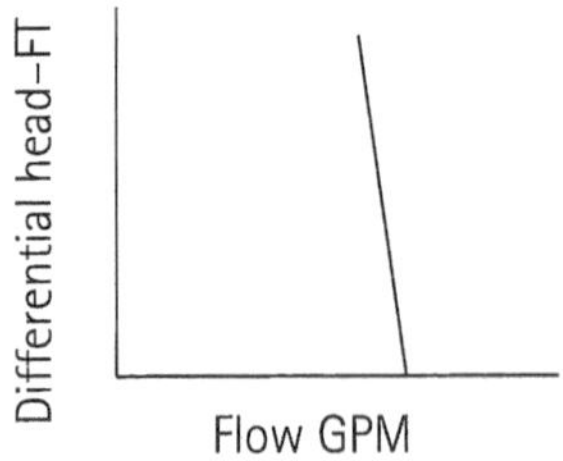 	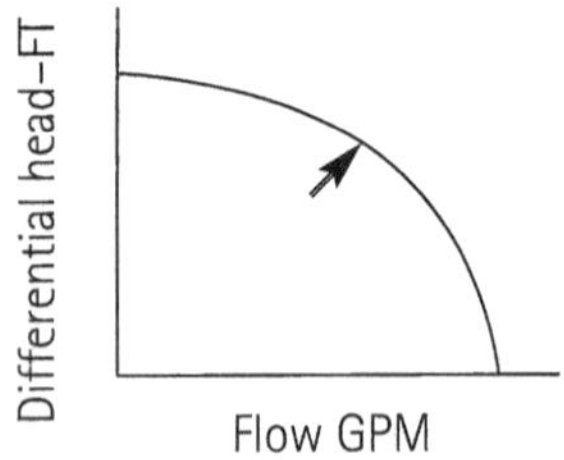

Figure 4.2 Positive displacement – dynamic pump comparison

Regardless of whether the pumps moves the liquid by positive displacement or dynamic means, each pump is divided into a hydraulic and a mechanical end. Figure 4.3 identifies these features for a positive displacement pump.

Figure 4.3 Power pump (Courtesy of Union Pump Co)

One important fact to remember is that the liquid doesn't care how it is moved; that is, the performance relationships of head, horsepower and efficiency will remain the same for all pumps regardless of whether they are positive displacement or dynamic. These relationships will be discussed.

As far as the mechanical end is concerned, the mechanical components of shafts, bearings, seals, couplings and casings perform the same function regardless of the pump type. Figure 4.4 is a chart that shows the similarities between the hydraulic and the mechanical ends of pumps regardless of their type.

Pump similarities – hydraulic and mechanical ends regardless of pump type

Hydraulic end

Head (energy) required

$$HD_{REQ} = \frac{2.31 \times \Delta P}{S.G.}$$

$\Delta P = P_2 - P_1$ (PSIG)

S.G. = specific gravity

$$BHP = \frac{HD \times GPM \times S.G.}{3960 \times \text{pump efficiency}}$$

GPM = U.S. Gallons/minute

S.G. = Specific gravity

Mechanical end

Mechanical components

- Casing (cylinder)
- Seals (mechanical or packing)
- Radial bearings
- Thrust bearing

Figure 4.4 Pump similarities – hydraulic and mechanical ends regardless of pump type

Again it can be seen from this chart that the relationships for pump performance are identical regardless of pump type. The only difference being the efficiency of one pump type relative to another. Secondly, it can be shown that the mechanical ends, housings, bearings, seals, etc. are very similar in each type of pump so that the function of a bearing or seal remains the same regardless of the type of pump.

Positive displacement pumps

As can be seen from Figure 4.2 a positive displacement pump is a constant flow variable head device. Refer to Figure 4.5 which shows a schematic of a double acting piston pump.

As the piston moves from left to right the pressure of the liquid will be increased and the pump will displace the liquid regardless of its specific gravity and viscosity as long as sufficient power is available from the pump driver. The types of positive displacement pumps which can be found in any petrochemical plant, refinery or gas plant are noted below. Refer to Table 4.1 for the application limits of positive displacement pumps.

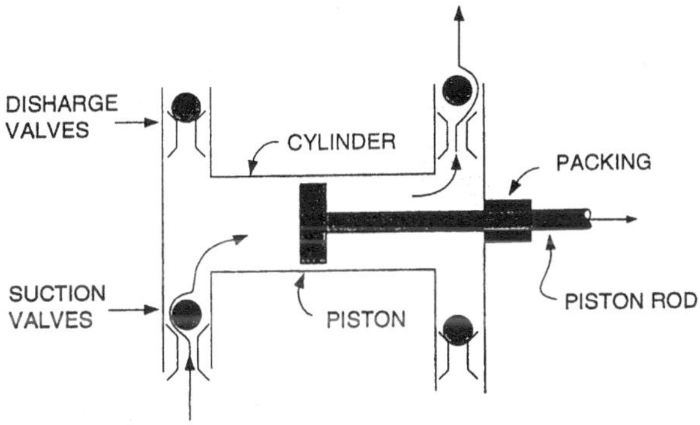

Figure 4.5 Double acting piston pump

Table 4.1 Application envelope positive displacement pumps

Pump type	Pressure-PSIG	Flow rate-GPM	Horsepower (max)
Power pumps			
■ Plunger-(horizontal)	1,000–30,000	10–250	200
■ Plunger (vertical)	1,000–30,000	10–500	1,500
■ Piston (horizontal)	up to 1,000	up to 2,000	2,000
Direct acting			
■ Piston	up to 350	up to 1,000	500
■ Plunger	up to 2,000	50–300	500
Metering	up to 7,500	up to 1,000 gal/hr	10*
Rotary			
■ Gear	up to 200	up to 300	50 HP
■ Screw	up to 5,000	up to 5,000	1,000 HP

* horsepower per cylinder, some applications use multiple cylinders

Reciprocating pumps

Reciprocating pumps are those types of positive displacement pumps that increase liquid energy by a pulsating action. The types are power pumps, direct acting steam pumps, diaphragm pumps and metering pumps. All reciprocating pumps produce pulsations that can cause damage to the pumps and/or process system if the system is not properly analyzed and designed. Anti pulsation devices (volume bottles, orifices or pulsation bottles) are usually required.

Figure 4.6 Power pump (Courtesy of Union Pump Co)

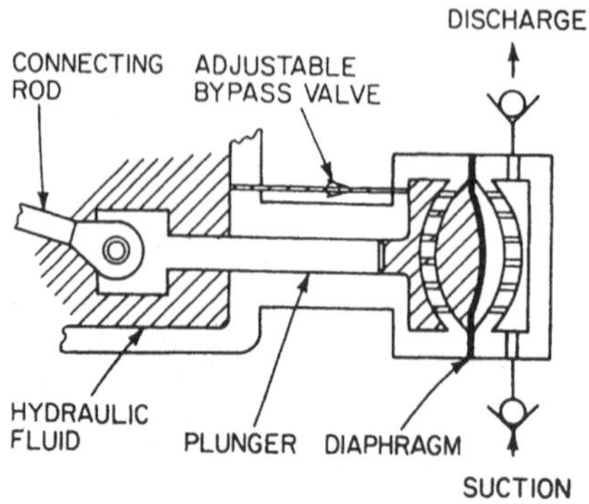

Figure 4.7 Diaphragm pump

Power pumps

A picture of a power pump is shown in Figure 4.6. Power pumps are used normally for high pressure, low flow applications, typically carbonate, amine service or high pressure water or oil services. They can either be horizontal or vertical. The major parts of a power pump as shown in Figure 4.6 include the liquid cylinder with pistons and rods, the valves and power end. The power end consists of the crankshaft with bearings, connecting rod and crosshead assembly. It is termed, 'Power Pump' because it is driven by an external power source, such as an electric motor, or internal combustion engine, instead of steam cylinders as in direct-acting pumps.

Diaphragm pumps

A schematic of a diaphragm pump is shown in Figure 4.7. In this type of pump the power end and liquid end areas are approximately the same. This results in the pump being capable of pumping against pressures no greater than that of the motive fluid. This pump has limited use in the refining and petrochemical industry and is used primarily for metering services.

Metering pumps

A diaphragm type metering pump or 'proportioning' pump is shown in Figure 4.8.

This type of pump is most commonly used for chemical injection service when it is required to precisely control the amount of chemical or inhibitor being injected into a flowing process stream. Volume control is provided by varying the effective stroke length. There are two basic types of metering pumps:

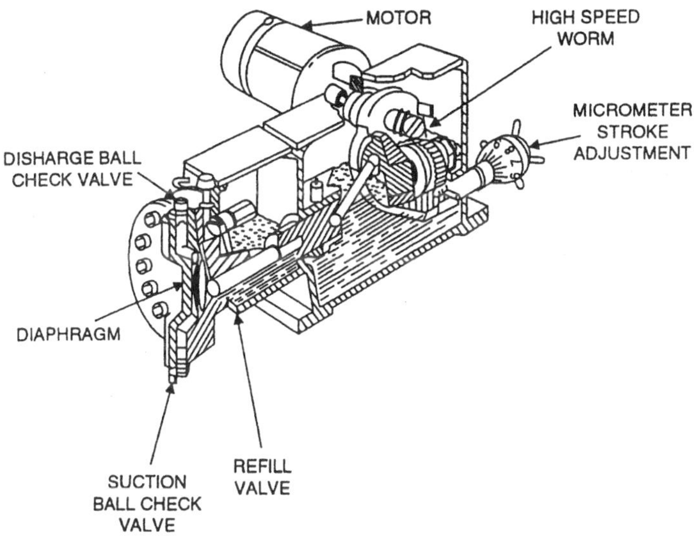

Figure 4.8 Diaphragm type of metering pump

1 Packed plunger pump – the process fluid is in direct contact with the plunger and is used for higher flow applications.

2. Diaphragm pump – process fluid is isolated from the plunger by means of a hydraulically actuated flat or shaped diaphragm and is used for lower flow applications or where escape of the pumped liquid to atmosphere is not acceptable.

Metering pumps can be furnished with either single or multiple pumping elements.

When the pumped liquid is toxic or flammable, diaphragm pumps can be provided with double diaphragms with a leak detector to alarm on failure of either diaphragm. The American Petroleum Institute has published standard 675 which covers the minimum requirements for controlled volume pumps for use in refinery service.

Rotary pumps

There are a number of different types of pumps which are classified as 'rotaries'. Rotary pumps are positive displacement pumps that do not cause pulsation. The inherent high efficiency and versatility of the 'rotary' (screw, gear and others) makes this design very suitable for use in lube oil, seal oil and other high viscosity oil services. They can handle capacities from a fraction of a gallon to more than 5,000 GPM, with pressures ranging up through 5,000 PSI when pumping liquids with viscosities from les than one (1) centistoke to more than 1,000,000 SSU.

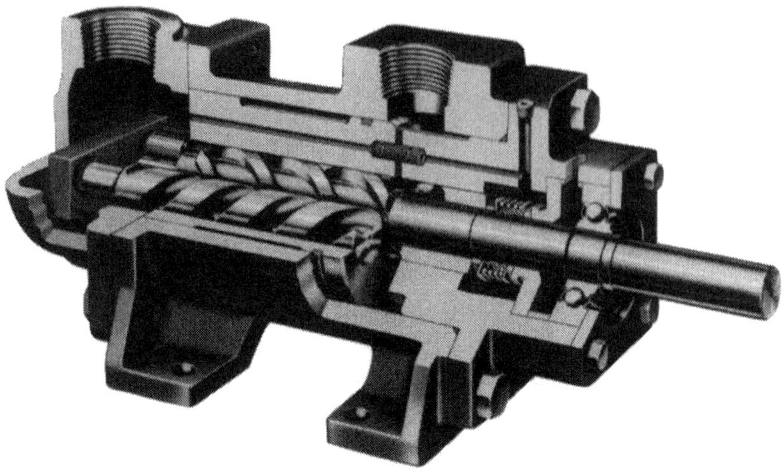

Figure 4.9 Screw pump (Courtesy of IMO Industries)

Screw pumps

Figure 4.9 illustrates the screw pump design. Fluid flow is carried axially between the threads of two or more close clearance rotors so that a fixed volume of fluid is displaced with each revolution. This design is frequently used for lube and seal service.

Gear pumps

A picture of a commonly used gear pump is shown in Figure 4.10. With this type of pump, fluid is carried between the teeth of two external gears and displaced as they mesh. Gear pumps are used for small volume lube oil services and liquids of very high viscosity (asphalt, polyethylene, etc).

Figure 4.10 Gear pump (Courtesy of IMO Industries)

Dynamic pumps

Centrifugal pumps can be referred to as 'dynamic' machines. That is to say they use centrifugal force for pumping liquids from one level of pressure to a higher level of pressure. Liquid enters the center of the rotating impeller, which imparts energy to the liquid. Centrifugal force then discharges the liquid through a volute as shown in Figure 4.11.

The centrifugal pump is one of the most widely used fluid handling devices in the refining and petrochemical industry. Every plant has a multitude of these types of pumps operating. A brief description of the various designs found in operating plants follows. Refer to Table 4.2 for the application limits of dynamic pumps.

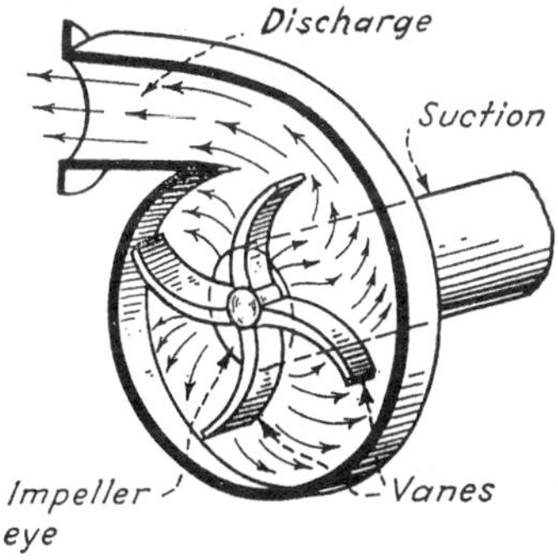

Figure 4.11 Dynamic pump principle

Table 4.2 Application limits – Dynamic pumps

Pump type	Pressure PSIA	Head (FT)	Flowrate-GPM	Horsepower-BHP
■ Single stage overhung	300	800	7,000	2,000
■ Single stage double Flow between bearing	300	800	>70,000	>8,000
■ Single stage inline	300	800	7,000	200
■ Integral gear centrifugal	1,000	2,500	2,000	400
■ Multistage horizontal Split	2,000	6,000	3,000	500
■ Multistage barrel	3,000	8,000	2,000	>5,000
■ Vertical canned pump	1,500	4,000	>70,000	1,000
■ Sump pumps	100	300	7,000	250
■ Submersible	100	300	4,000	150
■ Magnetic drive pump	300	800	3,000	800

Single stage overhung pump

The single stage overhung pump design shown in Figure 4.12 is probably the most widely used in the industry. Its construction incorporates an impeller affixed to a shaft which has its center of gravity located outside the bearing support system.

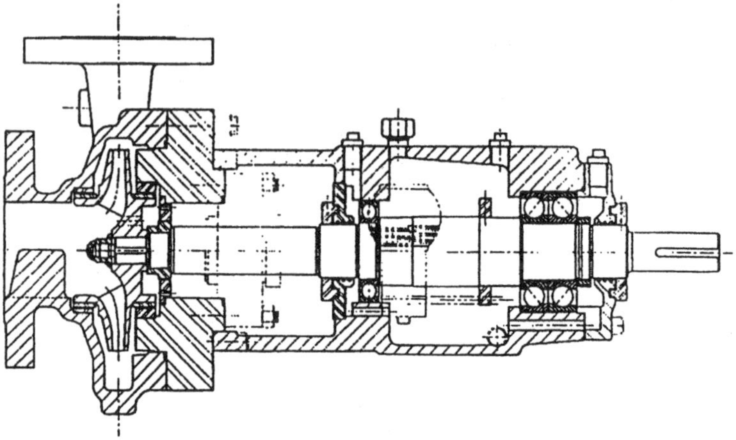

Figure 4.12 Single stage overhung pump (Courtesy of Union Pump Co)

Single stage inline

This type of pump is finding increased usage in applications of low head, flow and horsepower. Refer to Figure 4.13.

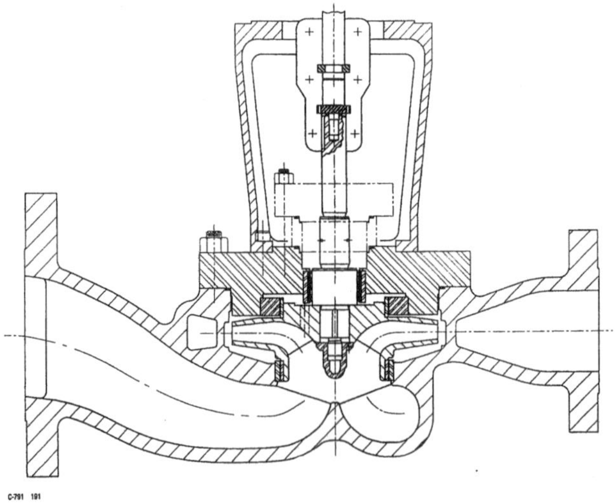

Figure 4.13 Single stage inline (Courtesy of Union Pump Co)

The advantage of this pump design is that it can be mounted vertically (inline) between pipe flanges and does not require a baseplate. A concrete, grouted support plate however, is strongly recommended. It should be noted that many inline designs do not incorporate bearings in the pump and rely on a rigid coupling to maintain pump and motor shaft alignment. Acceptable pump shaft assembled runout with these types of pumps should be limited to 0.001 ".

Integral gear centrifugal

This type of pump is used for low flow applications requiring high head. Refer to Figure 4.14. The pump case design is similar to the inline, but incorporates pump bearings and an integral gear to increase impeller speeds over 30,000.

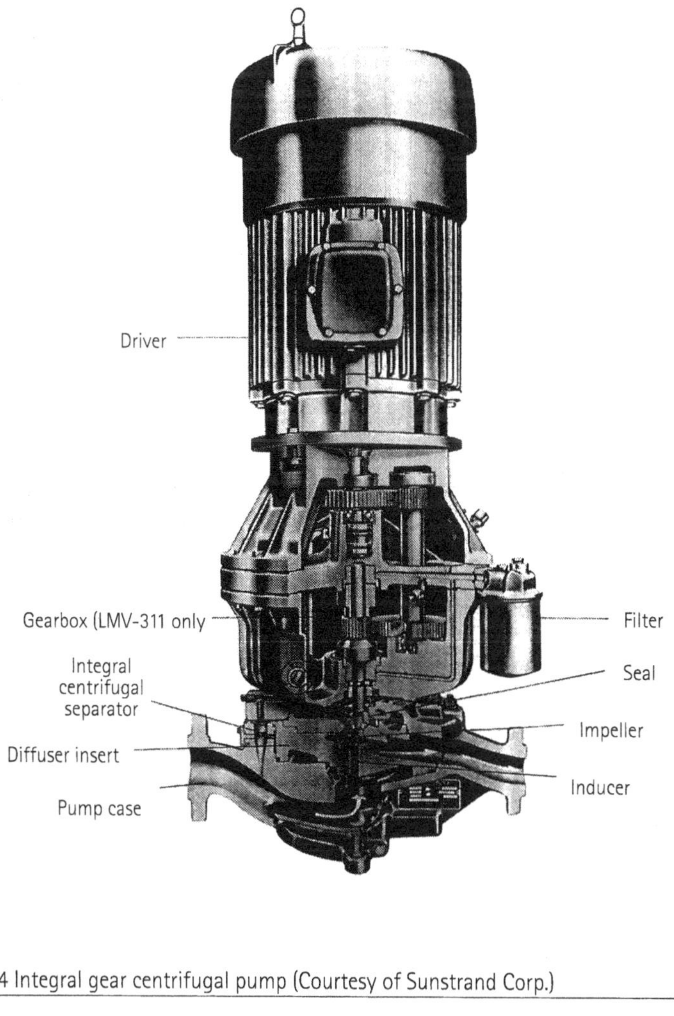

Figure 4.14 Integral gear centrifugal pump (Courtesy of Sunstrand Corp.)

Single stage double flow between bearing

As the name implies, double suction impellers are mounted on between-bearing rotors as shown in Figure 4.15.

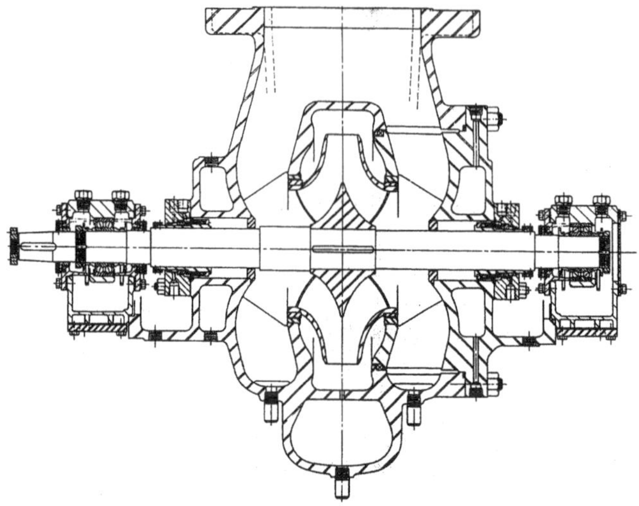

Figure 4.15 Single stage double flow between bearing (Courtesy of Union Pump Co)

This pump design is commonly used when flow and head requirements make it necessary to yield low values of NPSH required. When designing piping systems for this type of pump, care must be taken to assure equal flow distribution to each end of the impeller to prevent cavitation and vibration.

Multistage horizontal split

When the hydraulic limits of a single stage pump are exceeded, it is common practice to use a multistage pump shown in Figure 4.16.

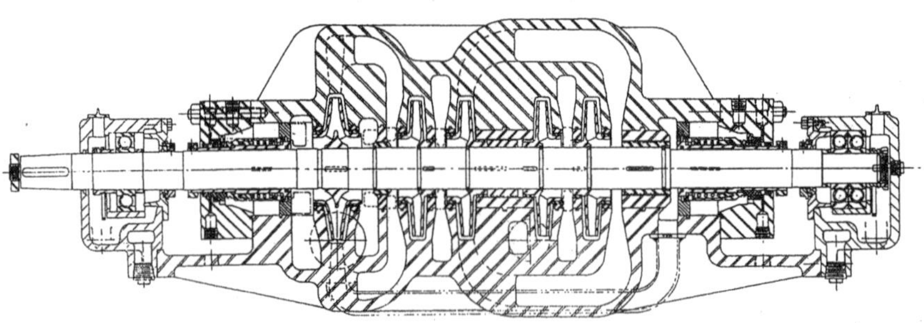

Figure 4.16 Multistage centifugal pump (Courtesy of Union Pump Co)

This figure illustrates a horizontal split casing design which allows the rotor to be removed vertically after the top half casing is unbolted. This type of pump is normally limited to working pressure of approximately 2000 PSI, temperatures to 600°F and S.G. of 0.7 or greater. Impeller configuration for this type of pump can be either 'inline' or 'opposed'. The 'opposed' impeller arrangement has the advantage of not requiring a thrust balancing device which is required for the 'inline' configuration.

Multistage barrel

The so called barrel casing design is shown in Figure 4.17.

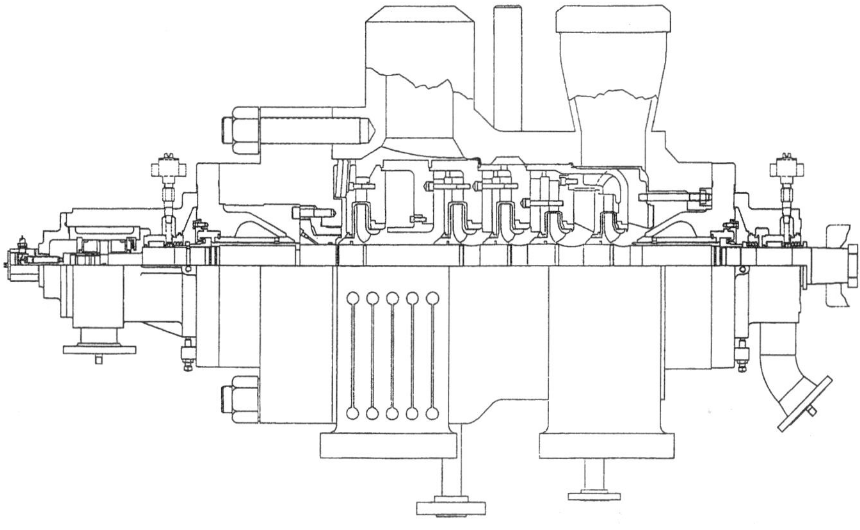

Figure 4.17 Multistage barrel (Courtesy of Demag Delaval)

It is used for service conditions exceeding those normally considered acceptable for a horizontal split case design. A thrust balance device is required since the impeller configuration is almost always 'inline'. The circular mounted end flange results in excellent repeatability for a tight joint as compared to a horizontal split case design.

Vertical canned pump

There are a variety of vertical pump types and designs. Use of the vertical pump is usually dictated by low available NPSH. To provide adequate NPSH characteristics, the first stage impeller can be lowered. The can-type design shown in Figure 4.18 is one that is widely used in the Petrochemical industry, particularly when pumping low specific

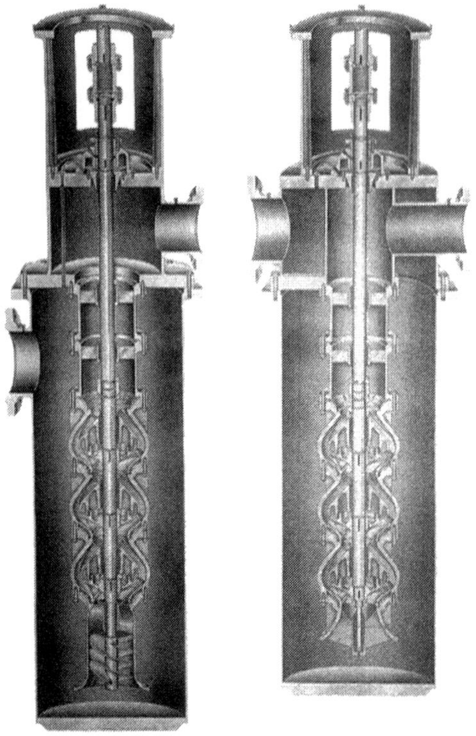

LEFT Figure 4.18 Vertical multistage pump (Courtesy of Dresser-Pacific Pumps)

gravity liquids from tank farm facilities. The multistage can pump is comprised of a number of bowl assemblies all contained within a can. This reduces the risk of hydrocarbon leakage to atmosphere. Lubrication to the sleeve bearings located throughout the length of the shaft is provided by the pumped liquid. Therefore, bearing material vs product compatibility must always be considered. It is necessary however to prevent leakage where the shaft passes through the can to connect to the driver. Sealing is normally provided using a mechanical seal.

Sump pump

The sump pump illustrated in Figure 4.19 is a popular design for handling run off streams of rain water, non corrosive or corrosive liquids. The setting limitation for this cantilever design is approximately 10 ft. This particular design incorporates an enclosed lineshaft with external lubrication to the bottom bearing. The pump shaft and impeller are coupled to the driver which is supported by a motor support bracket bolted to a cover plate.

Submersible pumps

This type of pump consists of an electric motor driver is coupled directly to the impeller/bowl assembly (see Figure 4.20). All components are designed to be submerged in the pumped fluid. In the past, this type of pump did not find widespread use in the refining and petrochemical industry.

However, with increasing environmental restrictions, this type of pump is being used more frequently in the refining and petrochemical industry.

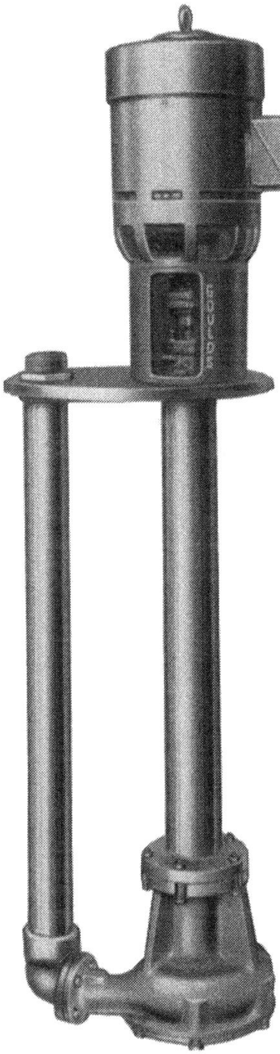

ABOVE Figure 4.20 Submersible Pump
(Courtesy of Goulds pumps)

LEFT Figure 4.19 Cantilever design sump pump
(Courtesy of Goulds pumps)

Magnetic drive pumps

As a result of more stringent environmental constraints and regulations, sealless pump technology has gained prominence. One such design is the magnetic drive pump (MDP), shown in Figure 4.21. This is a design such that the motor shaft is attached to the power frame of the magnetic drive pump by means of a flexible or rigid coupling. The outer magnet and shaft assembly is supported by its own bearings. Alignment requirements for this type pump are similar to that for horizontal mounted centrifugal pumps fitted with mechanical seals or packing. The sealless pump is generally applied when there is a need to contain toxic or hazardous fluids.

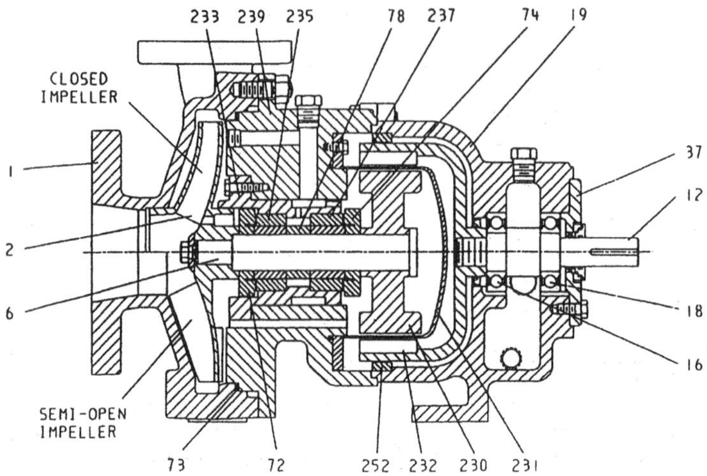

1 Casing	37 Collar, bearing outboard	232 Magnet assembly, outer
2 Impeller	72 Collar, thrust inboard	233 Housing, bearing
6 Shaft, pump	73 Gasket	235 Bushing, bearing inboard
12 Shaft, drive	74 Collar, thrust outboard	237 Bushing, bearing outboard
16 Bearing, inboard	78 Spacer, bearing	
18 Bearing, outboard	230 Magnet assembly, inner	239 Cover, casing
19 Frame	231 Shell containment	251 Ring, rub

Figure 4.21 Magnetic drive pump (Courtesy of Hydraulic Institute)

When to use positive displacement or centrifugal pumps

Coverage chart

Selecting the class of pump to use for a specific application can be somewhat confusing to engineers at times. When deciding on the class of pump to use, consideration must be given to hydraulic conditions (head, capacity, viscosity) and the fluid to be handled (corrosive, non-corrosive). The chart in Figure 4.22 provides a guideline for the range of coverage for different classes of pumps.

Obviously, there are overlapping areas and one must be familiar with the hydraulic and mechanical characteristics of each pump to properly apply it to the system. It can be seen that rotary and centrifugal pumps overlap in capacity range of 5,000 GPM and pressure range of 500 PSI. Centrifugal pumps can handle liquid viscosities up to 3,000 SSU, while rotary type pumps normally move materials from 60 SSU to millions of SSU.

PD pumps vs centrifugal pumps – advantages/disadvantages of each

The question of why use a PD pump in place of a centrifugal or vice

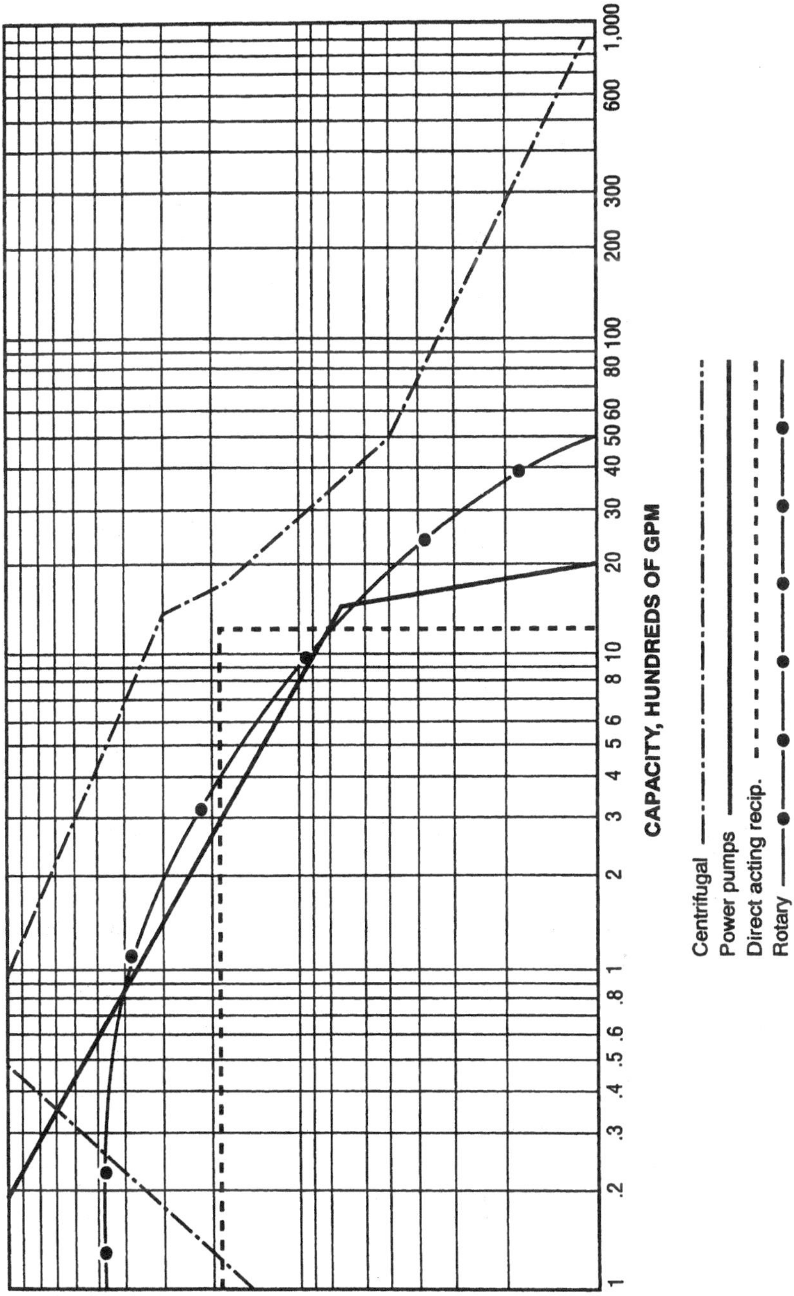

Figure 4.22 Operating range of typical pump types

versa, is often raised. Table 4.3 identifies some advantages and disadvantages associated with each class or equipment which may help make the decision less difficult.

Table 4.3 Advantages/disadvantages of PD pumps vs centrifugal pumps

Advantages	Disadvantages
Centrifugal pumps	
■ Variable capacity control over operating range at constant speed ■ Can handle liquids containing catalyst, dirt solids ■ Can pump liquids with poor lubricity ■ Weight, size, initial cost and installed cost is lower than PD pump with same hydraulic conditions ■ Liberal clearances, no rubbing parts, minimum wear – higher availability ■ Does not normally require overpressure protection over operating range at constant speed	■ Flow rate is effected by specific gravity ■ Viscosity affects performance ■ Requires priming ■ Develops limited head over operating range at constant speed ■ Low to moderate efficiencies
Positive displacement	
■ Not limited to delivery pressure for given capacity ■ Can handle high viscosity liquids efficiently ■ Higher efficiencies than centrifugal ■ Flow rate is not significantly affected by specific gravity	■ Requires over-pressure protection ■ Flow controls with bypass or speed ■ Pulsations associated with reciprocating PD pumps

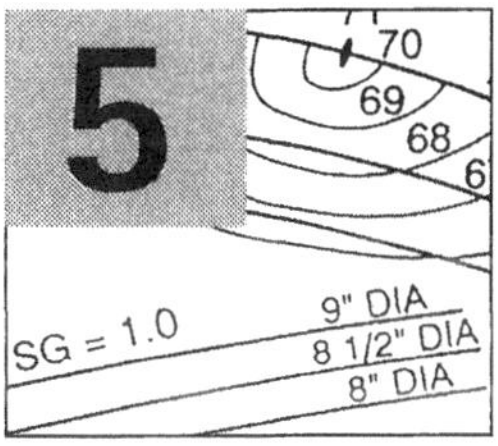

Pump performance data

- The pump curve
- The limits of the centrifugal pump curve
- Increasing head produced by a centrifugal pump

The pump curve

The pump characteristic curve defines the signature of a pump over its entire life. It identifies the range of flows and energy produced for a fixed speed, size, design and suction conditions of the pump. In order for a pump to move a liquid from a level of low pressure to a level of higher pressure, the head (energy) produced by the pump must equal or exceed the net head (energy) required by the system. Also, the net head (energy) available on the suction side of the pump system must be greater than the liquid vapor pressure to assure that liquid enters the pump without potential deterioration of performance or mechanical damage.

The positive displacement curve

Figure 5.1 illustrates the characteristics which defines the performance curve for a rotary type positive displacement pump.

- Capacity, GPM = Displacement – Slip
- Pressure, PSI
- Brake HP = Friction HP + Hydraulic HP
- EFF $= \dfrac{\text{GPM x PSID}}{\text{1714 x BHP}}$

$$= \frac{\text{GPM x HEAD x S.G.}}{\text{3960 x BHP}}$$

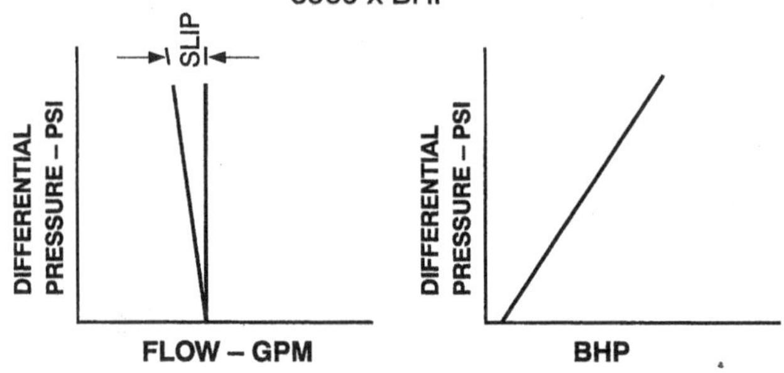

Figure 5.1 The positive displacement pump performance

Head

As shown in Figure 5.1, differential pressure is usually plotted vs flow instead of head to express energy produced by positive displacement pumps. Since the head (energy) produced is infinite for all types of P.D. pumps, produced differential pressure is not affected by specific gravity as in the case of centrifugal pumps, therefore differential pressure can be used to express the head (energy) produced by P.D. pumps.

Flow

As shown in Figure 5.1, flow reduces slightly with increased energy produced. This is a result of increased pump slip.

Efficiency

Positive displacement pump efficiencies are expressed as volumetric efficiency which is defined as:

$$\text{V.E.} = \frac{\text{actual capacity}}{\text{displacement}}$$

The displacement is defined as the capacity the pump would produce if there were no (slip) leakage losses. P.D. pump efficiencies vary from:

- 70% – 80% for rotary pumps
- 75% – 85% for plunger and reciprocating pumps

Horsepower

As shown in Figure 5.1 horsepower is directly proportional to pump differential pressure for a given liquid.

NPSH required

To overcome the continuous action of pulsating flow, which occurs in reciprocating type P.D. pumps, the acceleration energy must also be accounted for in the system to assure that the liquid to be pumped does not vaporize prior to entering the pump. It should be noted that "acceleration head loss" is the term used to express this energy. Refer to Figure 5.2 for the equations used to calculate the NPSH available for a system with and without pulsating flow characteristics.

NPSH available calculation for a PD pump

$$NPSH_{AVAIL} = (P_{S1} + P_A)\frac{2.31}{S.G.} + H_{Z1} - H_F - (P_V)\frac{2.31}{S.G.} \text{ (non pulsating flow)}$$

$$NPSH_{AVAIL} = (P_{S1} + P_A)\frac{2.31}{S.G.} + H - H_F - H_{LA} - (P_V)\frac{2.31}{S.G.} \text{ (pulsating flow)}$$

P_{S1} = suction press. PSIG

P_A = atmos. press. PSIG

S.G. = specific gravity

P_V = vapor pressure PSIA

H_{Z1} = suction side elevation change

H_F = suction side system friction (expressed in FT.)

H_{LA} = acceleration head loss (magnitude depends on pump type)

Figure 5.2 NPSH available calculation for a PD pump

The centrifugal curve

When a centrifugal pump is designed, its performance characteristics are defined for a range of flows and head (energy) produced for fixed impeller geometry and a variety of impeller diameters. Figures 5.3 and 5.4 show the relationships between the head (energy) produced and the flow through a single stage centrifugal pump. By the affinity laws, pumps of the same type, fitted with impellers of similar design, will have characteristics curves of the same shape.

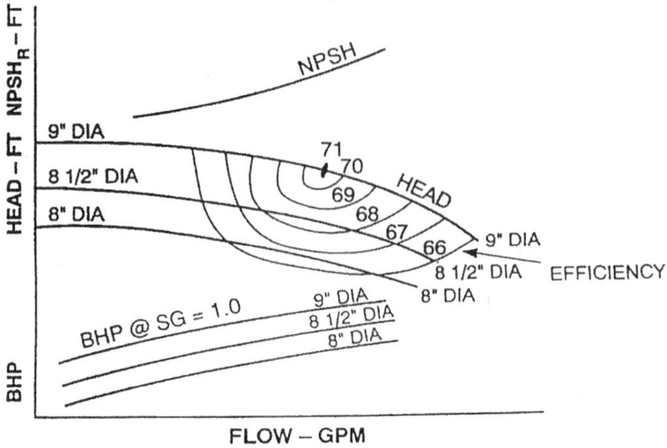

Figure 5.3 A typical centrifugal pump performance curve

Centrifugal pump performance definitions

$H = \frac{V^2}{2g}$ H = energy expressed in feet of liquid

V = velocity in feet per second

g = acceleration due to gravity in FT/SEC^2

$BHP = \frac{GPM \times H \times S.G.}{3960 \times EFF}$

EFF ratio of power output to power input

NPSHR energy which must be exceeded to avoid vaporization of liquid in pump suction passage

Figure 5.4 Centrifugal pump performance definitions

Head produced

The head produced by a centrifugal pump varies inversely with the flow rate. The curve head rise is a function of the impeller inlet and discharge blade angles. Typical centrifugal pump head rise values from design point to shutoff are 5% – 15%.

When the head required by the process exceeds the head produced by a single stage centrifugal pump, multistaging is used to produce the energy required by the system. Multistaging is nothing more than two or more impellers acting in series within a single casing to produce the total head (energy) required. It is common practice for each impeller to produce an equal amount of energy (refer to Figure 5.5).

Example – multistaging

- total net system energy (ft. head) – 1200 ft.
- number of impellers selected – 4
- energy (ft. head) produced for impeller – $\frac{1200}{4}$ = 300 ft/impeller

Figure 5.5 Example – multistaging

Flow

The flow rate of a centrifugal pump varies inversely with the head (energy) required by the process. For a given impeller design operating at a constant speed, increased process head requirements will reduce centrifugal pump flow rates. Since the typical head rise values for centrifugal pumps are 5% – 15%, a relatively small change in process head requirements can result in significant flow reductions and possible impeller recirculation on operation near zero flow (shutoff).

Efficiency

The pump efficiency is maximum at the pump design point using the maximum diameter impeller. Refer to Figure 5.3. The pump design point, often referred to as the B.E.P. – best efficiency point, is the flow where minimum losses occur in the pump stationary passages and the impeller. At off design flows, separation losses (low flows) and turbulence losses (high flows) increase internal produced head losses and reduce pump efficiency.

Horsepower

Horsepower required by a centrifugal pump varies directly with the specific gravity of the pumped liquid. Horsepower is the only parameter on a typical centrifugal pump curve that is effected by the specific gravity of the pumped liquid. Most pump curves present the horsepower curve based on water S.G. = 1.0. For pumped liquids of any other S.G. value, the horsepower on the pump curve must be multiplied by the actual specific gravity.

$NPSH_R$

The net positive suction head required by a centrifugal pump varies approximately with the flow rate squared since it is a measure of the pressure drop from the pump suction flange to the eye of the first impeller.

The $NPSH_R$ is also influenced by the pump rotational speed and varies somewhat less than the rotational speed squared.

The limits of the centrifugal pump curve

The centrifugal pump curve has high flow and low flow limits which can cause significant mechanical damage to the pump if not avoided. At the low flow end of the curve, flow recirculation can damage a pump while at the high flow end excessive $NPSH_{REQUIRED}$, horsepower and choke flow can result in mechanical damage to impellers, casing, shaft, bearings and seals. Each of these factors is discussed below.

Low flow operation

As we examine these factors we can see that oversizing a centrifugal pump will result in low flows through the impeller. A portion of the flow will reverse itself and set up turbulence as it reenters the impeller. The abrupt change in direction and very high acceleration can result in cavitation on the back side of the impeller vane (refer to Figure 5.6).

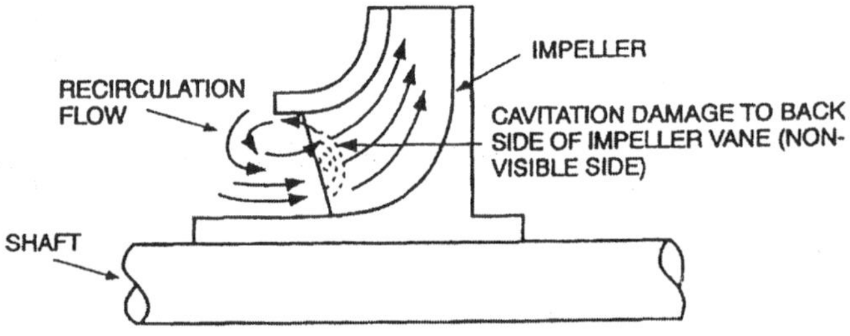

Figure 5.6 Recirculation flow pattern in impeller at low flows

As a result of oversizing an impeller, significant effects in performance and mechanical reliability can be experienced. These are outlined in Figure 5.7.

Effects of pump oversizing

Operation at low flows can result in:

- Internal recirculation damage to impeller
- Operation at less than best efficiency point
- High radial loads
- Bearing failures
- Seal failures
- High internal temperature rise and requirement for minimum flow bypass

Figure 5.7 Effects of pump oversizing

Pumps are designed to operate at minimum radial thrust loads at best efficiency point. Low flow operation results in high radial loads which can cause premature bearing failures unless bearings are selected to accept these higher loads in anticipation of operation at low flows. Pressure surges and flashing of the liquid can also occur at low flows. This can cause loading and unloading of the mechanical seal faces which can result in a seal failure. Depending on the fluid being pumped, low flow operation can result in a high temperature rise through the pump because the amount of energy absorbed by the liquid is low compared to that absorbed by friction losses. Refer to Figure 5.8 for the method to calculate the temperature rise through a pump.

Temperature rise through a pump

$$\text{RISE, DEG F} = \frac{H}{778 \times C_P}\left[\frac{1}{\text{EFF'Y}} - 1\right]$$

H = head in feet at pumping rate

eff'y = efficiency at pumping rate

CP = specific heat, BTU/LB–°F

778 = FT. LB/BTU

Figure 5.8 Temperature rise through a pump

The above relationship can also be used to determine the approximate flow rate of any centrifugal pump. By measuring the pipe temperature rise across any pump, the efficiency of the pump can be determined. Referring to the particular pump shop test curve for the calculated efficiency will allow the approximate pump flow rate to be determined. Note: This approach assumes the pump is in new condition. A worn condition will reduce the flow to a greater extent.

High flow operation

Selecting a pump to operate far to the right of best efficiency point can also result in potential pump problems as highlighted in Figure 5.9.

Effects of pump operation at high flows

Operation at high flows can result in:

- High to overloading horsepower with reduced system resistance
- Operation in the "break" of head capacity curve (significant changes in head with no change in flows)
- Higher NPSH required than available
- Recirculation cavitation at impeller tips

Figure 5.9 Effects of pump operation at high flows

Pump curve shapes

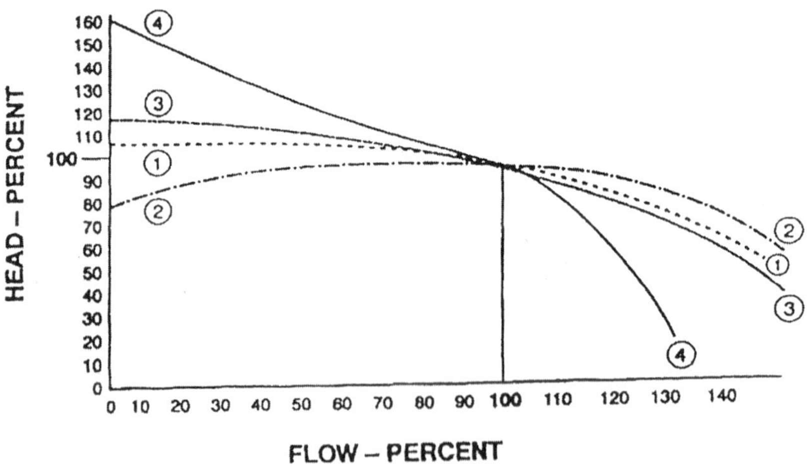

Figure 5.10 Centrifugal pump performance curve shapes

The various types of characteristic curves normally associated with centrifugal pumps include flat, drooping, rising, stable and unstable depending upon their shape. Figure 5.10 illustrates the different curve shapes and Figure 5.11 defines each type. The pump curve shape can play a significant role in determining if stable operation in a given process system is possible. Flat or drooping head characteristic curves (Figure 5.10 – curves 1 & 2) can result in unstable operation (varying flow rates). Pumps should be selected with a rising head curve or controlled such that they always operate in the rising region of their curve.

Definition of characteristic curve shape

Flat curves (1 and 2), show little variation in head for all flows between design point and shut-off

Drooping curve (2), head at shut-off is less than head developed at some flows between design point to shut-off

Rising curve (1,3,4), head rises continuously as flow decreases from design point to shut-off

Steep curve (4), large increase between head developed at design flow and that developed at shut-off

Stable curve (1,3,4), one flow rate for any one head

Unstable curve (2), same head can be developed at more than one flow rate

Figure 5.11 Definition of characteristic curve shape

Increasing head produced by a centrifugal pump

The affinity laws can be used to increase the head available from a centrifugal pump. Head produced by a centrifugal pump is a function of impeller tip speed. Since tip speed is a function of impeller diameter and rotational speed, two options are available. The characteristic curve can be affected by either a speed change or a change in impeller diameter with speed held constant. Figures 5.12 A&B show this relationship.

The affinity laws

Increasing head produced by a pump

$$H_1 = \left(\frac{D_1}{D}\right)^2 H \text{ or } H_1 = \left(\frac{N_1}{N}\right)^2 H$$

where: D_1 = diameter of changed impeller

D = diameter of existing impeller

H_1 = head in feet changed impeller

H = head in feet existing impeller

N_1 = Speed (RPM) changed condition

N = speed (RPM) existing condition

Figure 5.12A The affinity laws

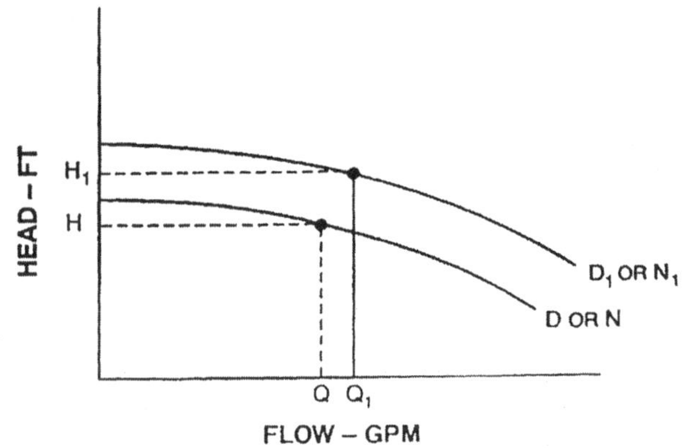

Figure 5.12B The affinity laws

The affinity laws

In actual practice, the affinity laws provide an approximation between flow, head and horsepower as pump impeller diameter or speed is varied. The actual values vary somewhat less than predicted by the affinity laws. That is, the actual exponents in the affinity equations are slightly less than their stated values and are different for each pump. This fact is a result of friction in hydraulic passages and impellers, leakage losses and variation of impeller discharge vane angles when diameters are changed. Pump manufacturers should be contacted to confirm actual impeller diameters and speed changes to meet new duty requirements.

Centrifugal pump hydraulic disturbances

- Introduction
- Maintaining a liquid inside a pump
- Causes of damage
- Preventing hydraulic disturbances
- The project design phase
- Field operation

Introduction

Liquid disturbances in centrifugal pumps are the major cause of reduction in pump reliability. The pressures generated by cavitation can exceed 100,000 PSI. Cavitation caused by many different factors is responsible for pump lost service time as a result of various pump component failures shown in Figure 6.1

Cavitation caused pump component failures

- Bearings
- Seals
- Impellers
- Shaft
- Wear rings

Figure 6.1 Cavitation caused pump component failures

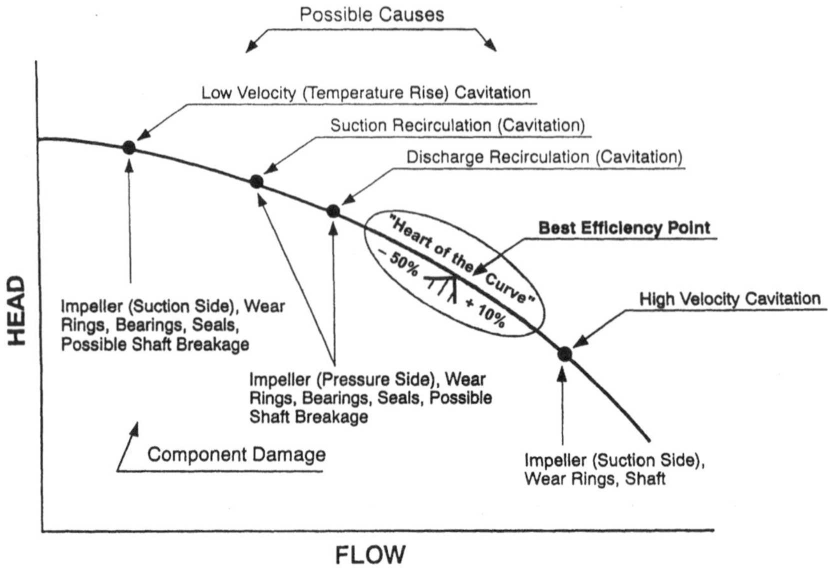

Figure 6.2 Centrifugal pump component damage and causes as a function of operating point

In the previous chapter we discussed proper pump selection guidelines that require a centrifugal pump to be selected in the 'heart of the curve' and that sufficient NPSH available be present to avoid mechanical damage (See Figure 6.2). The objective of the this chapter is to explain the reason for these requirements in detail and provide useful guidelines for preventing liquid disturbances in centrifugal pumps and for solving persistent, costly field pump problems caused by liquid disturbances.

Maintaining a liquid inside a pump

Design objectives

All pumps are designed to increase the energy of a fluid while, maintaining the fluid in its liquid state. Each centrifugal pump is designed to produce a specific amount of head at a specific flow rate based on a specified fluid density. Once the pump is designed, any reduction in the fluid density will result in a reduction in flow rate since the pump now requires greater head (energy) and can produce this value only at a reduced flow rate (See Figure 6.3).

One way to rapidly reduce the density of a fluid is to change its phase. If a liquid suddenly changes to a vapor, the impeller head required to meet the same process differential pressure requirements increases by a factor of 300 or more (See Figure 6.4). As a result, the pump will not be able to move the product since the maximum head produced by the

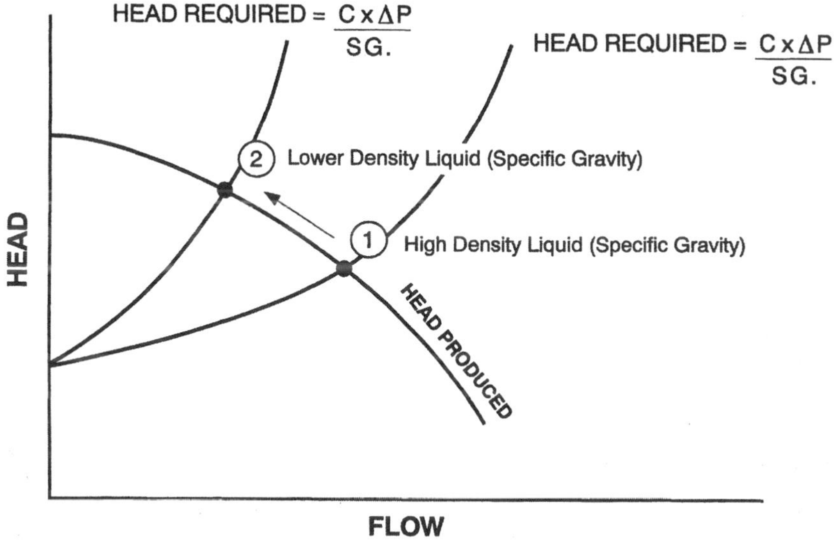

Figure 6.3 Head required by lower density fluid

pump will be much less than the head now required. The discharge pressure gauge will drop in pressure. This is commonly known as 'vapor lock' or 'loss of suction'.

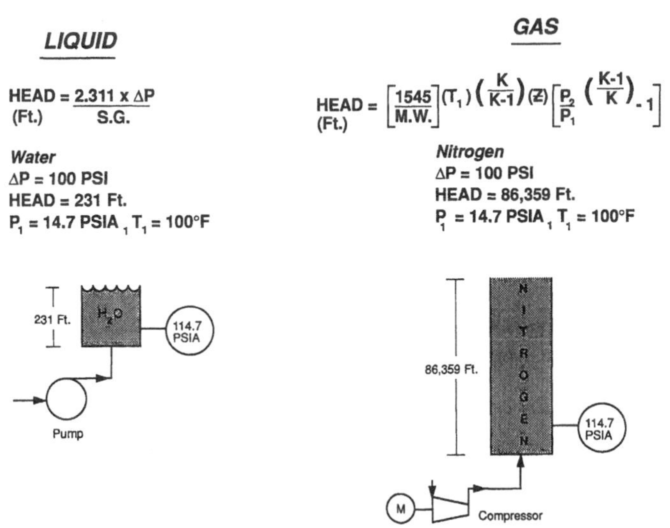

Figure 6.4 Fluid head

Vapor pressure

How can a liquid change phase inside a pump? The vapor pressure for any fluid is that pressure at which the fluid changes from its liquid to vapor phase. Vapor pressure changes with fluid temperature. The vapor pressure for any fluid can be obtained from its Mollier diagram. For a given fluid temperature, the vapor pressure is the intersection of that temperature and the saturated liquid line. Figure 6.5 is the Mollier diagram for ethylene.

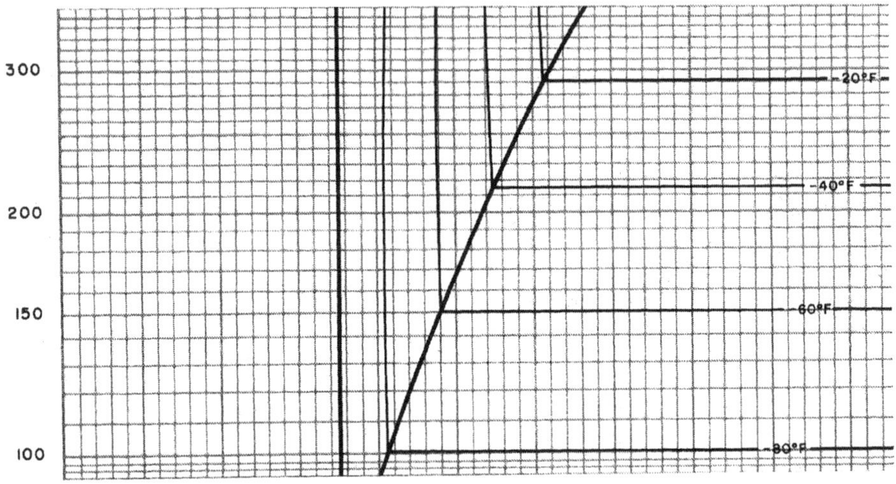

Figure 6.5 Ethelyne Mollier diagram

As can be seen from this figure, either a reduction in pressure (holding fluid temperature constant) or an increase in temperature (holding pressure constant) will cause a phase change. Another way of stating these facts is presented in Figure 6.6.

Maintaining a liquid

A fluid will remain a *liquid* as long as its vapor pressure is less than the pressure acting on the liquid

Figure 6.6 Maintaining a liquid

Before we leave this subject, let's examine how water can change phase (refer to Figure 6.7).

The most common cause of phase change for water is to increase its temperature at a constant atmospheric pressure. When the vapor pressure exceeds the pressure surrounding the liquid (14.7 PSIA) the liquid will change phase or 'boil'. However, water can also change phase 'boil' if its pressure is decreased when temperature is held constant (See Figure 6.7). When the pressure of the water is less than its vapor pressure, water vapor will form. Therefore, any action inside a pump that will either reduce the pressure of a liquid or significantly increase its temperature can cause the liquid to change to a vapor.

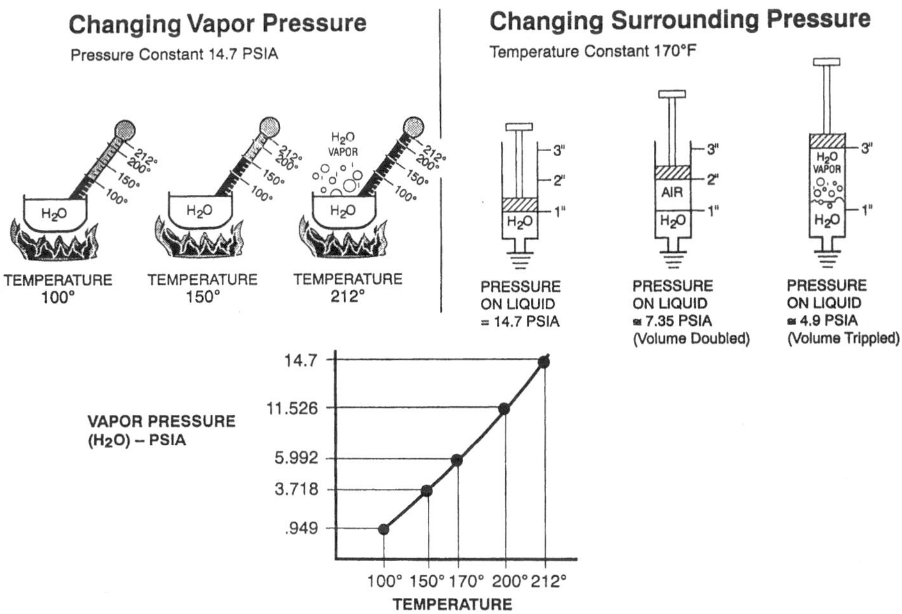

Figure 6.7 Causes of liquid phase change – H_2O

What can reduce the pressure of a liquid or significantly increase its temperature inside a pump? To answer this question, the entire cross section of a pump must be examined. In Figure 6.8, the flow path from the pump suction flange to the impeller vane is plotted vs pressure in the pump. It can be seen that the pressure decreases from the suction to the leading edge of the impeller vanes then increases rapidly once the liquid is inside the impeller vanes where vane tip speed (U) and liquid tangential velocity (V_T) increase head (energy). The total pressure reduction from the suction flange to the impeller vanes is the result of:

- Friction losses in the flow passage
- Liquid acceleration
- Entry shock losses at the impeller vane tips

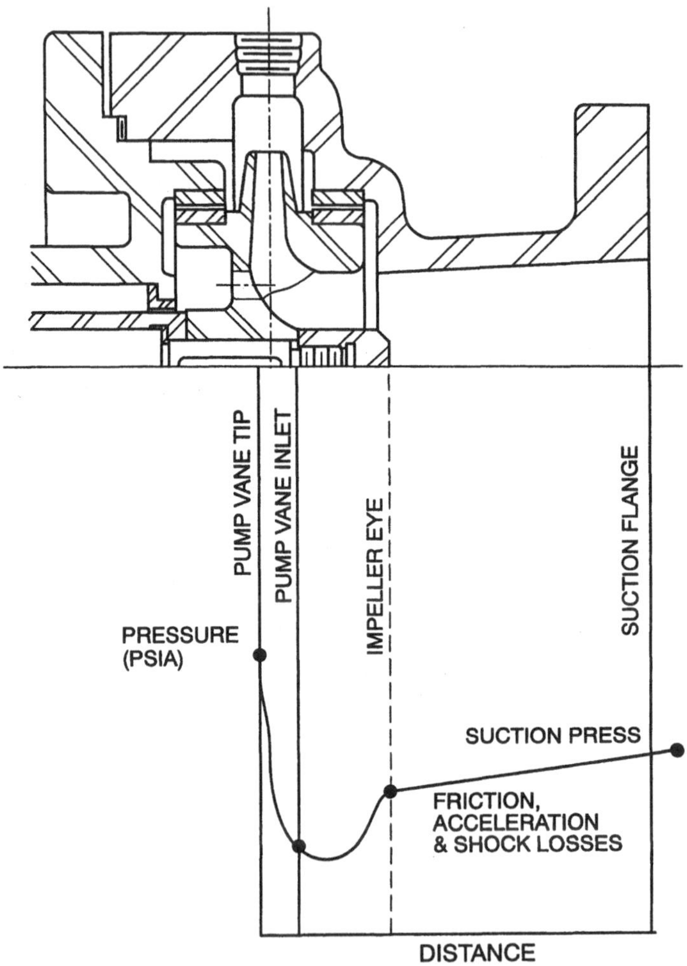

Figure 6.8 Flange to vane entrance losses

Therefore, each pump has a distinct pressure drop for a given flow which is the result of pump case, inlet volute and impeller design. If the pressure drop from the suction flange to the impeller vane reduces the pressure below the liquid's vapor pressure, vapor will be formed at the impeller vanes. There are also two (2) other causes of vapor formation within an impeller.

Low flow velocities at any location within the impeller can cause separation of flow stream lines and lead to low pressure areas (cells) inside the impeller. If the pressure within these cells is less than the fluids' vapor pressure, vapor will form (refer to Figures 6.9 and 6.10).

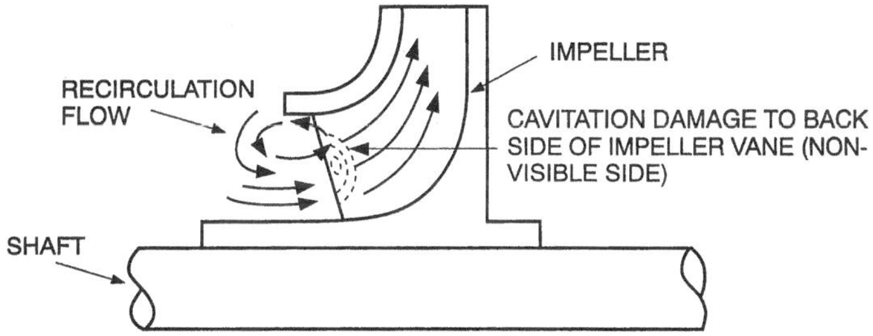

Figure 6.9 Recirculation flow pattern in impeller at low flows

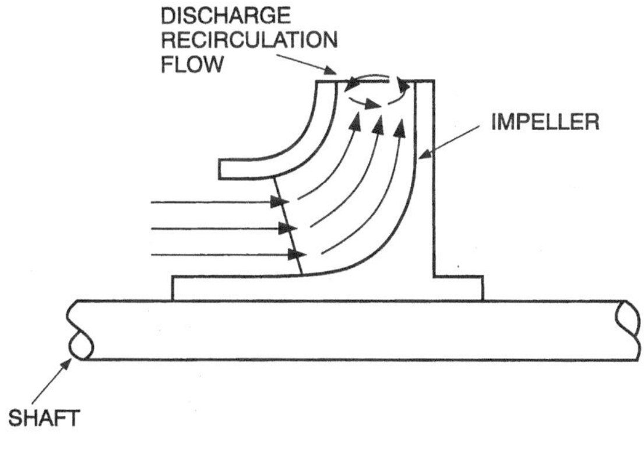

Figure 6.10 Discharge recirculation flow pattern

The curvature of the impeller vanes will always result in lower velocities on the pressure side of the vane (the side that cannot be readily observed). Therefore, vaporization caused by flow separation occurs on the pressure or 'backside' of the vane. This phenomena is commonly known as recirculation.

1. Pump efficiency is low at low flows

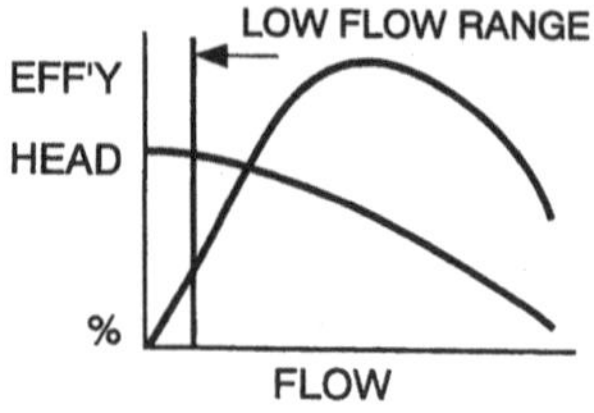

2. Liquid temperature rise increases by

$$\Delta T = \frac{\text{Pump head}}{778 \times Cp} \times \left[\frac{1}{\text{Pump efficiency}} - 1 \right]$$

Where: Pump head is calculated from data in $\frac{FT-LB_F}{LB_M}$

Cp is specific heat of the fluid in $\frac{BTU}{°F-LBmass}$

778 is conversion factor $\frac{FT-LB_F}{BTU}$

Pump efficiency is expressed as a decimal

3. If the temperature rise increases the fluid's vapor pressure above the surrounding pressure, the fluid will vaporize

Figure 6.11 Low flow temperature rise can cause vapor formation

The other cause of vapor formation within a pump is the temperature rise associated with low flows (Refer to Figure 6.11).

When operating at low flows, the impeller efficiency reduces significantly, thus increasing the liquid temperature rise. As previously mentioned, the vapor pressure of any liquid increases with temperature. Increased temperature can cause the liquid to vaporize at the impeller vanes. Referring to Figure 6.11, it can be seen that low specific gravity liquids with high vapor pressures are most susceptible to vaporization caused by low flow operation. Note that increased wear ring clearances can worsen this situation since the higher temperature liquid will mix with the cooler liquid entering the impeller. Figure 6.12 summarizes the causes of vaporization within a centrifugal pump.

Causes of vaporization within a centrifugal pump

- Internal inlet pressure losses
- Formation of low pressure cells at low flows
- Liquid temperature rise at low flows

Figure 6.12 Causes of vaporization within a centrifugal pump

Causes of damage

In the above section the causes of vapor formation within a pump were described. In this section the causes of damage to pump components will be discussed.

Cavitation

Figure 6.13 presents the definition of cavitation.

Cavitation definition

Cavitation is the result of released energy when an increase of pressure surrounding the fluid causes the saturated vapor to change back to a liquid

Figure 6.13 Cavitation definition

From the definition of above, vapor must be present before cavitation can take place. The sources of vapor formation were discussed and are summarized in Figure 6.12. Referring back to Figure 6.8, it can be seen that as soon as the liquid enters the impeller vane area, energy and pressure rapidly increase. When the pressure of the liquid exceeds the fluid's vapor pressure, the vapor bubbles will collapse and cavitation will occur. The solution to preventing cavitation is shown in Figure 6.14.

Preventing cavitation

Cavitation is prevented by preventing vapor formation within a pump

Figure 8.14 Preventing cavitation

Methods to prevent cavitation will be discussed later on in this chapter.

The effects of fluids on component damage

The energy released during cavitation caused by inlet pressure losses, recirculation or low flow temperature rise varies as a function of the fluid type and the amount of vaporization. In Figure 6.15, we have drawn a generic Mollier diagram to show that the fluid type (latent heat of vaporization for a given temperature) and degree of vaporization determine the energy released when the vapor is recompressed to a liquid. Note that the abcissa is BTU/LB. When cavitation occurs, a given amount of energy (BTU/LB) will be transferred from the fluid to the impeller. Energy can also be expressed in $FT\text{-}LB_F/LB_M$ by multiplying

$$\frac{BTU}{LB_M} \times \frac{778\ FT\text{-}LB_F}{BTU}$$

It can therefore be seen that in cavitation the energy transferred by the fluid to the vanes can greatly exceed the head produced by the impeller. Refer back to Figure 6.5 and determine the maximum amount of energy transferred to the impeller vanes when operating at 20 PSIG.

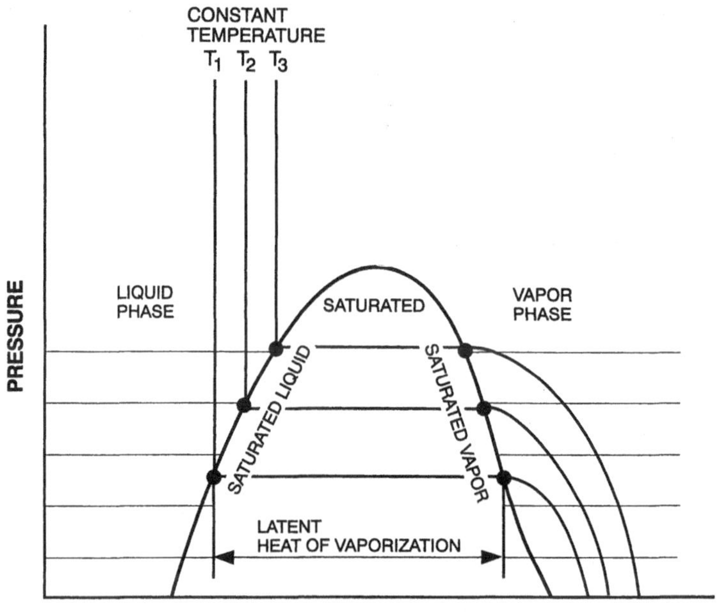

Figure 6.15 Mollier diagram

In general, single component liquids produce the highest energy values during cavitation and are therefore the most damaging fluids. Hydrocarbon mixtures produce lower energy values and have higher viscosities which reduce damage. Regardless of composition, all liquids produce high noise levels during cavitation which typically sounds like solids 'rocks' are passing through the pump. Always remember that there are different causes of vaporization which result in cavitation.

Preventing hydraulic disturbances

In the previous section, the causes of liquid disturbances in centrifugal pumps were discussed. In this section, practical advice on how to prevent purchasing troublesome pumps in the design phase and practical solutions on how to solve existing field problems caused by liquid disturbances will be presented.

The project design phase

Action taken during the early stages of a project can significantly increase pump reliability and safety by eliminating all sources of vaporization within a pump. Sources of vaporization exist both in the process and in the pump. Before presenting solutions, a number of important concepts must be reviewed and presented.

Concepts

Specific speed

Specific speed is a non-dimensional value that is a function of pump speed, flow and head

$$N_S = \frac{N\sqrt{Q}}{HD^{3/4}}$$

Where: N_S = specific speed
N = pump speed rpm
Q = pump flow gpm
Hd = pump produced head $\frac{FT/LB_F}{LB_M}$

Note: For double suction impellers, Q = Q/2

Specific speed is used extensively in both pump and compressor design to optimize stage efficiency for a given value of flow and head required. In pump design, specific speed is used to optimize the following design parameters:

- Impeller discharge flow velocity
- Impeller tip speed
- Impeller inlet and discharge blade angles
- Discharge throat velocity

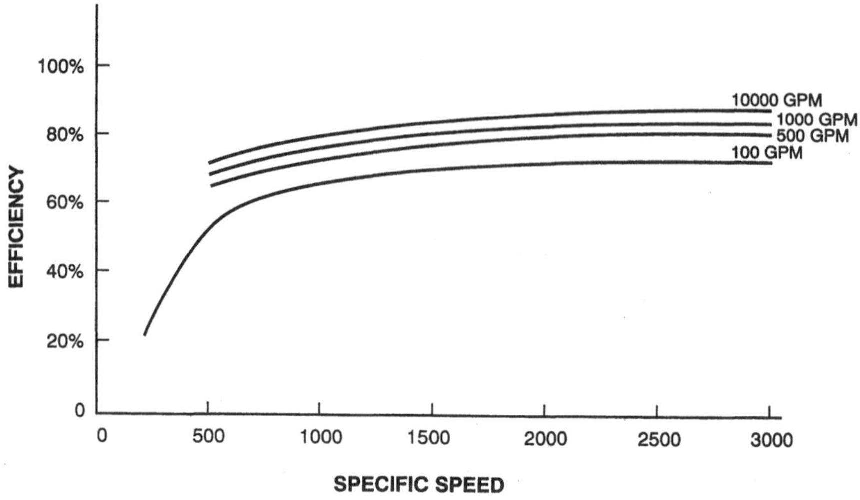

Figure 6.16 Efficiency as a function of pump specific speed and flow

Figure 6.16 presents a plot of pump specific speed vs efficiency for various flow rates. This is a typical chart and will vary slightly from pump vendor to pump vendor based on the specifics of pump design. It can be used for estimating purposes in determining pump efficiency to obtain pump required horsepower.

Net positive suction head available

$NPSH_A$ has been previously discussed. As we have learned, it must be greater than the $NPSH_R$ to prevent cavitation. Methods for determining $NPSH_A$ have been presented.

Net Positive Suction Head Required

Figure 6.17 shows the NPSH required within a typical centrifugal pump. It can be seen that this value is actually the pressure drop from the suction flange to the impeller vane inlet expressed in energy terms $\frac{(FT/LB_F)}{LB_M}$.

Perhaps now, we can truly understand why $NPSH_{AVAILABLE}$ must be $\geq NPSH_{REQUIRED}$. As we have learned,

- *If* $NPSH_A \geq NPSH_R$
- *Then*, the Fluid will not Vaporize
- *Therefore*, no vaporization, no cavitation, no damage

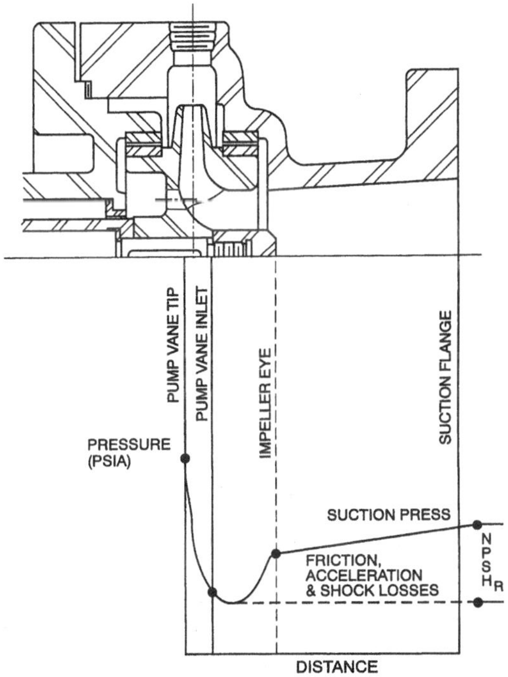

Figure 6.17 Flange to vane entrance losses

However, it must be remembered that $NPSH_A \geq NPSH_R$ is only one of the requirements that must be met to prevent vaporization. The following causes of vaporization must also be prevented:

- Low velocity stall
- Low flow temperature rise

Low flow temperature rise and fluid vaporization can be determined by the relationship shown in Figure 6.11 and the pumped fluid characteristics. However, the determination of low velocity stall or recirculation requires the understanding of the concept of suction specific speed.

Suction specific speed

N_{ss}, known as suction specific speed is determined by the same equation used for specific speed N_s but substitutes $NPSH_R$ for H (pump head). As the name implies, N_{ss} considers the inlet of the impeller and is related to the impeller inlet velocity. The relationship for N_{ss} is:

$$N_{SS} = \frac{N\sqrt{Q}}{(NPSH_R)^{3/4}}$$

Where: N = speed

Q = flow-GPM

$NPSH_R$ = Net Positive Suction Head Required

As previously explained, $NPSH_R$ is related to the pressure drop from the inlet flange to the impeller. The higher the $NPSH_R$, the greater the pressure drop. The lower the $NPSH_R$, the less the pressure drop. From the equation above, we can show the relationships between $NPSH_R$, N_{SS}, inlet velocity, inlet pressure drop and the probability of flow separation in Figure 6.18.

NSS related to flow separation probability

N_{SS}	$NPSH_R$	Inlet velocity	Inlet passageΔP	Probability of flow separation
14,000 (High)	Low	Low	Low	High probability
8,000 (Low)	High	High	High	Low probability

Figure 6.18 N_{SS} related to flow separation probability

Based on the information presented in Figure 6.18, it can be seen that flow separation will occur for high specific speeds resulting from low inlet velocity. The critical question the pump user needs answered is 'At what flow does the disturbance and resulting cavitation occur'? This is not an easy answer because the unstable flow range is a function of the impeller inlet design as well as the inlet velocity. A general answer to this question is shown in Figure 6.19.

Recirculation as a function of N_{SS}

The onset flow of recirculation increases with increasing suction specific speed

Figure 6.19 Recirculation as a function of N_{SS}

Another way of describing the statement in Figure 6.19 is 'The higher the value of N_{SS}, the sooner the pump will cavitate when operating at flows below the BEP'. Therefore, before an acceptable value of N_{SS} can be determined, the process system and pumped liquid characteristics must be defined.

Defining the process system

Reviewing the proposed process system prior to the purchase of a pump, as previously discussed, is strongly recommended. Figure 6.20 presents a typical process system with various control alternatives (flow, level, pressure).

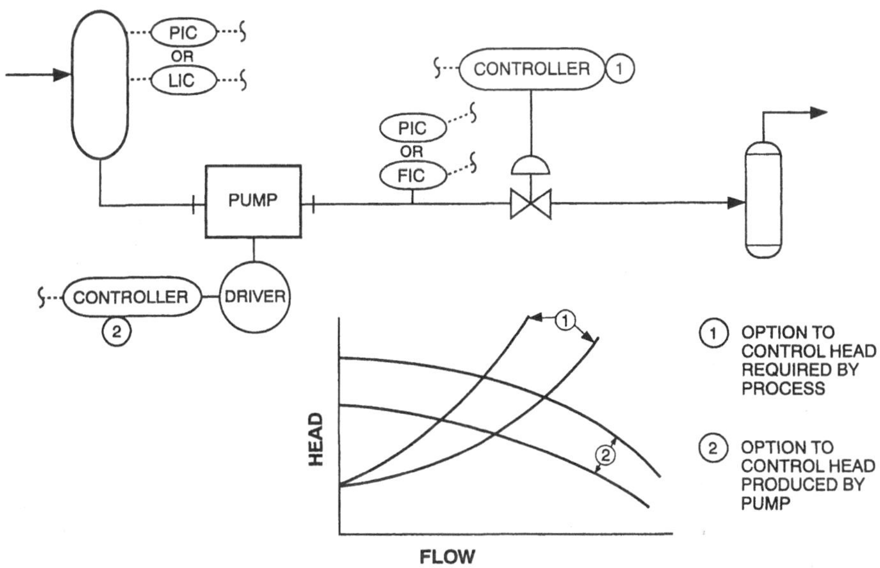

Figure 6.20 Centrifugal pump control options

The approach that should be followed when purchasing a pump is to define the required operating range of the pump based on the process system design and process requirements.

Once the operating range is defined, hydraulic calculations will determine the required flows, heads and $NPSH_{AVAILABLE}$. Care should be taken to define liquid composition and temperature as accurately as possible since these items will determine the vapor pressure which defines the $NPSH_{AVAILABLE}$. Steps in defining the process system are summarized in Figure 6.21.

Preventing liquid disturbances by accurately defining process requirements

- Define operating range of application
- Accurately define liquid characteristics
  - Vapor pressure
  - Pumping temperature
  - Viscosity
- Perform hydraulic calculations for all required flow rates to determine:
  - Head required
  - $NPSH_{AVAILABLE}$

Figure 6.21 Preventing liquid disturbances by accurately defining process requirements

Once the process requirements are accurately defined, a pump can be selected that will meet these requirements without the risk of hydraulic disturbances.

Selecting a pump for hydraulic disturbance free service

Having discussed the concepts used to prevent liquid disturbances and the requirements for accurately defining the process system requirements, this information can be used to select a pump free of hydraulic disturbances.

Refer to Figure 6.23 which shows a typical pump performance curve.

Based on previous discussions, there are three areas of concern to assure trouble-free operation (See Figure 6.22).

1. NPSH margin at maximum operating flow
2. Approximate recirculation margin at minimum operating flow
3. NPSH margin at minimum operating flow

Figure 6.22 Hydraulic disturbances – areas of concern

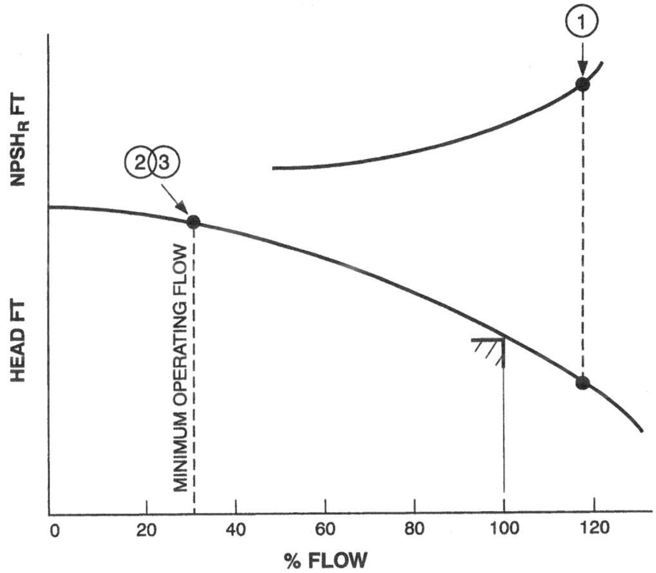

Figure 6.23 Hydraulic disturbance – areas of concern

The practical approach is to select a type of pump that will enable operation under all conditions in Figure 6.22 if possible. Figure 6.24 presents guidelines for selecting a pump free of hydraulic disturbances.

Guidelines for selecting pumps free of hydraulic disturbances

Step	Action
1.	Confirm $NPSH_A \geq NPSH_R$ at maximum operating flow. If margin less than two (2) feet, require witnessed $NPSH_R$ test. If $NPSH_R > NPSH_A$ ■ Increase $NPSH_A$ by: ● Increasing suction drum level ● Decreasing pumping temperature ● Decreasing suction line losses ■ Reselect pump (if possible) ■ Select canned pump
2.	For the pump selected calculate N_{ss} based on pump BEP conditions. Note: if double suction first stage impeller, use 1 /2 of bep flow
3.	If $N_{ss} > 8000$ contact pump vendor and require following data for actual pump fluid and conditions

- Predicted onset flow of cavitation caused by recirculation for actual fluid conditions
- Reference list of proposed impeller (field experience)

4. Compare cavitation flow to minimum operating flow. If this value is within 10% of minimum operating flow:
   - Reselect pump if possible
   - Install minimum flow bypass
   - Consider parallel pump operation
5. Calculate liquid temperature rise at minimum operating flow. If value is greater than 5% of pumping temperature:
   - Calculate $NPSH_A$ based on vapor pressure at calculated pumping temperature. If $NPSH_A < NPSH_R$
     - Install minimum flow bypass
     - Consider parallel pump operation

Figure 6.24 Guidelines for selecting pumps free of hydraulic disturbances

The guidelines presented in Figure 6.24 attempt to cover all situations, however, technical discussions with the pump vendor is encouraged whenever necessary.

Before proceeding, an important question regarding the typical pump performance curve needs to be asked. Why is the $NPSH_R$ curve not drawn to zero flow like the head curve? Based on the information presented in this course, you should be able to answer this question. Consider the following facts:

- The standard shop test fluid for all pumps is water
- The causes of vaporization at low flows

Hopefully your answer took the following 'form:

1. 'Liquid disturbances can occur at low flows if the vapor pressure of the pumped liquid exceeds the surrounding pressure of the liquid'
2. 'Flow separation and/or liquid temperature rise which can occur at low flows will either reduce the surrounding pressure on a liquid or increase its vapor pressure'
3. 'Since the actual liquid characteristics are not known when the standard pump curve (tested on water) is drawn, the vendor stops the $NPSH_R$ curve where flow separation and liquid temperature rise can cause liquid disturbances'

Therefore, trouble free operation to the left of this point is dependent on the pumped liquid and must be discussed with the pump vendor.

In conclusion, preventing liquid disturbances in the project design phase requires a thorough, accurate investigation of both the process and pump characteristics and some serious decisions on required action.

Justification of required action will be easier if the operating company looks beyond the project costs and examines what the actual cost (lost in production) will be if a pump experiences hydraulic disturbances. Most operating companies have documented Case Histories of problem pumps that will provide proven facts relating to the 'Total Cost' of operating problem pumps for the life of a project (Refer to Figure 6.25).

Determine the cost effectiveness of pump selection not only on the project (capital investment) costs, but on the cost to the operating company of unreliable pumps

Figure 6.25 Justifying the selection of troublefree pumps

Field operation

The preceding section described requirements to assure optimum pump reliability during the project design phase. This section will describe how to detect and correct hydraulic disturbances in the field.

Determining the potential for damage

Hydraulic disturbances are detected by monitoring the conditions noted in Figure 6.26.

Indicators of hydraulic disturbances

- Loud noise – continuous or varying
- Suction and discharge pressure pulsations
- High values of overall vibration
- Drop in produced head of >3%
- High values of vane passing frequency in overall spectrum
- Pressure pulsations in inlet and discharge piping
- Possible high bearing temperatures

Figure 6.26 Indicators of hydraulic disturbances

Once hydraulic disturbances are detected, the root cause of the disturbance must be defined. As previously discussed, vaporization of liquid must be present for cavitation to occur. Therefore, the requirement is to determine the root cause of vaporization. As previously discussed, there are three (3) primary root causes of vaporization:

- Internal inlet pressure losses
- Formation of low pressure cells at low flows
- Liquid temperature rise at low flows

Confirmation of process conditions and specific tests are required to determine the root cause of the problem.

Determining the cause of hydraulic disturbances

Confirmation of stated value of $NPSH_A$

Refer to the Pump Data Sheet and/or the Hydraulic Calculation Sheet for that pump service to determine the stated $NPSH_A$. Proceed to check the $NPSH_A$ at field operating conditions. Using the relationship:

$$NPSH_A = \frac{(P_{suction-PSIA} - P_{vapor\ pressure-PSIA}) \times 2{,}311}{S.G.\ at\ pumping\ temperature}$$

Substituting in the equation the actual values as follows:

- Pump suction pressure (down stream of suction streiner)
- Actual vapor pressure at measured pumping temperature
- Actual S.G. at measured pumping temperature

Compare calculated $NPSH_A$ to predicted $NPSH_A$. If actual $NPSH_A$ < predicted $NPSH_A$ modify operation if possible to attain predicted value. If this is not possible, the following alternatives exist:

- Operate pump at lower flow rate to reduce $NPSH_R$
- Modify pump to reduce $NPSH_R$ at operating flow rate

A summary of this discussion is presented in Figure 6.27.

Confirmation of $NPSH_A$ and recommended action

- Obtain predicted value of $NPSH_A$
- Calculate $NPSH_A$ using actual conditions
- If $NPSH_A$ actual < $NPSH_A$ predicted
  - Increase $NPSH_A$ if possible
  - Operate pump at lower flow rate (if cost effective)
  - Modify pump to reduce $NPSH_R$

Figure 6.27 Confirmation of $NPSH_A$ and recommended action

Internal pressure loss test

The test outlined in Figure 6.28 will confirm if the liquid disturbance is caused by pump inlet pressure losses resulting from high liquid velocity.

Test to confirm high velocity cavitation

- Close pump discharge control valve to reduce flow
- If pump noise significantly reduces and conditions noted in Figure 6.26 become stable, cause is confirmed

Figure 6.28 Test to confirm high velocity cavitation

If high velocity cavitation is confirmed and the stated value of $NPSH_A$ is confirmed, possible solutions are presented in Figure 6.29.

- Increase $NPSH_A$ until quiet operation is achieved
- Reduce pump throughout if cost effective
- Operate two pumps in parallel
- Increase impeller eye area
- Impeller material change
- Purchase new pump with acceptable $NPSH_R$

Figure 6.29 High velocity cavitation solutions

Modification of impeller eye area is not always possible and the pump vendor must be consulted to confirm it is possible and that satisfactory results have been achieved.

Before leaving this subject, mention of the inlet piping arrangement is required. Suggested inlet piping arrangements are presented in Figure 6.30.

Failure to conform with the guidelines presented in Figure 6.30 can lead to hydraulic disturbances caused by:

- Entrained vapor
- Additional internal ΔP caused by turbulence
- Liquid separation from impeller vanes

Inlet piping arrangements to avoid hydraulic disturbances caused by process piping

- Piping runs directly vertically or horizontally into pump without high pockets that can cause vapor formation
- Minimum straight suction pipe runs of:
  - Three (3) pipe diameters – single suction
  - Five (5) pipe diameters – double suction
- Double suction pumps should have pipe elbows perpendicular to the pump shaft
- The 'belly' of an eccentric reducer should be in the bottom location

Figure 6.30 Inlet piping arrangements to avoid hydraulic disturbances caused by process piping

Low flow hydraulic disturbance test

The test outlined in Figure 6.31 will determine if the liquid disturbance is caused by low flow circulation or temperature rise.

Test to confirm low flow hydraulic disturbances

- Open pump discharge control valve or bypass to increase flow
- If pump noise significantly reduces and conditions noted in Figure 6.26 become stable, cause of either low flow recirculation or temperature rise cavitation is confirmed
- Calculate liquid flow temperature rise to confirm if recirculation is the root cause

Figure 6.31 Test to confirm low flow hydraulic disturbances

Possible solutions are presented in Figure 6.32.

Solutions – low flow hydraulic disturbances

- Increase pump flow rate, if possible, until quiet operation is achieved
- Modify inlet volute to increase impeller inlet velocity (if sufficient $NPSH_A$ exists)
- Install impeller with reduced eye area (assuming-sufficient $NPSH_A$ exists) *
- Install minimum flow bypass to increase flow rate (to eliminate low flow temperature rise)

* Note: wear ring modifications are required

Figure 6.32 Solutions – low flow hydraulic disturbances

Internal volute and/or impeller eye modifications are not always possible. The pump vendor must be consulted to confirm these modifications are acceptable and satisfactory results have been achieved.

Justification of proposed action plan

Regardless of the cause of hydraulic disturbances, the problem cannot be resolved without management endorsement of a cost effective action plan. As previously mentioned, all action plans must be justified by cost savings. In the field, the largest revenue loss is usually lost product revenue resulting from unplanned critical equipment downtime.

Regardless of the proposed action, be sure to show lost product revenue against capital investment for problem solution.

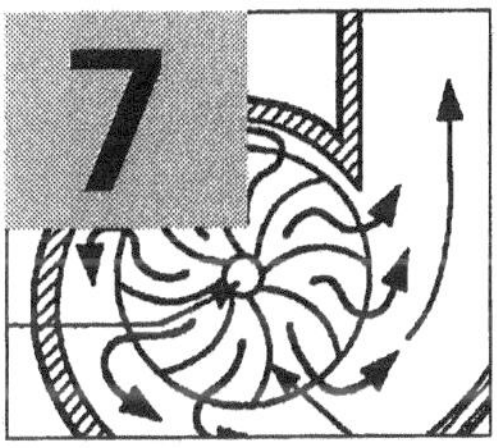

7 Pump mechanical design

- Introduction
- Basic elements of centrifugal pumps
- Volutes
- Wear rings
- Impellers
- Bearings
- Anti-friction bearings
- Hydrodynamic bearings
- Balancing devices
- Shaft and key stress

Introduction

Regardless of the degree of pump performance optimization, the availability of any pump depends on the quality of its mechanical design and manufacture. Understanding the function of each mechanical component is essential in properly specifying, maintaining and operating any pump.

Basic elements of centrifugal pumps

Each type of centrifugal pump is made up of two (2) basic elements, stationary and rotating (Refer to Figure 7.1).

Basic elements of centrifugal pumps

I. Casing
- Provides nozzles to connect suction and discharge piping
- Directs flow into and out of impeller and converts kinetic energy into pressure energy
- Provides support to the bearing housing

II. Bearings and stuffing box
- Provides support and enclosure for the rotating element

Figure 7.1 Basic elements of centrifugal pumps

In addition, any pump is comprised of hydraulic and mechanical components. The first prt of this chapter will cover the hydraulic components (casing, volute, diffusers, wear rings and impellers. The remainder of this chapter will cover the mechanical components (bearings and seals). The major components which make up these elements, and their functions, will be described.

Volutes

The volute is that portion of the pump casing where the liquid is collected and discharged by centrifugal force when it leaves the impeller (Refer to Figure 7.2).

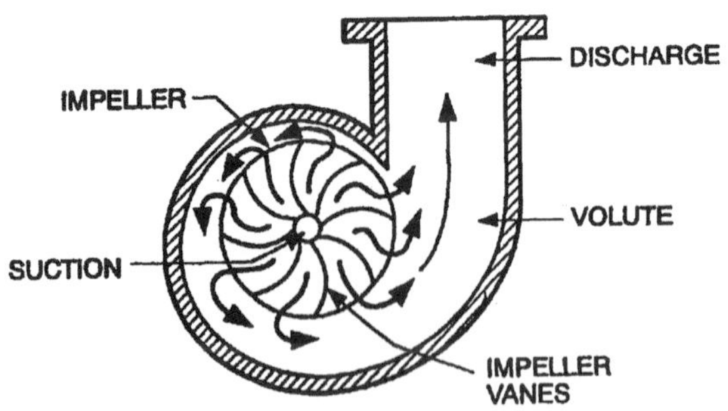

Figure 7.2 Single volute casing form

As the liquid leaves the rotating impeller the volute continually accumulates more liquid as it progresses around the casing. Because we want to keep the liquid velocity reasonably constant as the volume increases, the volute area between the impeller tip and the casing wall must be steadily increased since:

$$Q = AV$$

Where Q = Flow (gpm)

A = Area

V = Velocity

Figure 7.2 shows this relationship.

The form of volute casing shown in Figure 7.2 results in uneven pressure distribution around the periphery of the impeller. In a single stage overhung impeller type pump with cantilever shaft designed for high heads, the unbalanced pressure may result in increased shaft deflection and bearing loadings at off-design conditions.

The double volute casing design shown in Figure 7.3 equalizes the pressure around the impeller and reduces the unbalanced loading on the bearings. This is accomplished with two (2) similar flow channels which have outlets 180 degrees apart.

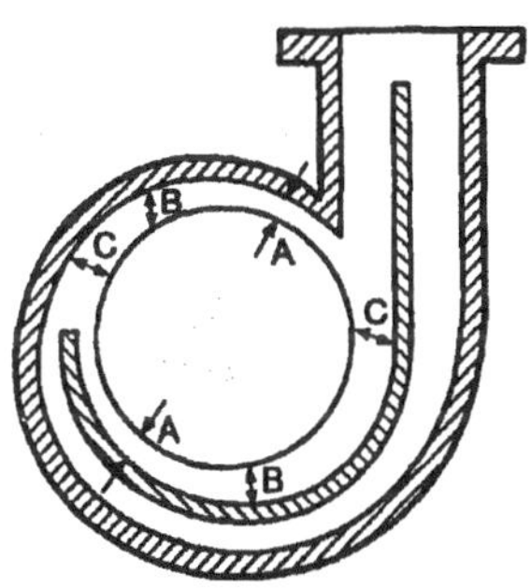

Figure 7.3 Dual volute casing form

Figure 7.4 shows the relationship between the radial reaction force acting on an impeller vs capacity for single and double volute pumps. The radial reaction force is directly proportional to the following parameters:

- Impeller produced head
- Impeller diameter
- Impeller discharge width (b_2)
- Specific gravity

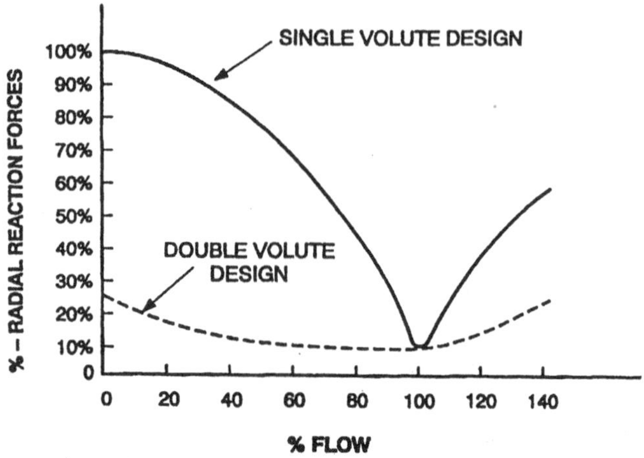

Figure 7.4 Volute radial reaction forces versus capacity

To balance the pressure around the periphery of the impellers of a multistage volute casing type pump, staggered volutes or double volute designs are available, when size permits.

Some multistage pumps, particularly the radial split barrel type casing design, utilize vaned diffusers to convert kinetic energy to pressure energy. The vanes are designed such that the liquid flow area gradually increases to effect a gradual decrease in velocity from inlet to outlet (Refer to Figure 7.5).

Figure 7.5 Vane type diffuser

Wear rings

To minimize the amount of liquid leakage from the high pressure side to the suction side of a centrifugal pump, wear rings are usually fitted to the casing and also to the suction side of the impeller eye outside diameter. To further minimize the leakage, wear rings can be fitted to the casing cover inside diameter and to the back side of the impeller (Refer to Figure 7.6).

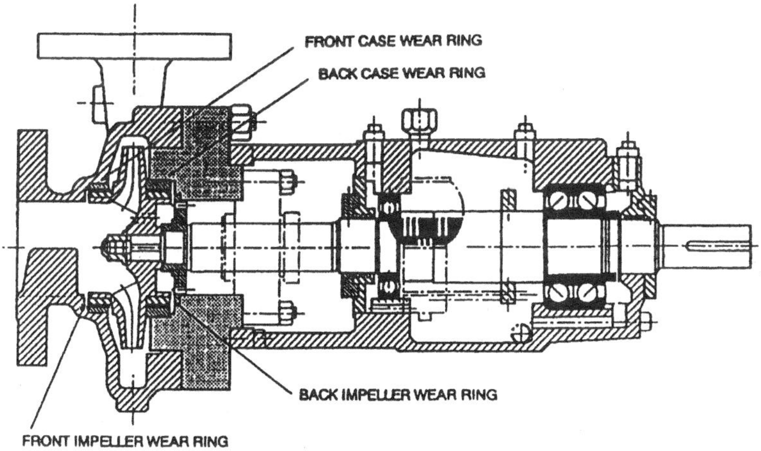

Figure 7.6 Wear ring locations

The clearance provided is small and the material hardness of the case and impeller wear rings is slightly different to avoid potential galling of the parts. The wear rings are lubricated by the pumped liquid. As wear takes place over time, clearance will increase and more flow will pass back to suction, resulting in some degradation of efficiency. API Standard 610 recommends clearances for these parts which are slightly more liberal than would otherwise be provided by most manufacturers. When a pump is furnished with API clearances, a slight decrease in efficiency is accounted for in pump performance.

Impellers

The impeller can be considered the heart of the centrifugal pump. It is the only component which produces head (energy). Therefore, they should be well designed to handle the liquid pumped with minimum loss. There are two (2) basic impeller designs available; the enclosed type, and the open type (Refer to Figure 7.7).

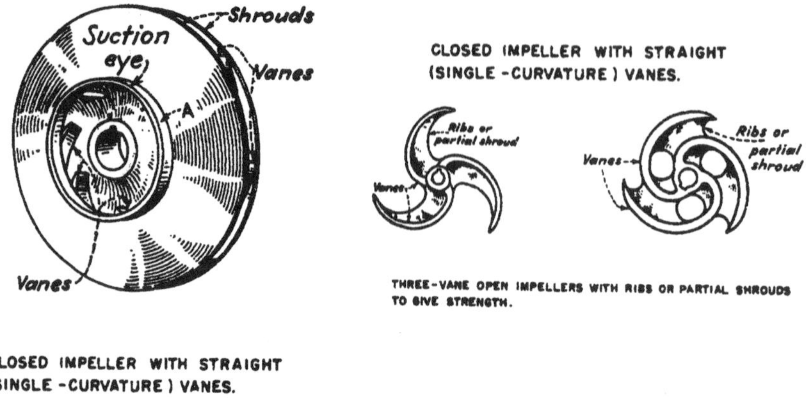

Figure 7.7 Impeller designs

In the enclosed impeller design, liquid flows into the impeller where it is contained between front and back shrouds affixed on each side of the vanes. There is no flow along the walls of the casing and suction head as there is with the open impeller. As noted in the previous Chapter, flow circulation from the high pressure side to the suction side is minimized by the use of close clearance wear rings.

The open impeller design may have vanes cast integral with the hub, usually without side walls. This type of design requires that the impeller be positioned with close clearance between it and the casing wall. This close clearance operation minimizes recirculation of liquid from the area under discharge pressure to the area under suction pressure. The efficiency of the open impeller is lower than that of the enclosed type. If the clearance between the impeller vane and the casing wall increases, some adjustment, to close the clearance, is possible. The open impeller does not find widespread use in hydrocarbon service.

The impeller may be further classified as single suction or double suction type. The term single and double suction defines the number of inlets to the impeller. A single suction type has one (1) inlet and a double suction type has two (2) inlets (Refer to Figure 7.8).

The larger the suction area in a double suction impeller allows the pump to operate with less positive suction head for a given capacity as compared to a single suction impeller because the flow area to the impeller vanes is increased. This results in a reduced pressure drop from the pump suction flange to the vanes in the eye of the impeller and thus reduces the NPSH required.

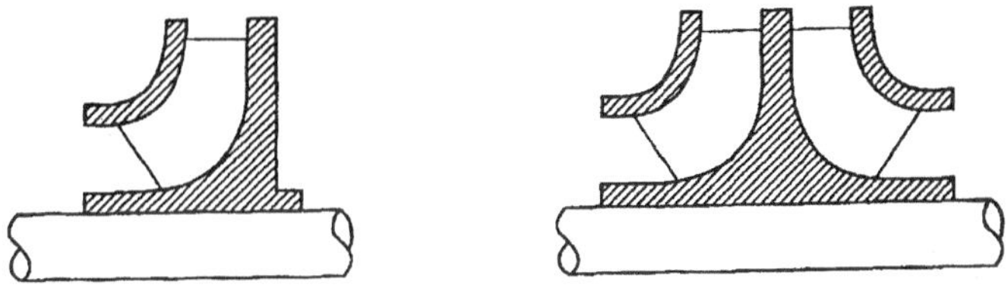

Figure 7.8 Impeller suction types

Specific speed

Specific speed is a non-dimensional value that is a function of pump speed, flow and head.

$$Ns = \frac{N\sqrt{Q}}{Hd^{(3/4)}}$$

Where: N_S = specific speed

N = pump speed RPM

Q = pump flow GPM

Hd = pump produced head $-\frac{FT\text{-}lb_f}{lb_m}$

Specific speed is use extensively in both pump and compressor design to optimize stage efficiency for a given value of flow and head required. In pump design, specific speed is used to optimize the following design parameters:

- Impeller discharge flow velocity
- Impeller tip speed
- Impeller inlet and discharge blade angles
- Discharge throat velocity

Figure 7.9 presents a plot of pump specific speed vs efficiency for various flow rates. This is a typical chart and will vary slightly from pump vendor to pump vendor based on the specifics of pump design. It can be used for estimating purposes in determining pump required horsepower.

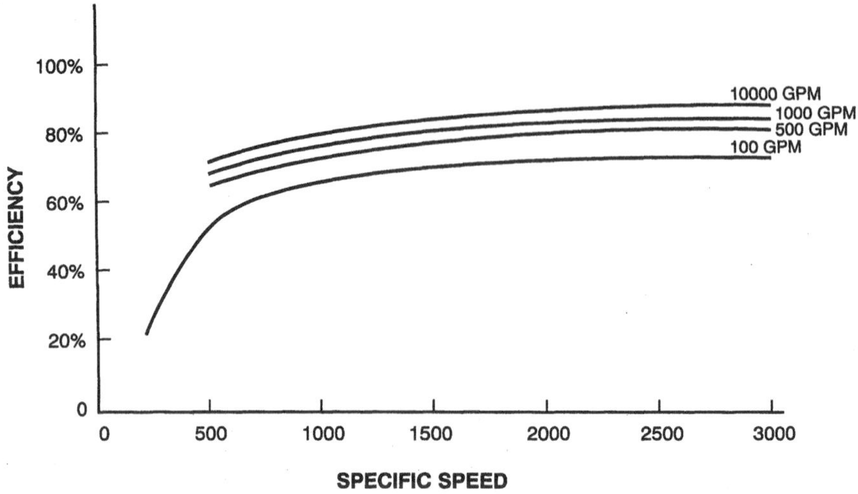

Figure 7.9 Efficiency as a function of pump specific speed and flow

Bearings

Functions of bearings

Bearings provide support for the rotating element. They are required to carry radial and axial loads. The basic relationship for any bearing design is:

$$P = \frac{F}{A}$$

Where: P = Pressure on the supporting oil film (hydrodynamic) or element (anti-friction) in P.S.I.

F = The total of all static and dynamic forces acting on the bearing in LB force

A = The bearing area in IN^2

In many cases, a pump bearing may perform for years and suddenly fail. This abrupt change in performance characteristics is usually due to a change in the forces acting on the bearing. Factors that can increase bearing load are shown in Figure 7.10.

Sources of forces

- Increased process pipe forces and moments
- Foundation forces ('soft' foot, differential settlement)
- Fouling or plugging of impeller
- Misalignment
- Unbalance
- Rubs
- Improper assembly clearances
- Thermal expansion of components (loss of cooling medium, excessive operating temperature)
- Radial forces (single volute – off design operation)
- Poor piping layouts (causing unequal flow distribution to the pump)

Figure 7.10 Sources of forces

In this section we will discuss the functions and various types of bearings used in pumps.

There are three (3) basic types of bearings used in centrifugal pumps (Refer to Figure 7.11).

Basic types of bearings

- Anti-friction
- Hydrodynamic ring oil lubricated
- Hydrodynamic pressure lubricated

Figure 7.11 Basic types of bearings

Application guidelines

Centrifugal pumps in the process industry are normally fitted with bearings which are appropriate for the application and pump design (Refer to Figure 7.12).

The L-10 rating life for anti-friction bearings is defined as the number

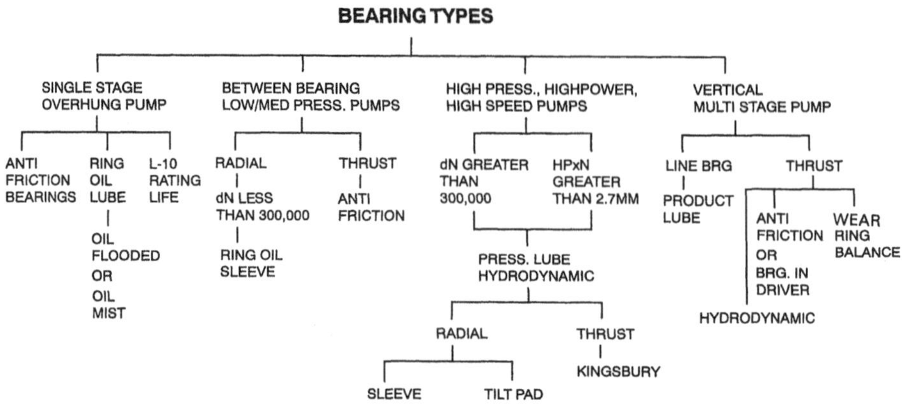

Figure 7.12 Bearing application guidelines

of hours at rated bearing load that 90% of a group of identical bearings will complete or exceed (25,000 hours of operation) before the evidence of failure. Failure evidence is generally defined as a 100% increase in measured vibration.

API Standard 610 recommends applying pressurized hydrodynamic bearings when the product of pump rated horsepower and rated speed in revolutions per minute exceeds 2.7 million. Pressure lubrication systems may be integral or separate, but should include as a minimum an oil pump, reservoir, filter, cooler, controls and instrumentation.

The dN number is the product of bearing size (bore) in millimeters and the rated speed in revolutions per minute. It is used as a measure of generated frictional heat and to determine the type of bearing to be used and type of lubrication required. Figures 7.13 and 7.14 present the relationships for these two important factors.

L-10 Life
Used for all anti-friction bearings

$$L-10 = \frac{16700}{N} \bullet \left[\frac{C}{F}\right]^3$$

where: L-10 = hours of operation 90% of a group of identical bearings will complete or exceed

N = pump speed

C = the total force required to fail the bearing after 1,000,000 revolutions (failure defined as 100% increase in measured vibration)

F = the total of all actual forces acting on the bearing

Figure 7.13 L-10 Life

D-N Number

- Is a measure of the rotational speed of the anti-friction bearing elements
- D-N number = bearing bore (millimeters) X speed (RPM)
  Is used to determine bearing type and lubrication requirements

D-N range	bearing type	lubrication type
Below 100,000	anti-friction	sealed
100,000 - 300,000	anti-friction	regreasable
Below 300,000	anti-friction	oil lube (unpressurized)
Above 300,000	sleeve, multi-lobe or tilt pad	oil lube (pressurized)

Figure 7.14 D-N Number

Anti-friction bearings

Ball Bearings

The anti-friction bearing most commonly used in centrifugal pumps for carrying radial and thrust loads is the ball bearing. The design of each type of bearing has it's individual advantages for specific load carrying requirements (Refer to Figure 7.15).

TYPE	LOAD CAPACITY	DESCRIPTION
• SINGLE-ROW DEEP GROOVE	EQUAL THRUST CAPACITY IN EACH DIRECTION. MODERATE TO HEAVY RADIAL LOADS	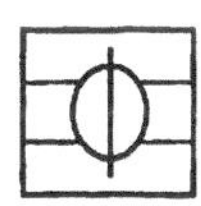
• SINGLE-ROW ANGULAR CONTACT	HIGHER RADIAL LOAD THAN DEEP GROOVE. HEAVY THRUST LOAD IN ONE DIRECTION	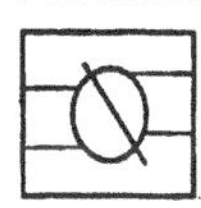
• DUPLEX, SINGLE-ROW BACK TO BACK (DB)	HEAVY RADIAL LOADS, COMBINED RADIAL/THRUST LOADS, REVERSE THRUST	

Figure 7.15 Typical ball bearing load capacity guidelines

Roller bearings

Anti-friction roller bearings are capable of accepting pure radial loads, pure thrust loads or various combinations. They have been applied as crankshaft bearings in power ends of some power pump designs. Refer to Figure 7.16. Specific uses of this type of bearing resides with the manufacture and is not usually a selective option for the user.

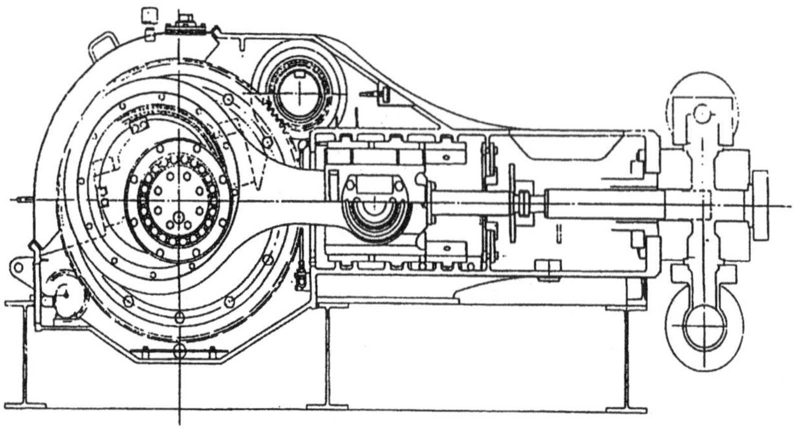

Figure 7.16 (Courtesy of Union Pump Co)

Lubrication

The purpose of anti-friction bearing lubrication is to increase bearing life, keep the balls or rollers separated, dissipate the heat generated in and conducted into the bearing and to prevent corrosion. There are various methods of oil lubrication for ball and roller bearings with the two most common methods being the ring oil and flooded method. Refer to Figures 7.17 and 7.18.

A word of caution – both types of lubrication usually incorporate constant level oilers to maintain oil level at a specified height in the bearing housing. The oil level in the constant level oiler is not the level of the oil in the bearing housing. The constant level oiler is a reservoir that provides oil to maintain a constant level in the bearing housing.

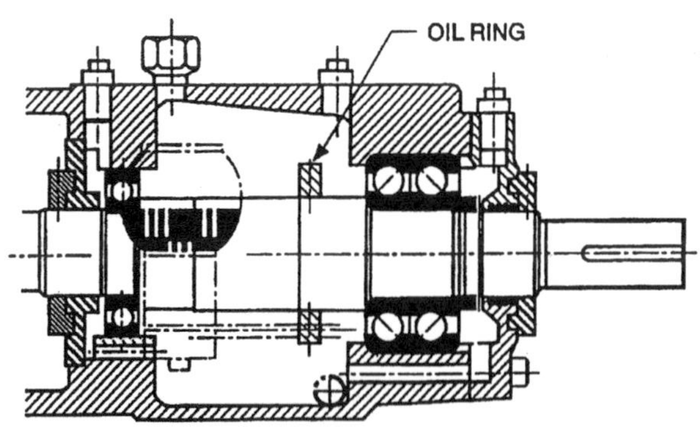

- IN RING OIL LUBRICATION, OIL IS RAISED FROM A RESERVOIR BY MEANS OF A RING WHICH RIDES LOOSELY ON THE SHAFT AND ROTATES WITH THE JOURNAL
- OIL LEVEL IS AT CENTER OF LOWEST BEARING

Figure 7.17 Ring oil lubrication

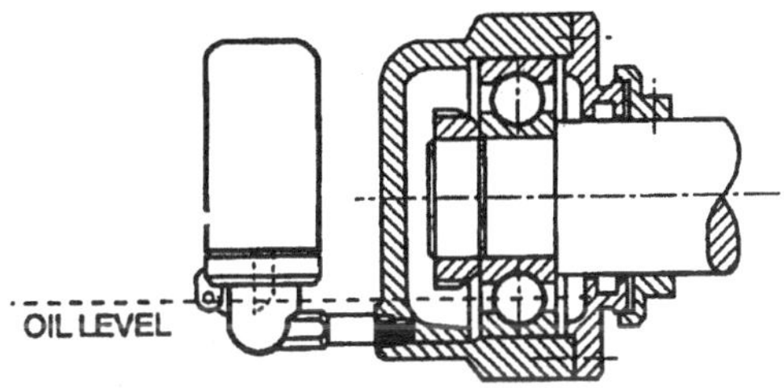

Figure 7.18 Flood type lubrication

Proper oil level must be confirmed by visual inspection or by using a dip stick.

Oil mist lubrication is another acceptable method of lubricating anti-friction bearings. It is gaining acceptance since it provides a controlled environment in the bearing housing in both operating and non operating pumps. A mist generator console is often used to provide services to many pumps simultaneously. Use of oil mist lubrication should be investigated in environments where sand or salts exist. (Refer to Figure 7.19).

Mist oil system features

Features of oil mist system include:

- Mist generator provides mixture of air and atomized oil under pressure
- Reduced operating temperature resulting from air flow preventing excessive Oil accumulation
- Oil consumption is low. controlled with instrumentation
- Entrance of grit and contaminants is prevented since air is under pressure
- Reliable continuous supply of lubricant is available
- Ring oil can be installed as back-up to mist system (called "wet sump system")

Figure 7.19 Mist oil system features

Hydrodynamic bearings

The function of the hydrodynamic bearings is to continuously support the rotor with an oil film that is less than one (1) thousandth of an inch. There are various types of bearings available (Refer to Figure 7.20).

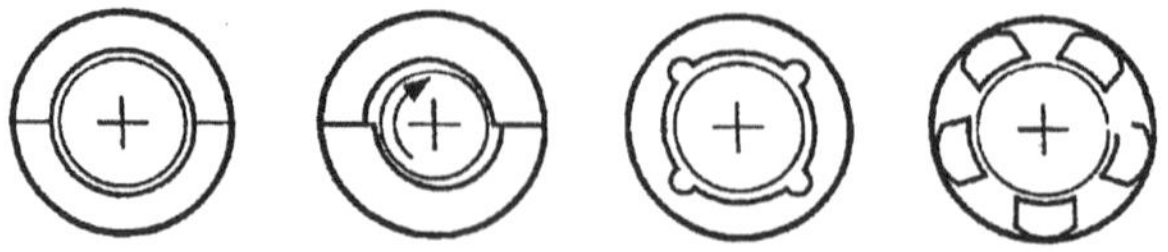

Figure 7.20 Journal bearing types

Sleeve bearings

Sleeve bearings are commonly used as radial bearings in centrifugal pumps. Lubrication is usually supplied by an external pressurized system, although some slow speed sleeve bearing applications (dN less than 400,000) can utilize ring oil lubrication.

Tilt Pad bearing

Tilt pad journal bearings are sometimes used for high horsepower, high speed centrifugal pumps where rotor stability may be of concern. They are always oil pressurized bearings.

Multilobe bearing

This type of bearing has found limited use in centrifugal pumps, but it can also be applied on high speed pumps where increased damping and stiffness is desired for lightly loaded bearings (most vendors will select tilt pad bearings).

Hydrodynamic bearing performance

When at rest, the journal settles down and rests at the bottom of the bearing (Refer to Figure 7.21A). As the journal begins to rotate, it rolls up the left side of the bearing, moving the point of contact to the left. There is then a thin film of oil between the contact surfaces, and fluid friction takes over for metal to metal contact (Refer to Figure 7.21B). The journal slides and begins to rotate, dragging more oil between the surfaces, forming a thicker film and raising the journal. As the speed of rotation increases, the oil drawn under the journal builds up pressure that forces the journal up and to the right in the direction of rotation

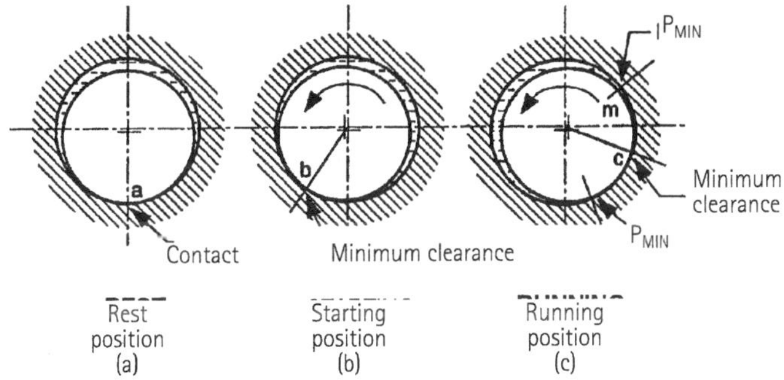

Figure 7.21 Hydrodynamic bearing performance

shown until a condition of equilibrium is reached resulting in a point of minimum clearance (Refer to Figure 7.21C).

As the oil film builds up under the journal, the center of the journal moves and the location of minimum clearance is away from the load line. The pressure distribution across the bearing varies and the maximum unit pressure reaches a value about twice the average pressure on the projected area of the bearing (Refer to Figure 7.22).

The permissible unit pressure (PSI) is a function of the static and dynamic forces and the dimensions of the bearing (projected area). Industry guidelines for continuously loaded bearings range from 50 to 300 PSI. Pressures exceeding 300 PSI can potentially result in breakdown of the oil film and subsequent metal to metal contact.

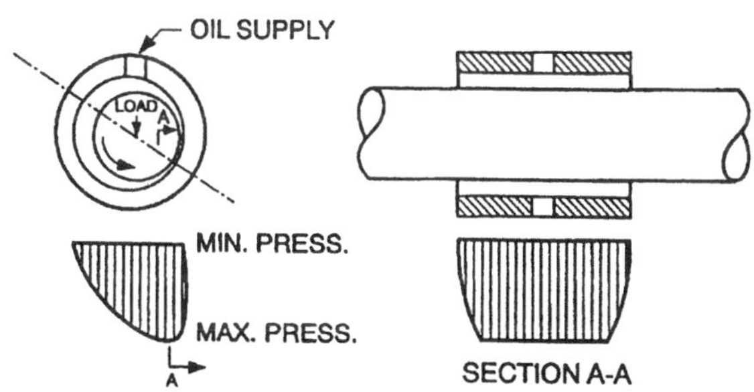

THE AVERAGE UNIT PRESSURE ON THE PROJECTED AREA (BEARING LENGTH X DIAMETER) IS GIVEN BY THE EQUATION

$$P = \frac{6 Z V d}{2C^2} K'$$

Z = ABSOLUTE VISCOSITY OF LUBRICANT, CP
V = JOURNAL SURFACE VELOCITY, FPS
d = JOURNAL DIAMETER, INCHES
C = DIAMETRAL CLEARANCE BETWEEN JOURNAL AND BEARING
K′ = FACTOR DEPENDING ON BEARING CONSTRUCTION AND RATIO OF LENGTH TO JOURNAL DIAMETER

Figure 7.22 Bearing pressure distribution

Balancing devices

Single suction and multi-stage pumps require hydraulic balancing to limit thrust; whereas, double suction impellers are theoretically in hydraulic balance (Refer to Figure 7.23).

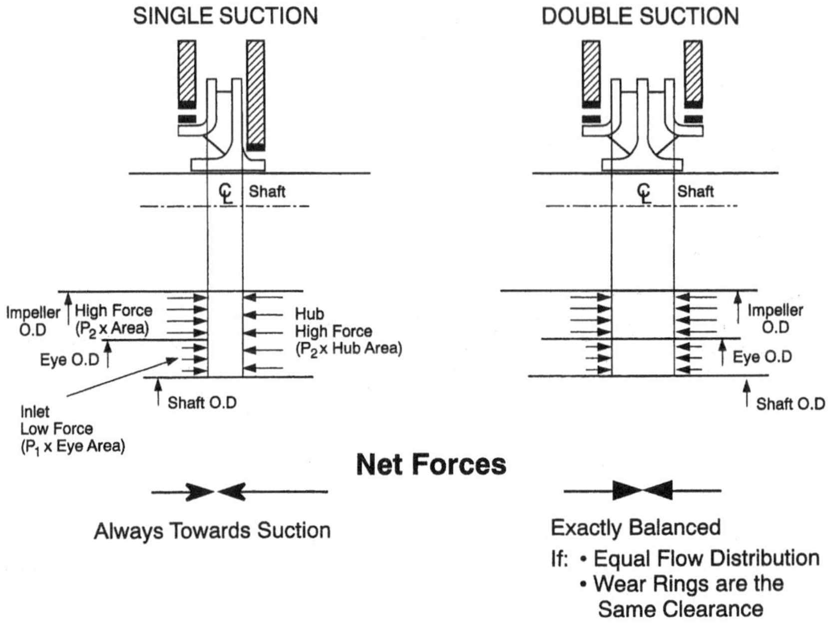

Figure 7.23 Impeller force

The thrust on a single suction impeller is always in the direction of suction. When this type of impeller is installed in a single stage overhung pump, holes are usually provided in the impeller to help maintain the stuffing box pressure in balance with the suction pressure of the pump. In some overhung pump designs pumping vanes are machined on the back hub of the impeller to reduce thrust and evacuate the area between the impeller and the stuffing box to allow flow from the stuffing box into the pump.

To minimize the impeller thrust of a multi-stage pump, two methods are used, either opposed impeller design or a hydraulic balance device. (Refer to Figure 7.24).

The claimed advantage for using opposed impeller arrangements for multi-stage pumps is its inherent zero thrust. However, there is some residual thrust resulting from leakage across bushings throughout the length of the shaft. A correctly sized thrust bearing is still required to handle the thrust load and the added thrust load resulting from

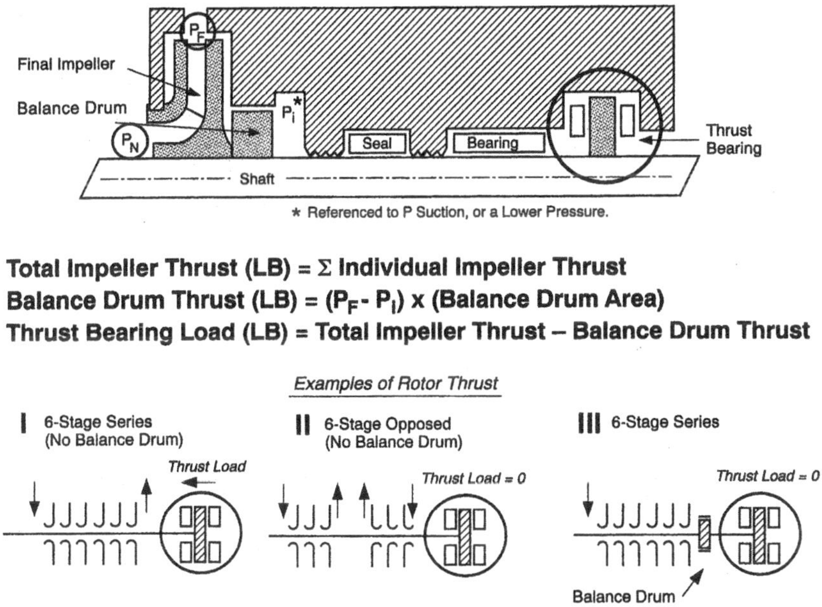

Figure 7.24 Pump thrust force

potential increased clearance of these bushings. In addition, for single stage double suction pumps, process piping design can unbalance the thrust load if turbulence is created in the suction piping as a result of unequal flow distribution to each impeller. As a rule of thumb all pumps utilizing double suction impellers should have 5–10 straight pipe diameters up stream of the suction flange.

The balance drum in Figure 7.25 is the conventional method for minimizing hydraulic axial thrust in multi-stage pumps with tandem impeller arrangements. It is not a self compensating device and thrust will increase as wear takes place and clearances open. To avoid having the rotor move back and forth in a 'shuttle' fashion, it is good practice to intentionally design the balance drum for some residual thrust. This is accomplished by sizing the drum diameter such that there is always some thrust force either in the direction of suction or discharge. The cavity behind the balance drum is routed to suction and maintained at a slight pressure above suction pressure due to the pressure drop in the balance line. Refer to Figure 7.26 for thrust bearing sizing criteria.

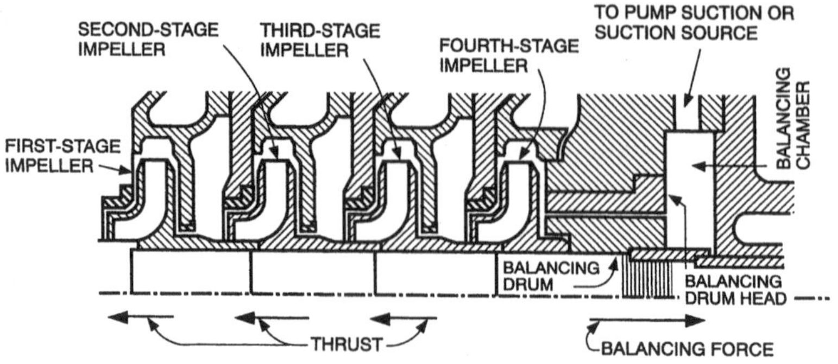

Figure 7.25 Thrust balancing drum

Thrust bearing sizing criteria

Thrust bearing is sized for:

- Net thrust load from impellers and leakage plus added thrust load resulting from twice the Increased clearance of the balance drum
- Thrust load from coupling

Figure 7.26 Thrust bearing sizing criteria

Shaft and key stress

Whenever a Root Cause Failure Analysis is required for a Shaft and/or Keyway change or the Equipment must operate under an increased power of greater than 25%, Shaft and Key Stresses should be checked. The Limiting Shaft and/or Key Stress is always the Shear Stress, which is the Stress produced by the Load (Torque) of the Equipment. Figure 7.27 presents these facts.

Always check shaft & key stress when:

- Shaft end and/or key stress damage is observed
- Equipment is being modified (upgraded) to transmit additional power

Figure 7.27

Shaft Shear Stress is calculated first and is a function of the Transmitted Torque (Proportional to Horsepower/Shaft Speed) and Shaft Diameter to the 3rd power. This relationship is shown in Figure 7.28.

Shaft shear stress

I. The relationship: $\text{Shear Stress} = \dfrac{16T}{\pi (D)^3}$

Where: $T = \text{torque} = \dfrac{\text{shaft horsepower}}{\text{shaft speed}} \times 63025 \text{ in-LBS/HP}$

D = shaft diameter (smallest diameter) inches

II. Guidelines: use maximum driver hp, rated speed typical limit for STD. (4140) shaft material = 8,500 PSI

Figure 7.28 Shaft shear stress

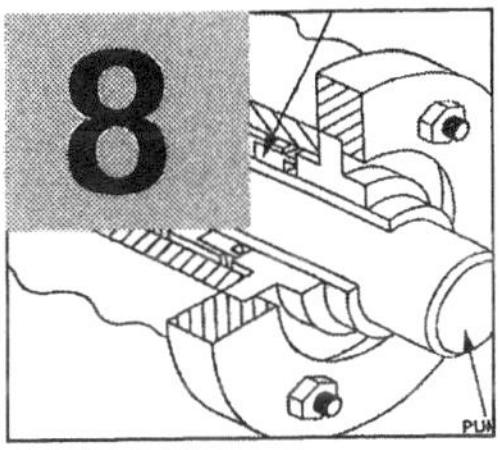

Mechanical seals

- Introduction
- Function of mechanical seals
- The seal system
- Controlling flush flow to the seal
- Examining some causes of seal failures
- Seal configurations
- Flush system types
- Auxiliary stuffing box and flush plans

Introduction

The part of the pump that is exposed to the atmosphere and through which passes the rotating shaft or reciprocating rod is called the stuffing box. A properly sealed stuffing box prevents the escape of pumped liquid. Mechanical seals are commonly specified for centrifugal pump applications (Refer to Figure 8.1).

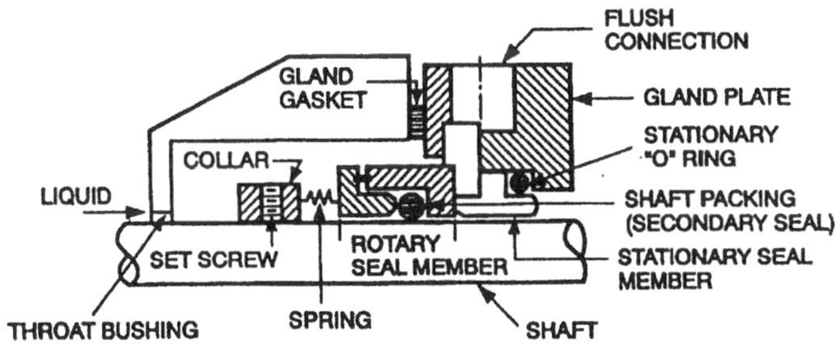

Figure 8.1 Typical single mechanical seal

Function of mechanical seals

The mechanical seal is comprised of two basic components (Refer to Figure 8.2).

Basic seal components

- Stationary member fastened to the casing
- Rotating member fastened to shaft, either direct or with shaft sleeve

Figure 8.2 Basic seal components

The mating faces of each member perform the sealing. The mating surface of each component is highly polished and they are held in contact with a spring or bellows which results in a net face loading closure force (Refer to Figure 8.1).

In order to meet the objective of seal design (prevent fluid escape to the atmosphere), additional seals are required. These seals are either 'O' rings, gaskets or packing (Refer to Figure 8.1). For high temperature applications (above 400°) the secondary seal is usually 'Graphoil' or 'Kalrez' material in a 'U' or chevron configuration. An attractive alternative is to eliminate the secondary seal entirely by using a bellows seal since the bellows replaces the springs and forms a leak tight element thus eliminating the requirement for a secondary seal (Refer to Figure 8.3).

To achieve satisfactory seal performance for extended periods of time, proper lubrication and cooling is required. The lubricant, usually the pumped product, is injected into the seal chamber and a small amount passes through the interface of the mating surfaces. Therefore, it can be stated that all seals leak and the amount of leakage depends on the pressure drop across the faces. This performance can be considered like flow through an equivalent orifice (Refer to Figure 8.4).

The amount of heat generated at the seal face is a function of the face loading and friction coefficient, which is related to material selection and lubrication. Figure 8.5 shows the equation for calculating the amount of heat which needs to be removed by the flush liquid.

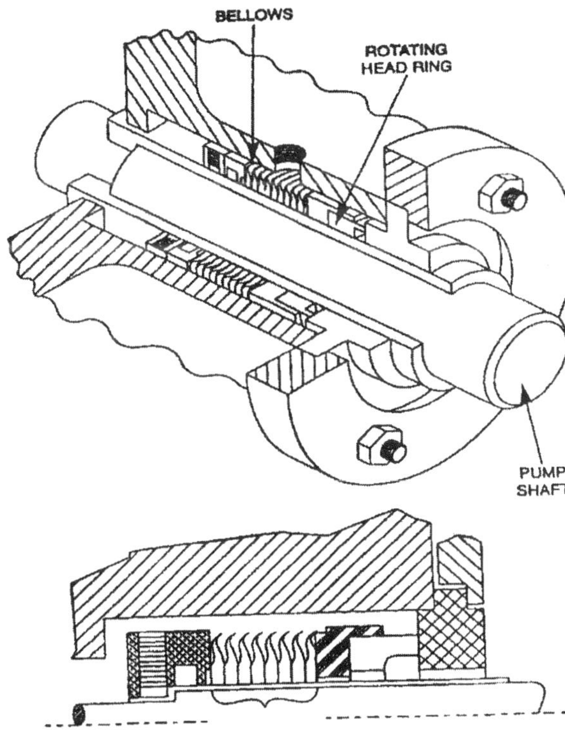

Figure 8.3 Metal bellows seal

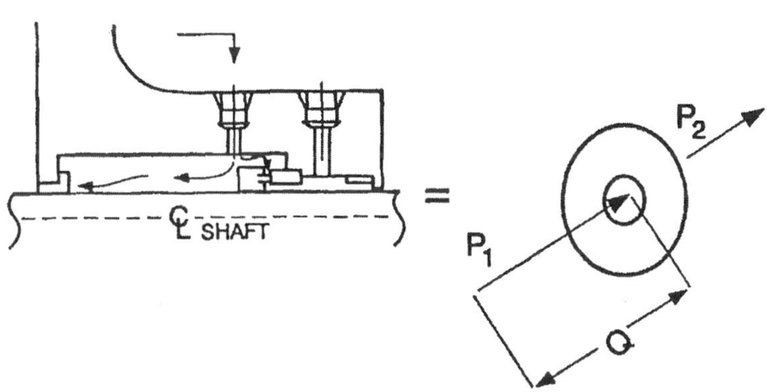

Figure 8.4 Equivalent orifice flow across seal faces

Heat generated by mechanical seal

$Q = 500 \bullet S.G. \bullet Q_{,NJ} \bullet Cp \bullet \Delta T$

where: Q = heat load (BTU/HR)

S.G. = specific gravity of injection liquid

C_p= specific heat of injection liquid $\left(\frac{BTU}{LB-°F}\right)$

ΔT = temperature rise of injection liquid (°F)

Q_{INJ}= injection liquid flow rate (G.P.M.)

Figure 8.5 Heat generated by mechanical seal

As the lubricant flows across the interface, it is prone to vaporization. The initiation point of this vaporization is dependent upon the flush liquid pressure and its relationship to the margin of liquid vapor pressure at the liquid temperature. The closer the liquid flush pressure is to the vapor pressure of the liquid at the temperature of the liquid, the sooner vaporization will occur (Refer to Figure 8.6).

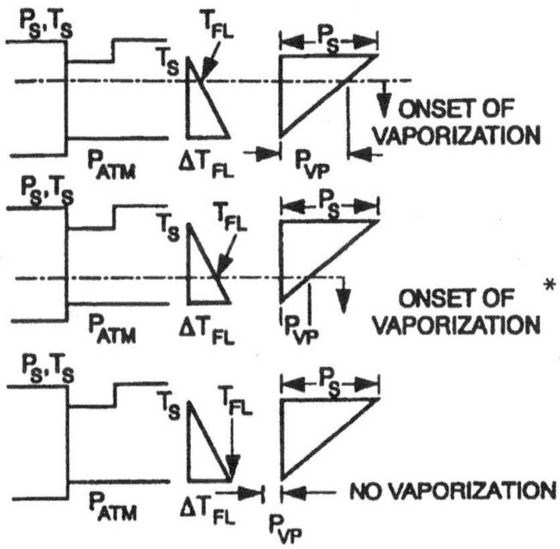

*This represents the design case, vaporization approximately 3/4 down the faces

Figure 8.6 Typical seal face pressure temperature relationship to vaporization

The seal system

To assure reliable trouble free operation for extended periods of time, the seal must operate in a properly controlled environment. This

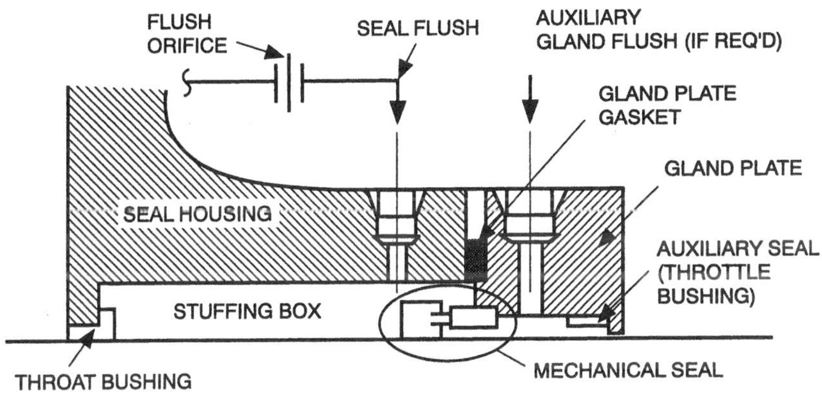

Figure 8.7 Simple seal system

requires that the seal be installed correctly so that the seal faces maintain perfect contact and alignment and that proper lubrication and cooling be provided. A typical seal system for a simple single mechanical seal is comprised of the seal, stuffing box throat bushing, liquid flush system, auxiliary seal and auxiliary flush or barrier fluid (when required) (Refer to Figure 8.7).

The purpose of the seal is to prevent leakage of pumped product from escaping to the atmosphere. The liquid flush (normally pumped product from the discharge) is injected into the seal chamber to provide lubrication and cooling. An auxiliary seal is sometimes fitted to the gland plate on the atmospheric side of the seal chamber. Its purpose is to create a secondary containment chamber when handling flammable or toxic fluids which would be considered a safety hazard to personnel if they were to leak to atmosphere. A liquid (non-toxic) flush or barrier fluid, complete with a liquid reservoir and appropriate alarm devices can be used to assure toxic fluid does not escape to the atmosphere.

Controlling flush flow to the seal

The simple seal system shown in Figure 8.8 incorporates an orifice in the flush line from the pump discharge to the mechanical seal. Its purpose is to limit the injection flow rate to the seal and to control pressure in the seal chamber. A minimum bore diameter or 1/8" is normally specified (to minimize potential of blockage) and the orifice can either be installed between flanges or in an orifice nipple.

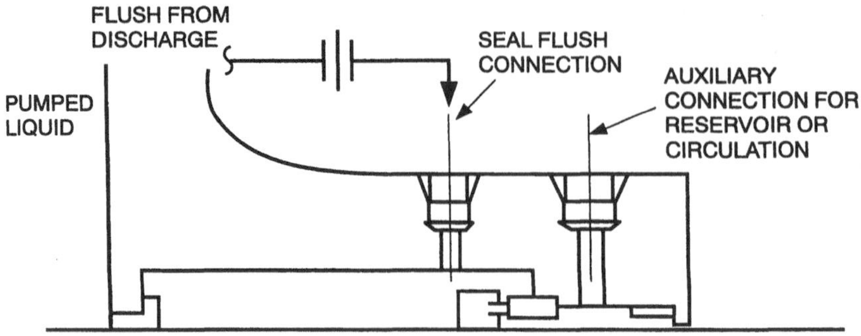

Figure 8.8 Seal flow control orifice

Examining some causes of seal failures

An indication of some causes of seal failures can be obtained while the seal is operating. When you consider the seal as an equivalent orifice, examination of 'tell tale' symptoms can indicate potential failure causes for which corrective action can be implemented or at least can provide direction of subsequent failure analysis (Refer to Figure 8.9). It should be noted that improper application, installation, and/or manufacturing errors can also result in mechanical seal failures.

Comments	Possible causes	Comments/recommendations
■ Seal squeal during operation	Insufficient amount of liquid to lubricate seal faces	Flush line may need to be enlarged and/or orifice size may need to be increased
■ Carbon dust accumulating on outside of seal area	Insufficient amount of liquid to lubricate seal faces	See above
	Liquid film vaporizing/ flashing between seal faces	Pressure in seal chamber may be too low for seal type
■ Seal spits and sputters in operation (popping)	Product vaporizing/flashing across seal faces	Corrective action is to provide proper liquid environment of the product at all times
		1. Increase seal chamber pressure if it can be achieved within operating parameters (maintain at a minimum of 25 Psig above suction pressure)

	2. Check for proper seal balance with manufacturer 3. Change seal design to one not requiring as much product temperature margin (ΔT) 4. Seal flush line and/or orifice may have to be enlarged 5. Increase cooling of seal faces Note: A review of seal balance requires accurate measurement of seal chamber pressure, temperature and product sample for vapor pressure determination

Figure 8.9 Possible causes of seal failure

Seal configurations

Mechanical seals are the predominant type of seals used today in centrifugal pumps. They are available in a variety of configurations, depending upon the application service conditions and/or the User's preference (Refer to Figures 8.10 to 8.13 for the most common arrangements used in refinery and petrochemical applications).

Single mechanical seal applications

Single mechanical seals (Refer to Figure 8.10) are the most widely used seal configuration and should be used in any application where the liquid- is non-toxic and non-flammable. As mentioned earlier in this section, many single mechanical seals are used with flammable and even toxic liquids and only rely on the auxiliary seal throttle bushing to prevent leakage to atmosphere. Since the throttle bushing does not positively contain leakage, State and Federal environmental regulations now require use of a tandem or double seal for these applications. In some plants, a dynamic type throttle bushing ('Impro' or equal) is used to virtually eliminate leakage of the pumped fluid to atmosphere in the event of a mechanical seal failure.

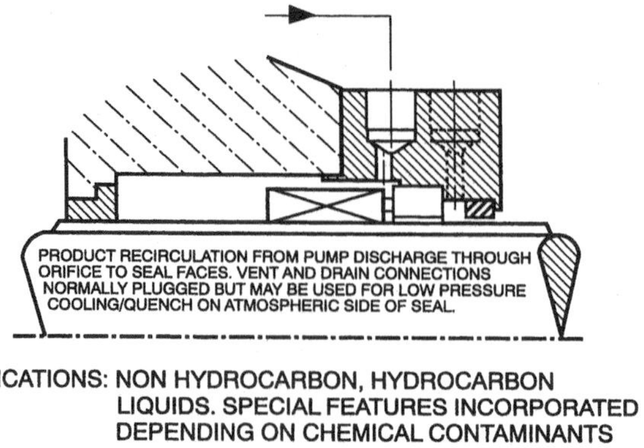

Figure 8.10 Single mechanical seal

Tandem mechanical seal applications

Tandem mechanical seals (Refer to Figure 8.11) are used in applications where the pumped fluid is toxic and/or flammable. They consist of two (2) mechanical seals (primary and back-up). The primary seal is flushed by any selected seal flush plan. The back-up seal is provided with a flush system incorporating a safe, low flash point liquid. A pressure alarm is provided to actuate on increasing stuffing box pressure between the primary and back-up seal thus indicating a primary seal failure. Since the pumped product now occupies the volume between the seals, failure of the back-up seal will result in leakage of the pumped fluid to atmosphere. In essence any time a tandem seal in alarm, it is actually a single seal and should be shut down immediately to assure that the toxic and/or flammable liquid does not leak to atmosphere.

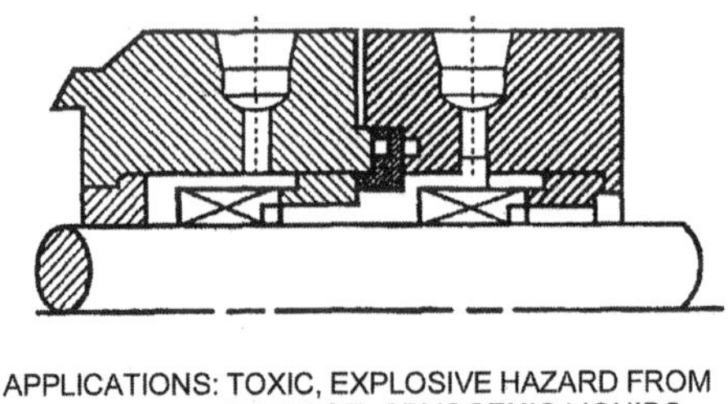

Figure 8.11 Tandem seal arrangement

Double mechanical seal applications

Double mechanical seals (Refer to Figure 8.12) are used in applications where the pumped fluid is flammable or toxic and leakage to atmosphere cannot be tolerated under any circumstances. Typical process applications for double seals are H_2S service, Hydrofluoric acid alkylation services or sulfuric acid services.

Leakage of the pumped fluid to the atmosphere is positively prevented by providing a seal system, whose liquid is compatible with the pumped liquid, that continuously provides a safe barrier liquid at a pressure higher than the pumped fluid. The seals are usually identical in design with the exception that one seal incorporates a pumping ring to provide a continuous flow of liquid to cool the seals. Typical double seal system components are: reservoir, cooler, pressure switch and control valve.

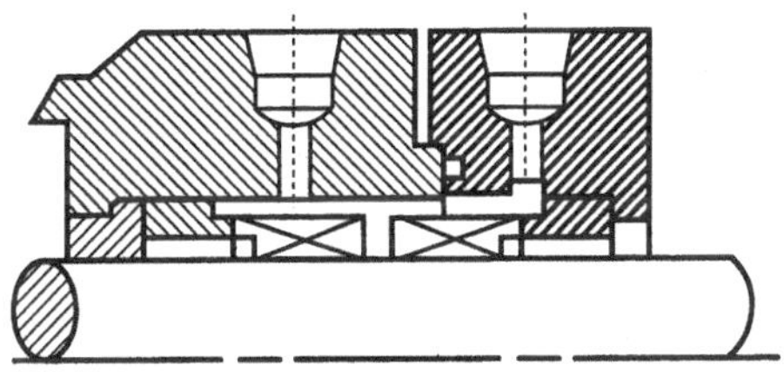

APPLICATIONS: SIMILAR TO THESE USED FOR TANDEM SEALS

Figure 8.12 Double mechanical seal

Liquid/Gas tandem mechanical seal applications

In this configuration (Refer to Figure 8.13) a conventional single liquid mechanical seal is used as the primary and a gas seal (non-contacting faces) that can temporarily act as a liquid seal in the event of primary seal failure serves as the back-up seal. This seal configuration is used in low specific gravity applications where the pumped fluid is easily vaporized. Using a gas seal as the back-up has the advantage of eliminating the vessel, cooler and pumping ring necessary for conventional tandem liquid seals.

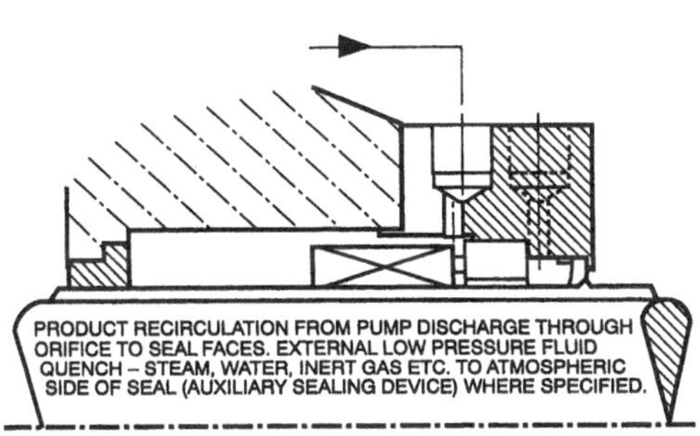

APPLICATIONS: CAN BE USED FOR LIQUIDS ABOVE AUTO IGNITION TEMP. TO PREVENT FIRE IF LEAKAGE EXPOSED TO ATMOSPHERE

Figure 8.13 Liquid/gas tandem seal combination

This application is well proven and has been used successfully for natural gas liquids, propane, ethylene, ethane and butane pump applications.

Double gas seal applications

Before leaving this subject, a relatively new application utilizes two (2) gas seals in a double seal configuration and uses N_2 or air as a buffer maintained at a higher pressure than the pumped fluid to positively prevent the leakage of pumped fluid to atmosphere. This configuration, like the tandem liquid/gas seal mentioned above eliminates the seal system required in a conventional liquid double seal arrangement. Note however, that the pumped product must be compatible with the small amount of gas introduced into the pumped fluid. This configuration cannot be used in recycle (closed loop) services.

An excellent resource for additional information covering design, selection and testing criteria, is API Standard 682.

Flush system types

Providing the proper environment for the seal is a key factor in achieving satisfactory seal operation. In conjunction with defining the seal design and materials of construction, it is necessary to decide which type of seal flush system will be selected to lubricate and cool the seal faces. The API industry has developed various systems to accommodate the requirements of almost every possible sealing arrangement (Refer to Figure 8.14 and 8.14A).

Clean product systems (Plan 11)

In this plan, product is routed from the pump discharge to the seal chamber for lubricating and cooling the seal faces. It will also vent air and/or vapors from the chamber as it passes to the pump suction through the throat bushing. NOTE: For single stage pumps without back hub pumping vanes, it is necessary to have holes in the impeller that reduce the pressure behind the impeller below the discharge pressure to allow flush flow to exit the stuffing box through the throat bushing. (Refer to Figure 8.15).

Clean product flush (Plan 13)

In this plan, the product is routed from behind the pump impeller, through the throat bushing into the stuffing box and out of the stuffing box, through an orifice back to the suction (Refer to Figure 8.16) This

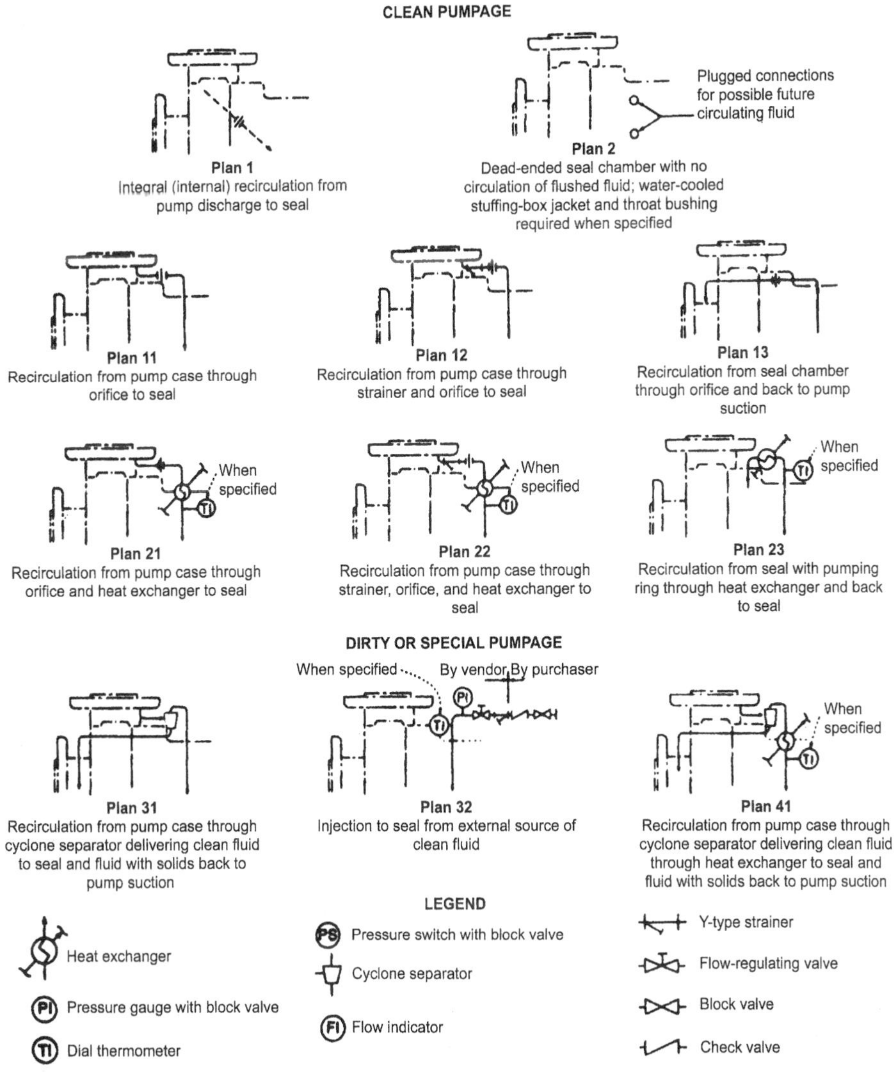

Figure 8.14 All flush plans reprinted with the permission of The American Petroleum Institute

plan is used primarily in vertical pump applications because it provides positive venting of the stuffing box. It is also used in single stage pump applications that do not employ pumping vanes or holes in the impellers.

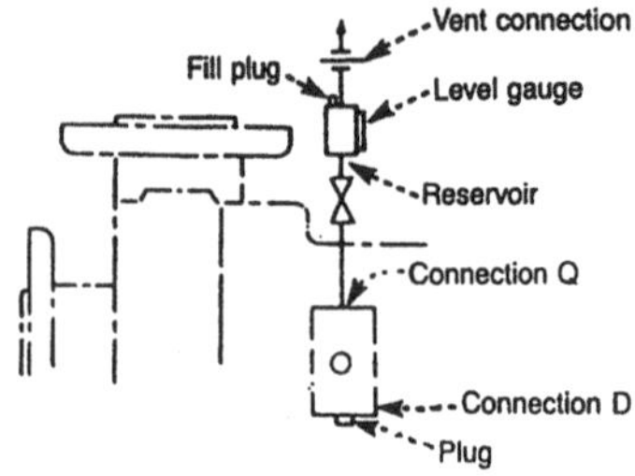

Plan 51
Dead-ended blanket (usually methanol, typically used with auxiliary sealing device (single- or double-seal arrangement

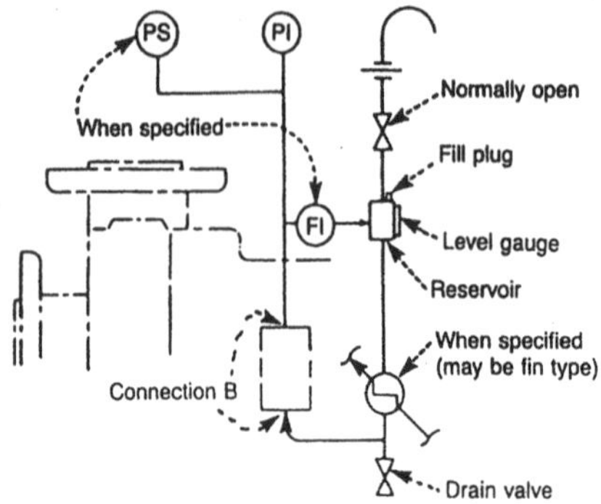

Plan 52
Nonpressurized external fluid reservoir with forced circulation; typically used with tandem-seal arrangement

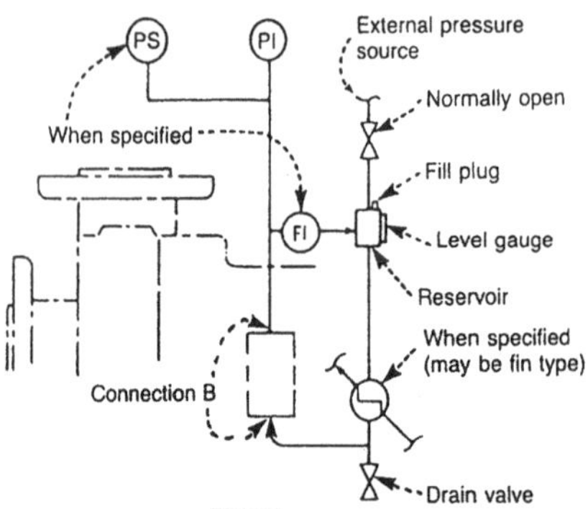

Plan 53
Pressurized external fluid reservoir with forced circulation; typically used with double-seal arrangement

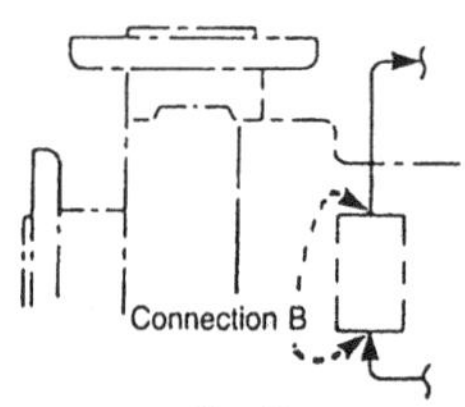

Plan 54
Circulation of clean fluid from external system typically used with double-seal arrangement

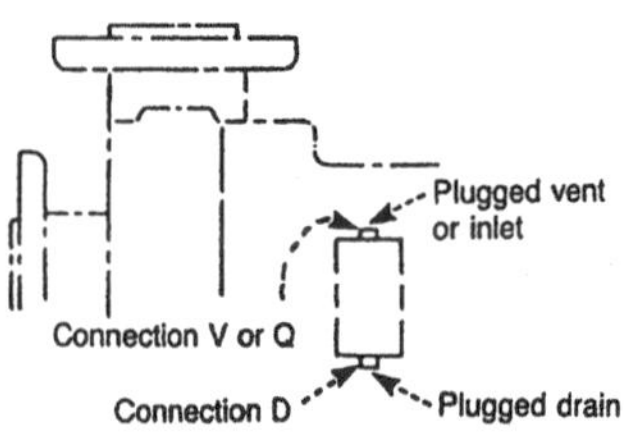

Plan 61
Tapped connections for purchaser's use; applies when purchaser is to supply fluid (steam, gas, water, etc.) to auxiliary sealing device (single- or double-seal arrangement

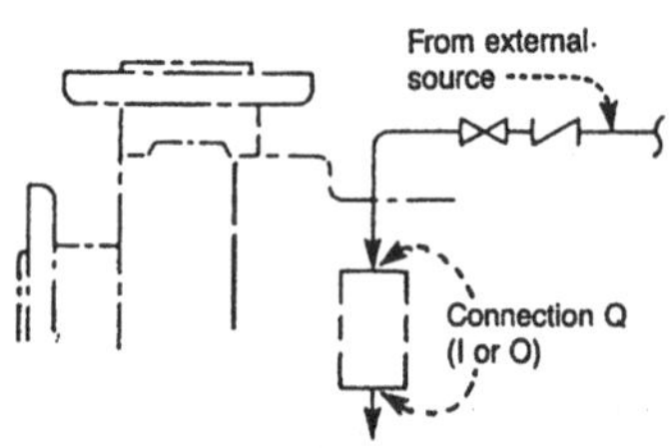

Plan 62
External fluid quench (steam, gas, water, etc.; typically used with throttle bushing or auxiliary sealing device (single- or double-seal arrangement

Figure 8.14A API flush plans continued

Dirty product system (Plan 31)

Liquid product from the pump discharge is routed to the seal chamber through a cyclone separator which is selected to optimize removal of solids across an individual pump stage (Refer to Figure 8.17 and to Figure 8.18 for guidelines covering the use of cyclones).

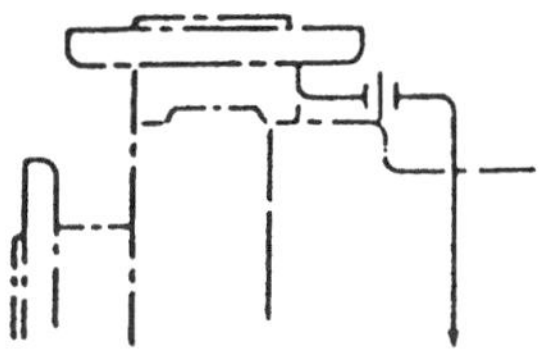

Plan 11
Recirculation from pump case through orifice to seal

Figure 8.15 API flush plan 11

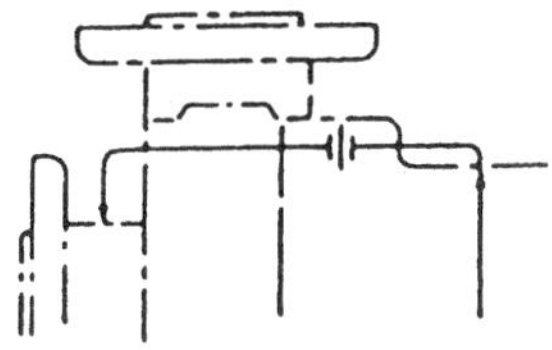

Plan 13
Recirculation from seal chamber through orifice and back to pump suction

Figure 8.16 API flush plan 13

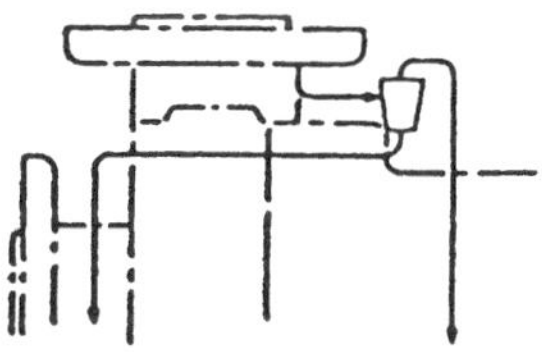

Plan 31
Recirculation from pump case through cyclone separator delivering clean fluid to seal and fluid with solids back to pump suction

Figure 8.17 API flush plan 31

Guidelines for use of cyclones

- Do not use cyclones when differential pressure is less that 25 Psi
- Consider using orifice when pressure differential exceeds cyclone design differential
- Solids to be removed should have density at least twice that of the fluid
- Efficiency of separation is reduced as differential pressure across cyclone varies from design differential
- Separation efficiency drops as particle size decreases

Figure 8.18 Guidelines for use of cyclones

When a clean, cool seal flush liquid is mandated for reasons of preventing solids accumulation in the seal chamber, an external liquid flush system (Plan 32) is used. The fluid pressure is higher than that behind the impeller so that flow is always towards the pump suction, thus preventing back flow of dirty product into the seal chamber (Refer

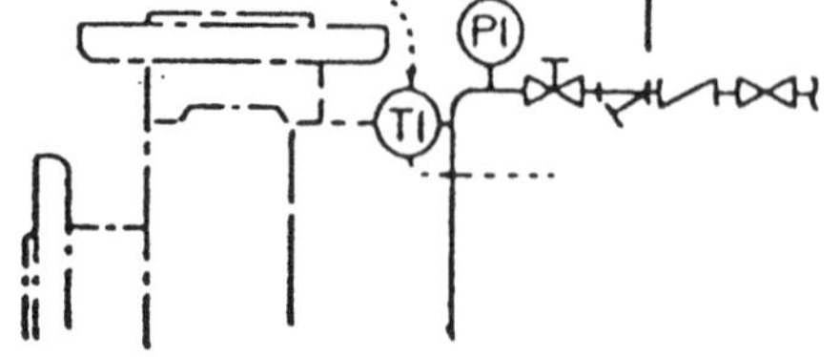

Figure 8.19 API flush plan 32

to Figure 8.19). When using Flush Plan 32, it is important to confirm that the flush fluid will not vaporize in the stuffing box.

High temperature product flush system (plan 23)

This flush plan is desirable when it is necessary to maintain the required margin between liquid vapor pressure (at seal chamber temperature) and seal chamber pressure. The feature about this plan is that the cooler only removes heat generated by the seal faces plus the heat soak through the shaft from the process. A throat bushing is installed in the seal chamber to isolate the product in this area from that in the impeller area of the pump. A circulating device (pumping ring) is mounted on the seal which circulates liquid in the seal chamber through a cooler and back to the seal chamber. It is more efficient than Plans 21 and 22 which incorporate a cooler to continuously cool the flush from the discharge. These plans are simply Flush Plan 1 1 with the addition of a cooler. Flush Plans 21, 22 or 23 should be used when the temperature of the pumped fluid is above 400°F and a bellows seal is not applied. It is important to confirm that the pumped fluid is clean when using Flush Plan 23 since there is not a constant external flush. This plan is typically used for boiler feed pump applications (Refer to Figure 8.20).

Low temperature product flush/buffer system (plan 52)

This system is well suited for low temperature applications such as ethylene, propylene and other low temperature liquids which are susceptible to forming ice on the seal faces when the atmospheric side of the seal is exposed to the atmosphere; thus separating the faces and resulting in excessive seal leakage. This plan consists of a tandem (dual) seal with a buffer liquid between them. A seal pot containing the buffer liquid (usually methanol – a drying agent) is vented to a lower pressure vent system. The seal pot system is usually equipped with a pressure

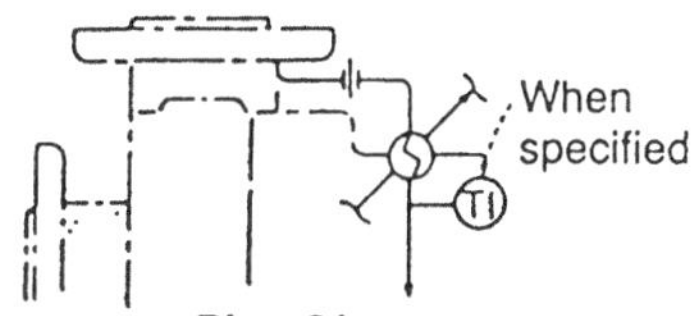

Plan 21
Recirculation from pump case through orifice and heat exchanger to seal

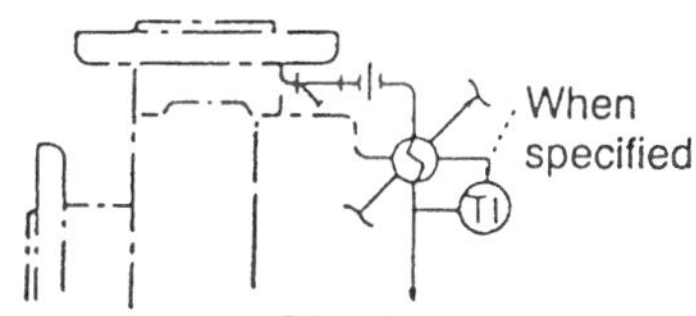

Plan 22
Recirculation from pump case through strainer, orifice, and heat exchanger to seal

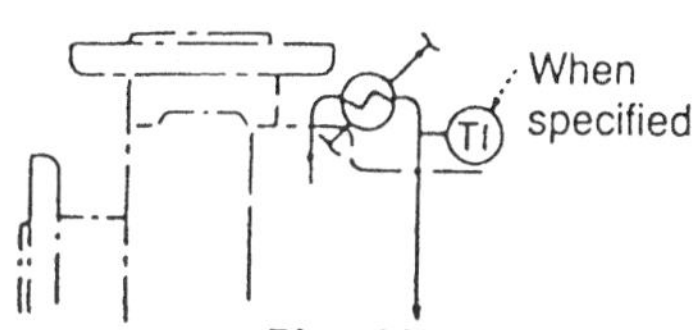

Plan 23
Recirculation from seal with pumping ring through heat exchanger and back to seal

Figure 8.20 API high temperature flush plans 21, 22, 23

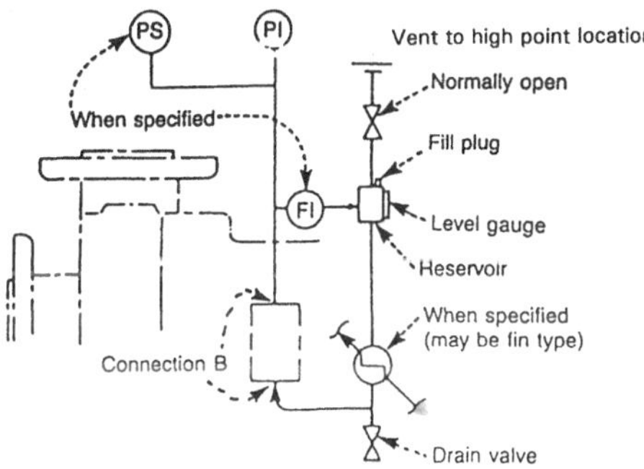

Plan 52
Nonpressurized external fluid reservoir with forced circulation; typically used with tandem-seal arrangement

Figure 8.21A plan 52

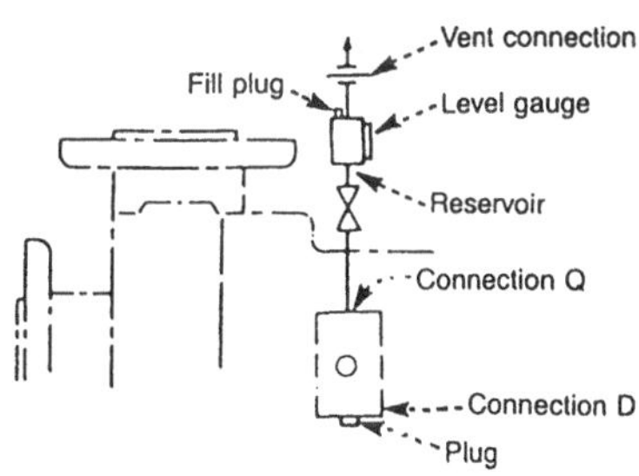

Plan 51
Dead-ended blanket (usually methanol, see Note 3); typically used with auxiliary sealing device (single- or double-seal arrangement in Figure D-1)

Figure 8.21B plan 51

switch to sound an alarm if the inner seal product leakage cannot be adequately carried away through the orifice vent system (Refer to Figure 8.21A). When the alarm sounds, the pump should be shut down as soon as possible since the back-up seal is now functioning as the primary seal and will leak the pumped fluid to atmosphere if it fails.

Seal flush Pan 51 may also be used when leakage to atmosphere cannot be tolerated. Plan 51 incorporates a dead ended system design and is shown in Figure 8.21B.

Toxic or flammable product flush system (plan 53)

This system is used when leakage to the atmosphere cannot be tolerated (see Figure 8.22A). It consists of a dual seal arrangement with a barrier liquid between them. A seal pot contains the barrier liquid at a pressure

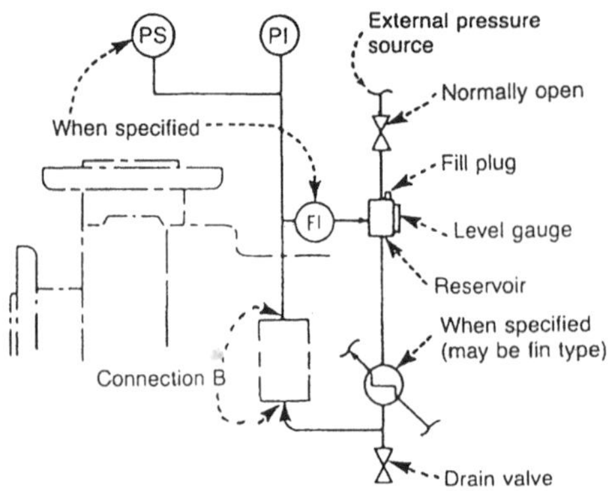

Plan 53
Pressurized external fluid reservoir (see Note 3) with forced circulation; typically used with double-seal arrangement in Figure D-1

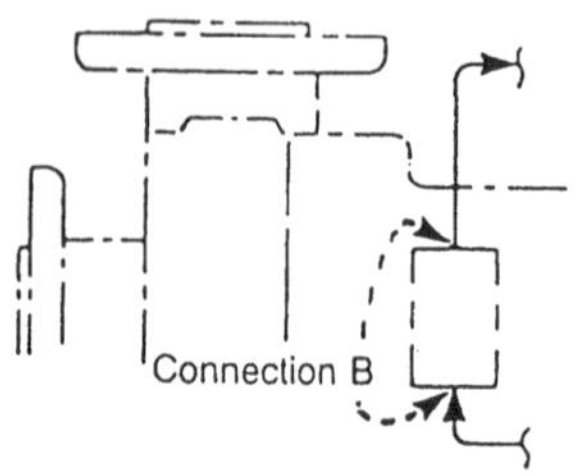

Plan 54
Circulation of clean fluid from external system typically used with double-seal arrangement

Figure 8.22 A & B plans 53 and 54

higher than seal chamber pressure (usually 20–25 PSI). Inner seal leakage will always be barrier liquid leakage into the product, resulting in some product contamination. The barrier liquid should be selected on the basis of its compatibility with the product. An internal pumping device (pumping ring) is used to circulate the barrier liquid into and out of the seal chamber through the seal pot. The integrity of the system always needs to be monitored to assure that seal pot pressure level is maintained with barrier liquid.

Plan 54 is also a dual system which utilizes a pressurized barrier liquid from an external reservoir or system to supply clean cool liquid to the seal chamber. As described in the previous plan, the barrier liquid pressure level is higher than the seal chamber pressure (usually 20–25 PSI) so that inner seal leakage is always into the pump. With this plan, it is also necessary to consider the compatibility of the barrier liquid and the pumped product. This system is considered to be one the most reliable systems available. However, it is more complex and more costly than other systems (Refer to Figure 8.22B).

Auxiliary stuffing box and flush plans

As mentioned previously, the auxiliary stuffing box can be used as an auxiliary sealing device in the event of seal failure. (Refer to Figure 8.23). It is important to note that the auxiliary stuffing box contains a restricted flow seal (packing or close fitting throttle bushing) and does not positively contain the pumped fluid. Therefore, the auxiliary

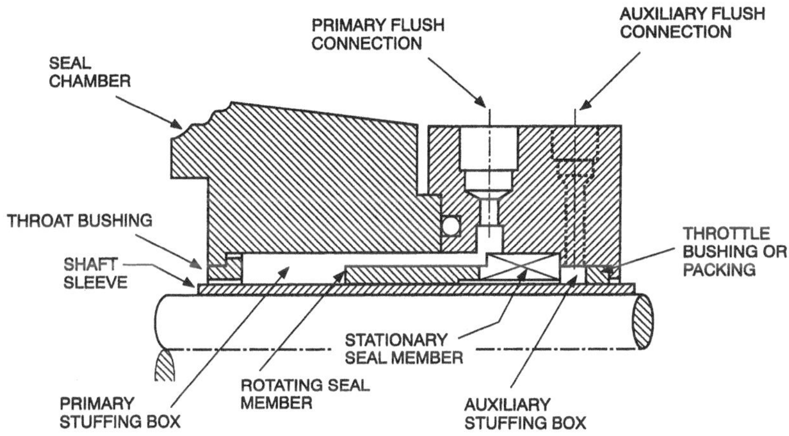

Figure 8.23 Auxiliary stuffing box

stuffing box seal device is for emergency containment of the pumped fluid only. The pump should be shut down immediately in the event of leakage observed from the auxiliary stuffing box if a quench is not supplied. It is always a good practice to require that the auxiliary stuffing box drain connection, which will come plugged, be piped to a drain system that meets environmental standards.

When the pumped fluid can vaporize and form hard deposits, the auxiliary stuffing box is used to contain a quench fluid that will dissolve (wash away) the hard deposits at the exit of the seal faces, thus eliminating seal face wear.

A typical refinery quench application is the use of low pressure (50 PSI) steam in and out of the auxiliary stuffing box to dissolve coke deposits on the seal face. This application also has the added advantage of keeping the standby pump seal warm which prevents thermal expansion problems in start-up. This arrangement uses a throttle bushing as the external seal.

Water is also used as an auxiliary stuffing box flush in caustic applications where solid deposits need to be flushed from the seal face. When a water flush is used, two (2) rows of packing are usually provided in the auxiliary stuffing box as an external seal to minimize external water leakage.

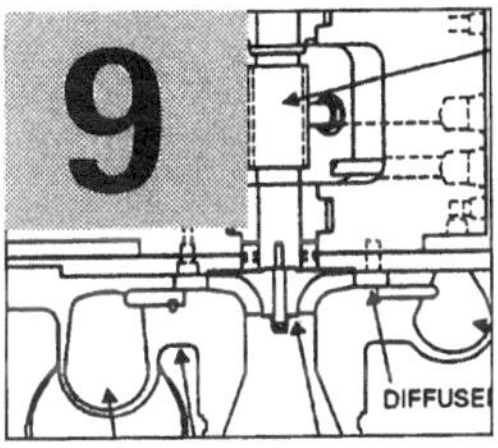

Compressor types and applications

- Introduction
- Positive displacement compressors
- Dynamic compressors

Introduction

In this chapter we will overview compressor types and their typical applications. The two basic classifications of compressors are positive displacement and dynamic compressors.

Positive displacement compressors are constant volume, variable energy (head) machines that are not affected by gas characteristics.

Dynamic compressors are variable volume, constant energy (head) machines that are significantly affected by gas characteristics.

The type of compressor that will be used for a specific application therefore depends on the flow rate and pressure required and the characteristic of the gas to be compressed.

In general, dynamic compressors are the first choice since their maintenance requirements are the lowest. The next choice are rotary type positive displacement compressors since they do not contain valves and are gas pulsation free. The last choice are reciprocating compressors since they are the highest maintenance compressor type and produce gas pulsations. However, the final selection depends upon the application requirements as discussed below.

Figure 9.1 presents a flow range chart showing the various types of compressor applications as a function of flow (ACFM) and discharge pressure (PSIG).

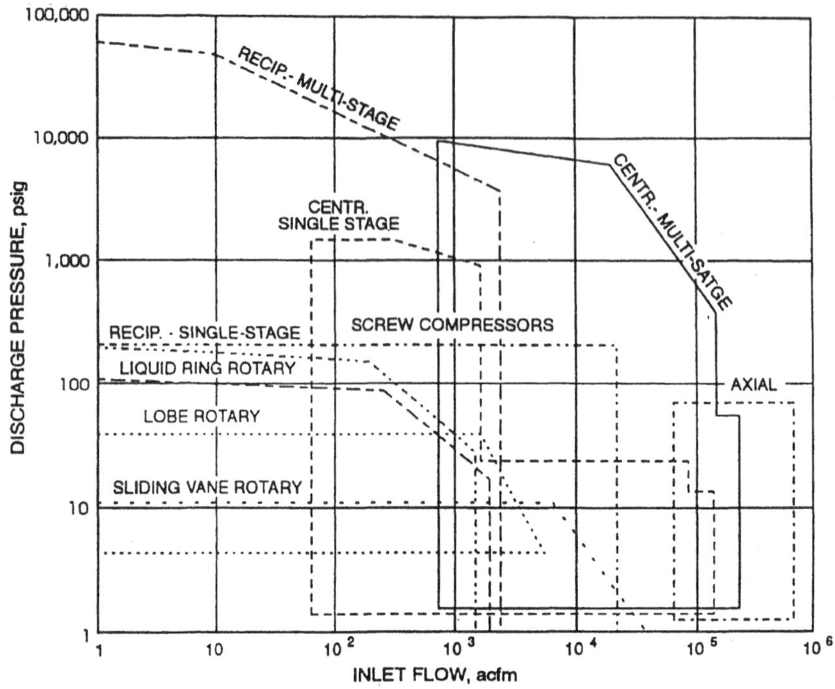

Figure 9.1 Compressor application range chart

Table 9.1 shows the typical operating ranges for the various types of compressors used in the refining, chemical and gas processing industries.

Table 9.1 Typical operating range of various types of gas compressors

Machine Type	Capacity ICFM Min	Max	T_2 Max °F	P_1 Max PSIA	P_2 Max PSIA	P/R MIN	P/R MAX
Rotary lobe		1 to 40,000	350	35	55	1.0+	2.4
Rotary vane		45 to 3,300	350	45	65	1.3	3.2
Rotary screw		50 to 20,000	350	150	615	2.0	6.0
Recip		1 to 10,000	800	1000	10,000	3.0	50.0
Liquid ring		10 to 10,000	N/A	100	140	1.0+	10.0
Centrifugal Single stage		700 to 150,000	500	1000	1,400	1.0+	3.4
Centrifugal Multi stage		300 to 150,000	800	2000	6,000	2.0	10.0
Axial		75,000 to 350,000	800	30	150	1.0	10.0

Table 9.2 describes the typical applications for the various types of compressors presented in Figure 9.1 and Table 9.1.

Table 9.2 Typical compressor applications

Compressor type	Application
Rotary lobe	Conveying – powder, polyethylene
Rotary vane air	Air blowers (low volume). Also used as gas turbine starters.
Rotary screw	Plant and instrument air, low flow process – off gas, recycle, sulphur blowers
Rotary liquid ring	Crude unit vacuum, various saturated gas applications
Reciprocating	Plant and instrument air off gas (low flow) recycle (low flow) H_2 make-up, gas reinjection (low flow)
Centrifugal single stage	Air blowers, recycle
Centrifugal single stage – Integral gear	Low flow recycle, off gas, plant air (can replace a recip in low flow, medium to high molecular weight applications).
Centrifugal multi-stage side load (Horizontal split or barrel)	(Propane, propylene, ethylene, Freon, mixed gas) refrigeration
Centrifugal multi-stage barrel	Recycle, reinjection, syn gas
Centrifugal multi-stage – integral gear	Plant and instrument air, *Process applications require proven field experience
Axial compressor	FCC blower, MTBE effluent, gas turbine air

Positive displacement compressors

Positive displacement compressors are used for low flow and/or low molecular weight (hydrogen mixture) applications. The various types are presented below.

Rotary lobe

A rotary lobe compressor consists of identically synchronized rotors. The rotors are synchronized through use of an external, oil-lubricated, timing gear, which positively prevents rotor contact and which minimizes meshing rotor clearance to optimize efficiency. This feature also allows the compressor to be oil free in the gas path. The rotors of the two-lobe compressor each have two lobes. When the rotor rotates, gas is trapped between the rotor lobes and the compressor casing. The rotating rotor forces the gas from the gas inlet port, along the casing, to the gas discharge port. Discharge begins as the edge of the leading

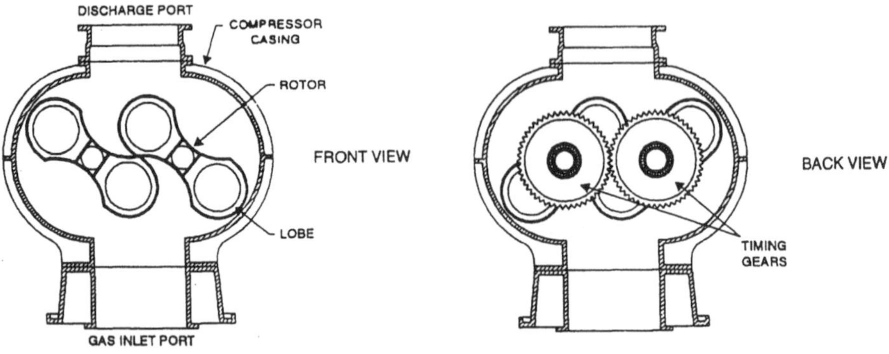

Figure 9.2 Rotary lobe

lobe passes the edge of the discharge port. The trailing lobe pushes the entrapped gas into the discharge port, which compresses the gas against the backpressure of the system. Rotary lobe compressors are usually supplied with noise enclosures or silencers to reduce their characteristic high noise level. A schematic of a two-lobe rotary compressor is shown in Figure 9.2.

Rotary vane

A sliding-vane rotary compressor uses a series of vanes that slide freely in longitudinal slots that are cut into the rotor. Centrifugal force causes the vanes to move outward against the casing wall. The chamber that is formed between the rotor, between any two vanes, and the casing is referred to as a cell. As the rotor turns, an individual vane passes the inlet port to form a cell between itself and the vane that precedes it. As an individual vane rotates toward the end of the inlet port, the volume

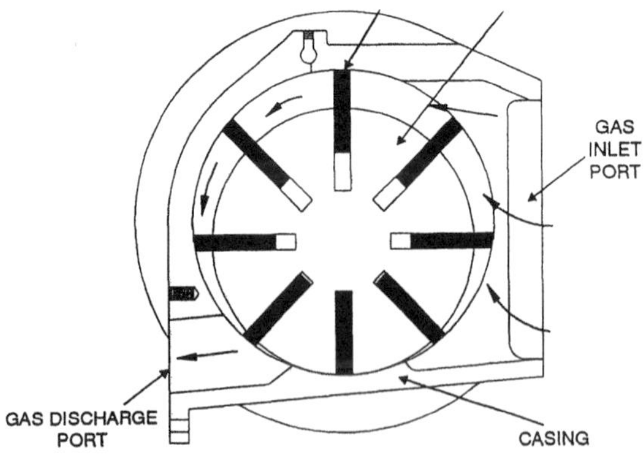

Figure 9.3 Rotary vane

of the cell increases. The increase in the cell volume draws a partial vacuum in the cell. The vacuum draws the gas in through the inlet port. When a vane passes the inlet port, the cell is closed, and the gas is trapped between the two vanes, the rotor and the casing. As rotation continues toward the discharge port, the volume of the cell decreases. The vanes ride against the casing and slide back into the rotor. The decrease in volume increases the gas pressure. The high pressure gas is discharged out of the compressor through the gas discharge port. Sliding-vane rotary compressors are characterized by a high noise level that results from the vane motion. A schematic of a sliding vane rotary compressor is shown in Figure 9.3.

Rotary screw

The single-stage design consists of a pair of rotors that mesh in a one-piece, dual-bore cylinder. The male rotor usually consists of four helical threads that are spaced 90 degrees apart. The female rotor usually consists of six helical grooves that are spaced 60 degrees apart. The rotor speed ratio is inversely proportional to the thread-groove ratio. In the four-thread, six-groove, screw compressor, when the male rotor rotates at 1800 rpm, the female rotor rotates at 1200 rpm.

The male rotor is usually the driven rotor, and the female rotor is usually driven by the male rotor. A film of foil is normally injected between the rotors to provide a seal between the rotor and to prevent metal-to-metal contact. An oil-mist eliminator, installed immediately downstream of the compressor, is required for plant and instrument air service. However, designs are available that do not require lubrication.

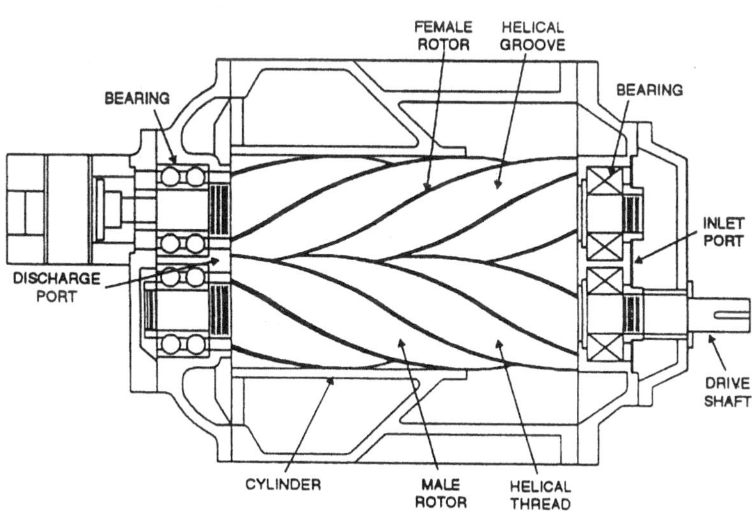

Figure 9.4 Rotary screw

Screw compressors that do not require lubrication are commonly referred to as "dry screw-type compressors".

The inlet port is located at the drive-shaft end of the cylinder. The discharge port is located at the opposite end of the cylinder. Compression begins as the rotors enmesh at the inlet port. Gas is drawn into the cavity between the male rotor threads and female rotor grooves. As rotation continues, the rotor threads pass the edges of the inlet ports and trap the gas in a cell that is formed by the rotor cavities and the cylinder wall. Further rotation causes the male rotor thread to roll into the female rotor groove and to decrease the volume of the cell. The decrease in the volume increases the cell pressure. Oil is normally injected after the cell is closed to the inlet port. The oil seals the clearances between the threads and the grooves, and it absorbs the heat of compression. Compression continues until the rotor threads pass the edge of the discharge port and release the compressed gas and oil mixture. A typical single stage screw compressor is shown in Figure 9.4.

Rotary liquid ring

Liquid ring rotary compressors consist of a round, multi-blade rotor that revolves in an elliptical casing. The elliptical casing is partially filled with a liquid, which is usually water. As the rotor turns, the blades form a series of buckets. As the rotor turns, the buckets carry the liquid

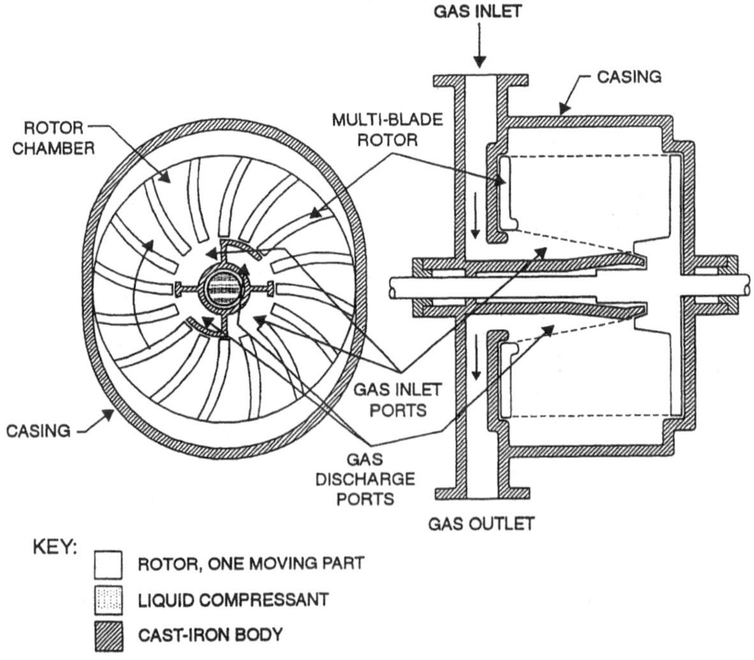

Figure 9.5 Rotary liquid ring (Courtesy of Nash Engineering Company)

around with the rotor. Because the liquid follows the contour of the casing, the liquid alternately leaves and returns to the space between the blades. The space between the blades serves as a rotor chamber. The gas inlet and discharge ports are located at the inner diameter of the rotor chamber. As the liquid leaves the rotor chamber, gas is drawn into the rotor chamber through the inlet ports. As the rotor continues to rotate, the liquid returns to the rotor chamber and decreases the volume in the chamber. As the volume decreases, the gas pressure increases. As the rotor chamber passes the discharge port, the compressed gas is discharged into a gas/liquid separator and then to the process. A typical liquid ring rotary compressor is shown in Figure 9.5.

Reciprocating

The basic components of a reciprocating compressor are a crankshaft, crossheads, piston rod packing, cylinders, pistons, suction valves, and discharge valves. Figure 9.6 is an illustration of a three-stage reciprocating compressor. Note that the third stage piston and cylinder are mounted on top of the second stage piston and cylinder. A prime mover (not shown) rotates the crankshaft. The crankshaft converts the rotary motion of the prime mover into reciprocating motion of the pistons.

The compression cycle of a reciprocating compressor consists of two strokes of the piston: the suction stroke and the compression stroke. The suction stroke begins when the piston moves away from the inlet

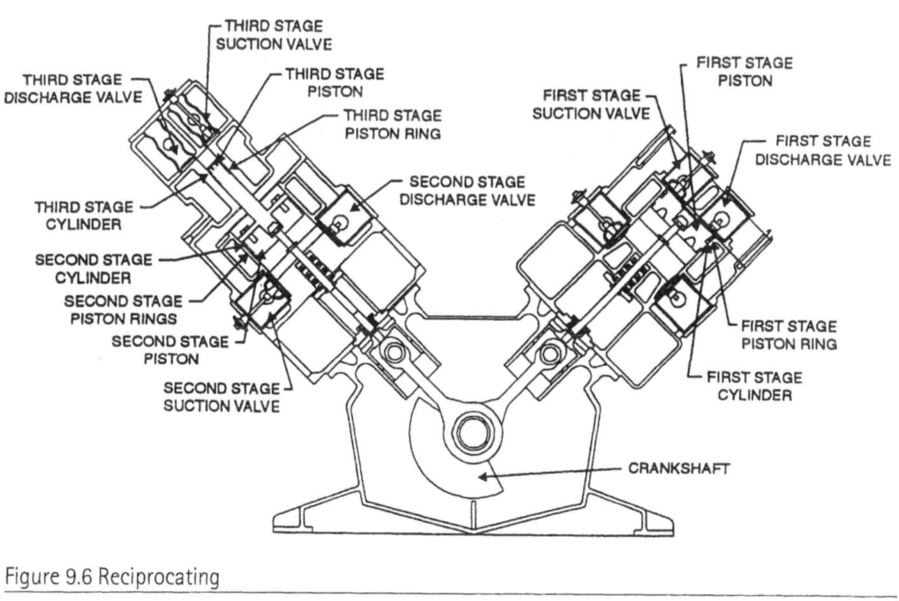

Figure 9.6 Reciprocating

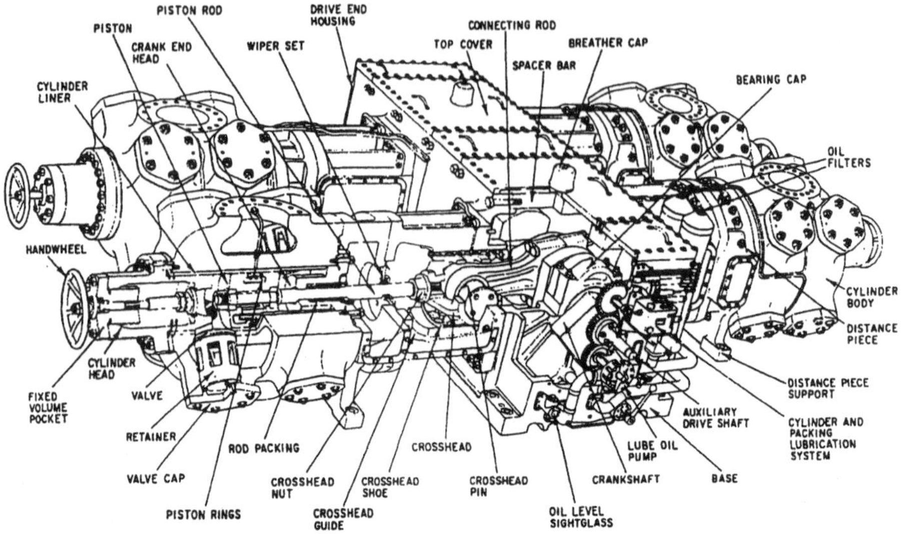

Figure 9.7 Reciprocating

port of the cylinder. The gas in the space between the piston and the inlet port expands rapidly until the pressure decreases below the pressure on the opposite side of the suction valve. The pressure difference across the suction valve causes the suction valve to open and admit gas into the cylinder. The gas flows into the cylinder until the piston reaches the end of its stroke. The compressor stroke starts when the piston starts its return movement. When the pressure in the cylinder increases above the pressure on the opposite side of the suction valve, the suction valve closes to trap the gas inside the cylinder. As the piston continues to move toward the end of the cylinder, the volume of the cylinder decreases and the pressure of the gas increases. When the pressure inside the cylinder reaches the design pressure of the stage, the discharge valve opens and discharges the contents of the cylinder to the suction of the second stage. The second stage takes a suction on the discharge of the first stage, further compresses the gas and discharges to the third stage. The third stage takes a suction on the discharge of the second stage and compresses the gas to the final discharge pressure. Please refer to Figure 9.6.

Figure 9.7 shows a balanced opposed, four throw reciprocating compressor typical of the type that would be used in refinery hydrogen make-up service.

Dynamic compressors

Dynamic compressors are used wherever possible s a result of their low maintenance requirements. The single stage integral gear centrifugal compressor allows the use of a dynamic compressor in many applications where positive displacement compressors have previously been used. The two types of dynamic compressors are centrifugal and axial.

Centrifugal compressors – general principles of operation

Centrifugal compressors increase the energy of a gas by increasing the tangential velocity (V_T) of a gas, as shown in Figure 9.8. The principle of operation of a centrifugal compressor is very similar to the principle of operation of a centrifugal pump. The gas enters the compressor through the inlet nozzle that is proportioned so the gas enters the impeller with a minimum of shock or turbulence. The impeller, which consists of a hub and blades, is mounted on a rotating shaft. The impeller receives the gas from the inlet nozzle and dynamically compresses it by increasing the energy of gas proportional to the product of the blade tip speed velocity (U_T) and the gas tangential velocity change (V_T) in the impeller. V_R represents the velocity of the gas relative to the blade. The resultant velocity (V) is the vector sum of the relative velocity (V_R) and the blade tip speed velocity (U_T).

A diffuser surrounds the impeller, and it serves to gradually reduce the velocity of the gas as the gas leaves the impeller. The diffuser converts the velocity energy to a higher pressure level. In a single stage compressor, the gas exits the diffuser through a volute casing that surrounds the diffuser. The volute casing collets the gas to further reduce the velocity of the gas and to recover additional velocity energy.

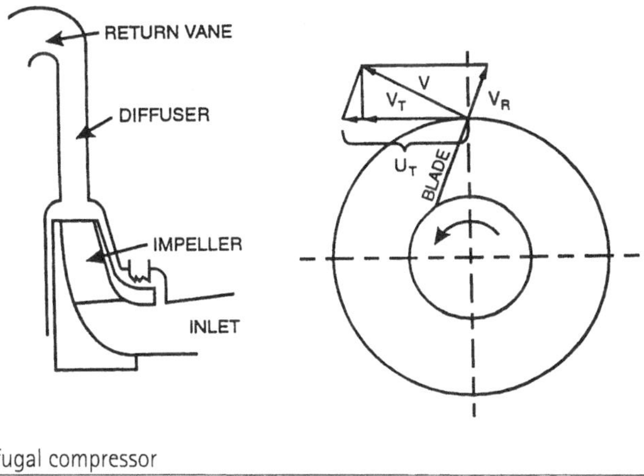

Figure 9.8 Centrifugal compressor

The gas exits through the discharge nozzle. In a multi-stage compressor, the gas exits the diffuser and enters return vanes. The return vanes direct the gas into the impeller of the next stage.

Centrifugal single stage (low ratio)

A typical single stage centrifugal low ratio compressor is shown in Figure 9.9. These types are known as single stage overhung compressors since the impeller is outboard of the radial bearings the cases are radially split.

Centrifugal single stage integral gear

Figure 9.10 shows a "Sundyne" single stage integral gear compressor section. This type of compressor is an in line type (similar to a pump) usually driven by motor through an integrally mounted gear box (not shown). These compressors are used for low flow high energy (head) applications and are being used in many applications previously serviced by positive displacement compressors. These compressors operate at high speeds, 8,000 – 34,000 rpm and are limited to approximately 400 horsepower.

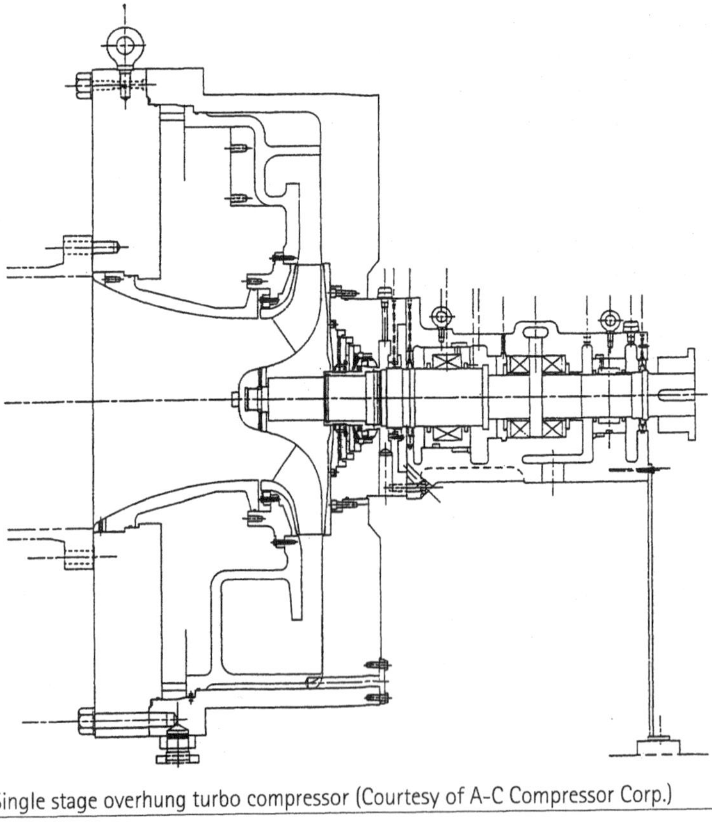

Figure 9.9 Single stage overhung turbo compressor (Courtesy of A-C Compressor Corp.)

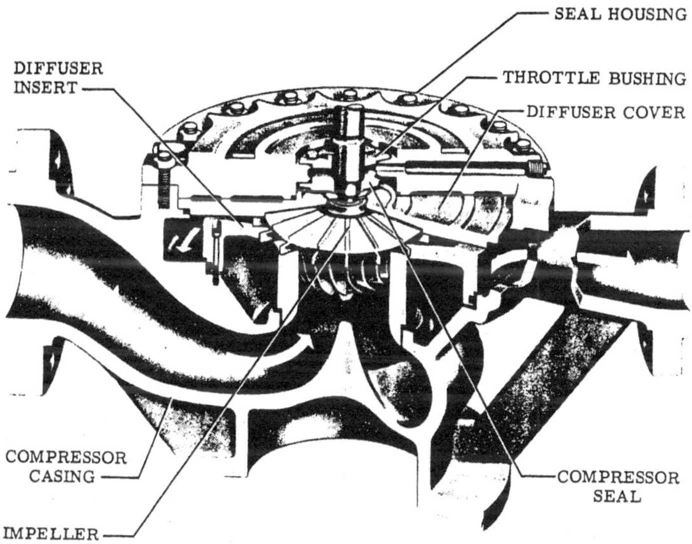

Figure 9.10 Single stage high speed compressor (Courtesy of Sundstrand Corp.)

Centrifugal multi-stage horizontal split

A typical multi-stage horizontally split centrifugal compressor is shown in Figure 9.11. The casing is divided into upper and lower halves along the horizontal centerline of the compressor. The horizontal split casing allows access to the internal components of the compressor without disturbing the rotor to casing clearances or bearing alignment. If possible, piping nozzles should be mounted on the lower half of the compressor casing to allow disassembly of the compressor without removal of the process piping.

Centrifugal multi-stage with side loads

This type of compressor is used exclusively for refrigeration services. The only difference from the compressor shown in Figure 9.11 is that gas is induced or removed from the compressor via side load nozzles. A typical refrigeration compressor is shown in Figure 9.12. Note that this type of compressor can be either horizontally or radially split.

Centrifugal multi-stage (barrel)

A typical multi-stage, radially-split, centrifugal compressor in shown in Figure 9.13. The compressor casing is constructed as a complete cylinder with one end of the compressor removable to allow access to the internal components. Multi-stage, radially-slit centrifugal compressors are commonly called barrel compressors.

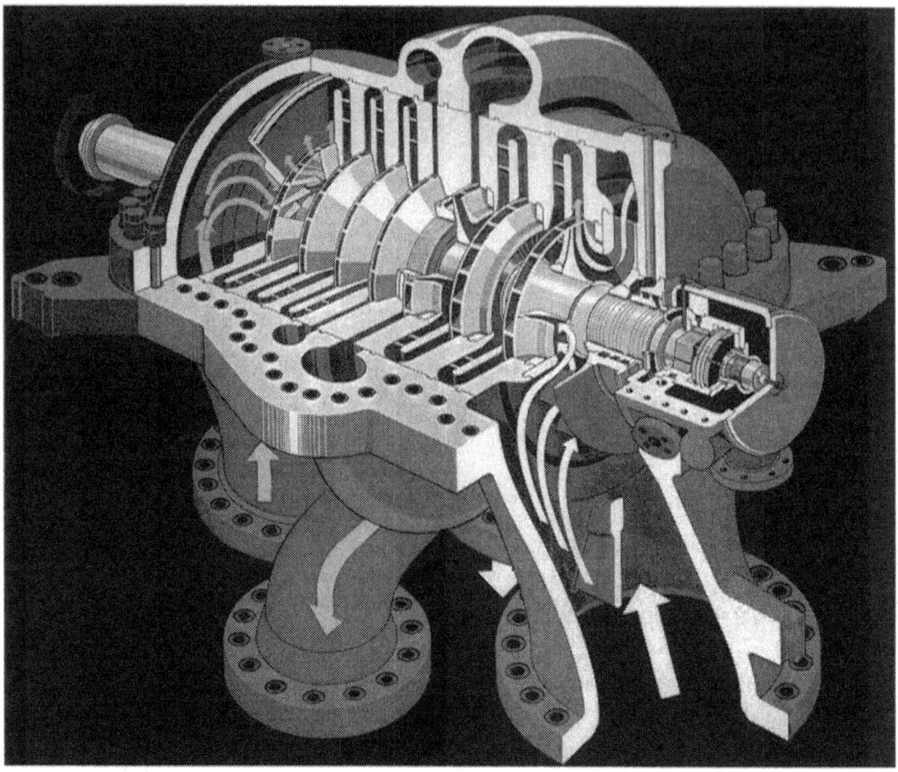

Figure 9.11 Centrifugal multi-stage horizontal split (Courtesy of Mannesmann Demag)

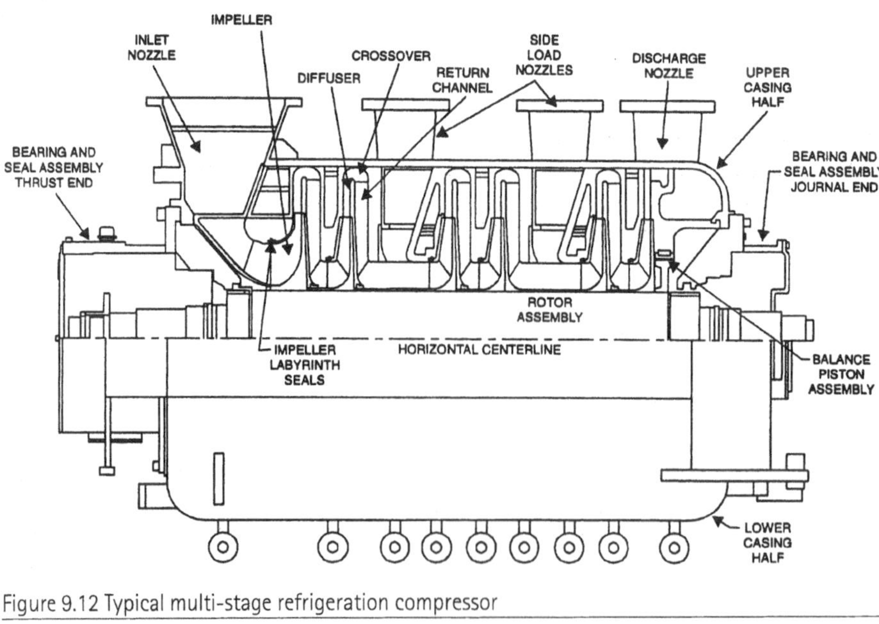

Figure 9.12 Typical multi-stage refrigeration compressor

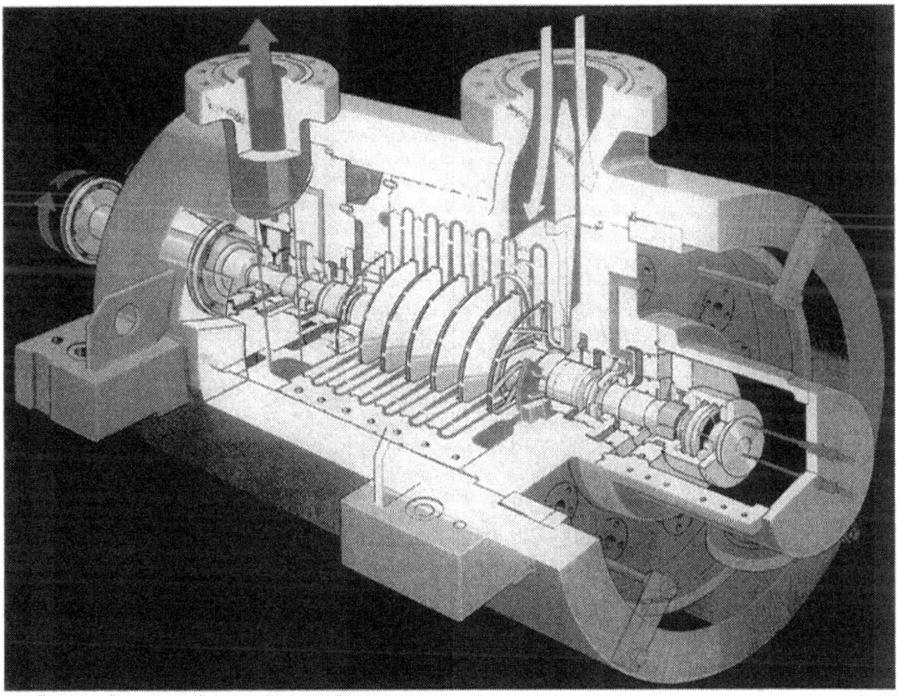

Figure 9.13 Typical multi-stage, radially-split centrifugal compressor (Courtesy of Mannesmann Demag)

Barrel compressors are used for the same types of applications as the multi-stage, horizontally-split centrifugal compressors. Because of the barrel design, however, barrel compressors are normally selected for higher pressure applications or certain low mol gas compositions (hydrogen gas mixtures).

Centrifugal multi-stage integral gear

A typical four-stage, integrally-geared, centrifugal compressor is shown in Figure 9.14. Integrally-geared, centrifugal compressors have a low-speed (bull) gear that drives two or more high-speed gears (pinions). Impellers are mounted at one end or both ends of each pinion. Each impeller has its own casing that is bolted to the gear casing. The gear casing may be horizontally split to allow access to the gears.

The gas enters the compressor through the first stage inlet nozzle to the impeller. The impeller receives the gas from the inlet nozzle, dynamically compresses it, and discharges to the diffuser. The diffuser surrounds the impeller, and it serves to gradually reduce the velocity of the gas as the gas leaves the impeller. The gas exits the diffuser through a volute casing. The volute casing collects the gas, further reduces the velocity of the gas, and recovers additional velocity energy. The gas exits the first stage through the first stage discharge nozzle, enters an

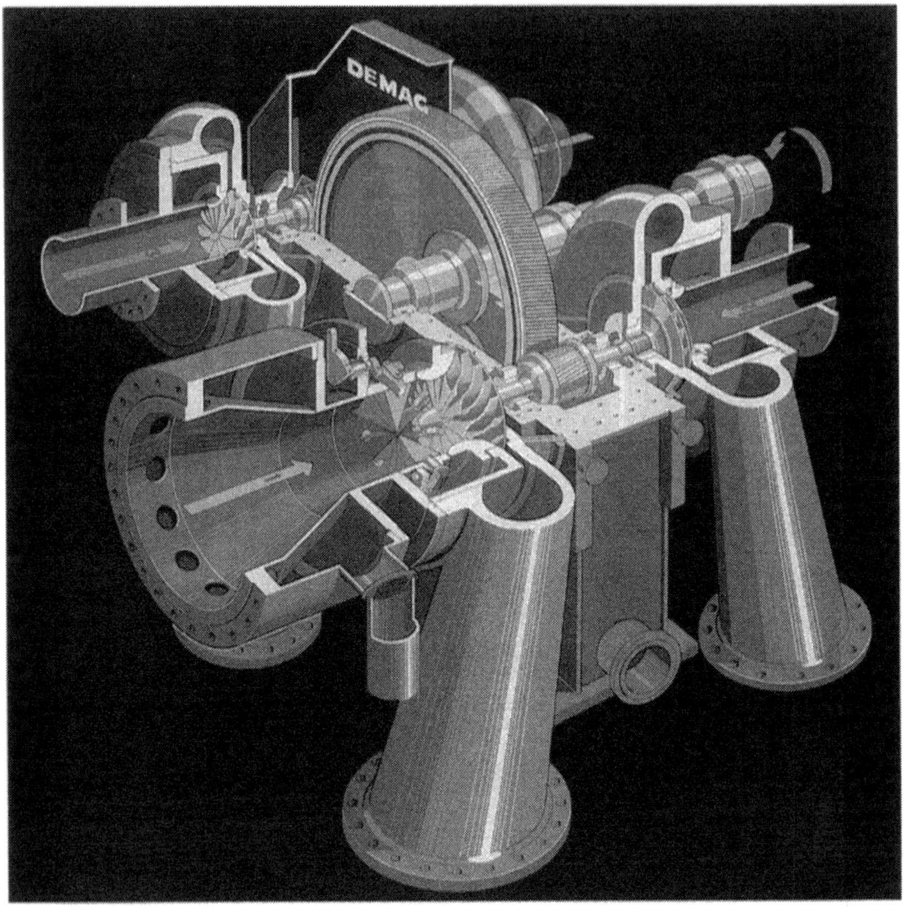

Figure 9.14 Typical integrally-geared, centrifugal compressor (Courtesy of Mannesmann Demag)

intercooler, and then it is piped to the second stage. The discharge from the second stage enters an intercooler, and then it is piped to the third stage.

Axial horizontal split

A typical axial compressor is shown in Figure 9.15. An axial compressor consists of a rotor shaft with a series of rotating blades and a tapered cylindrical casing with fixed stator vanes. Each set of blades is followed by a set of stator vanes. The gas enters the inlet nozzle, which guides the gas to the inlet volute. The inlet volute guides and accelerates the gas stream into the stator vanes. The stator vanes turn the gas stream to properly align the gas with the blades. The blades increase the energy of the gas by increasing the velocity of the gas. The stator vanes act as diffusers to provide resistance to the gas flow, and they cause the gas stream to decrease in velocity and to increase in pressure. The stator vanes also properly orient the gas stream for the next row of blades. Since blades and stator vanes alternate down the length of the casing, the gas is both accelerated and decelerated several times before it leaves the compressor. Pressure is increased each time the gas flow meets a set of stator vanes. The gas exits the compressor through the discharge volute and discharge nozzle.

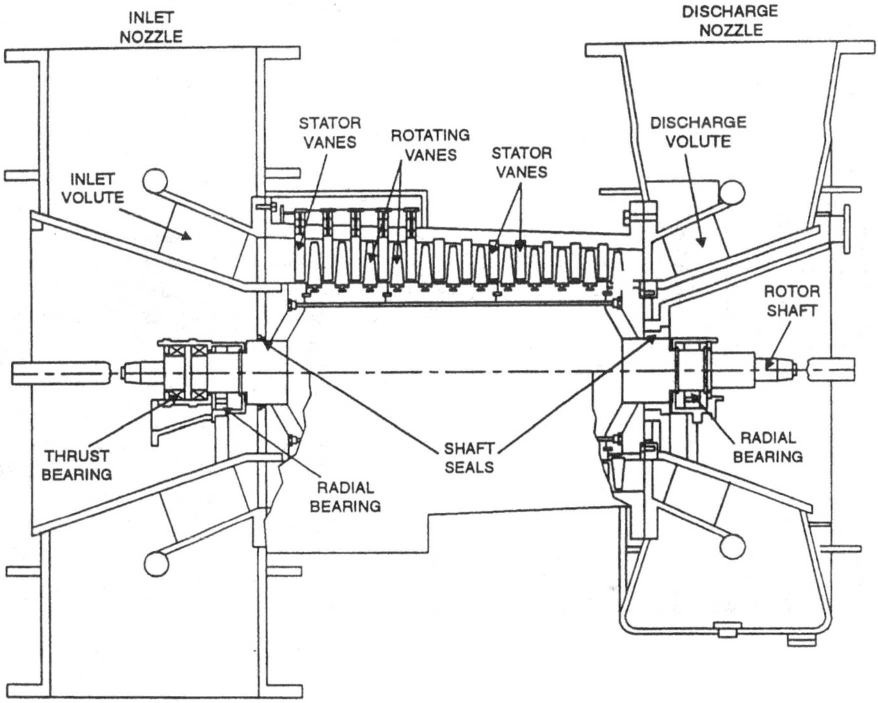

Figure 9.15 Typical axial compressor

10 The concept of fluid head

- Introduction
- Definition
- Paths of compression
- The different types of gas head
- Dynamic compressor curve format

Introduction

Without a doubt one of the most confused principles of turbo-compressor design in my experience has been that of fluid head. I have found that understanding this concept can best be achieved by recognizing fluid head is the energy required to achieve specific process requirements.

Head is the energy in foot-pounds force required to compress and deliver one pound of a given fluid from one energy level to another. One of the confusing things about this concept is that the industry persists in defining head in feet. Head should be expressed in foot-pounds force per pound mass or British Thermal Units per pound. A British Thermal Unit per pound of fluid is equal to exactly 778 foot-pounds force per pound mass of that fluid.

Remember, when we deal with fluid head, a fluid can be either a liquid or a gas depending upon the conditions of the fluid at that time. Ethylene for instance, can be either a liquid or a gas depending on it's pressure and temperature. If it is a liquid, an ethylene pump will be used and the energy required to increase the pressure of the liquid from P_1 to P_2 will be defined as head in foot-pound force per pound mass. Conversely, if the conditions render it a vapor, an ethylene compressor would be used to achieve the same purpose. We will see in this section that the amount of energy required to compress a liquid or a gas the same amount will be significantly higher in favor of the gas because the gas is at a much lower density than that of a liquid. Understanding a Mollier Diagram is an important aid to understanding the concept of fluid head. Every fluid can have a Mollier Diagram drawing which expresses energy on the X axis and pressure on the Y axis for various temperatures. Increasing the pressure of a vapor will result in increased energy required.

Having defined the concept of head as that of energy, one must now investigate the different types of ideal (reversible) gas heads, namely isothermal, isentropic or adiabatic and polytropic. All of these types of fluid head simply describe the path that the gas takes in being compressed. It must be remembered that any type of head can be used to describe a reversible compressor path as long as the Vendor uses the appropriate head and efficiency in his data reduction calculations. In this section we will show the assumptions for various kinds of heads and the relative difference in their values. In addition, the definitions of each type will be stated.

Definition

The definition of head required by any fluid compression process is presented in Figure 10.1.

HEAD REQUIRED *is the energy in foot pounds required to compress and deliver one pound of a given fluid from one energy level to another.*

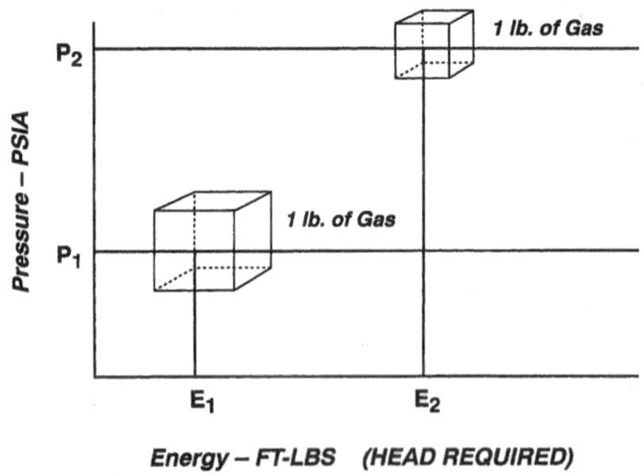

Figure 10.1 Head required definition

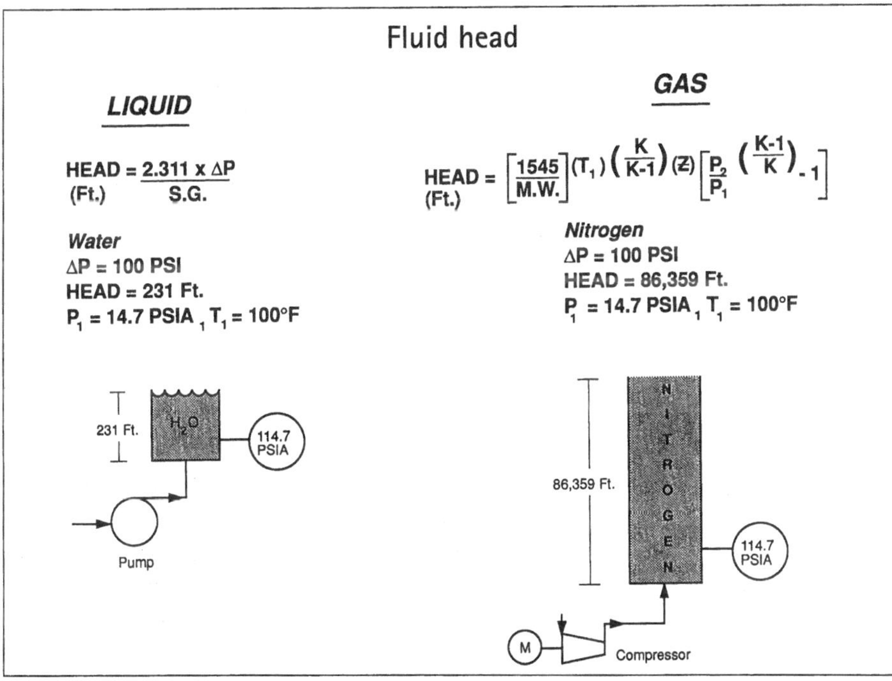

Figure 10.2 Fluid head definition

In any compression process, the amount of energy required to compress one pound of mass of a specific gas, at a given temperature from compressor suction flange (P_1) to the discharge flange (P_2) is defined as head required.

Figure 10.2 demonstrates how the density of a fluid significantly affects the amount of energy (head) required in a compression process.

Water with a density of 62.4 lbs/ft^3 requires only 231 ft-lb force per lb mass to compress the liquid 100 PSI. Note also, the equation for head required by a liquid is independent of temperature. On the other hand, nitrogen with a density of 0.07 lbs/ft^3 requires approximately 350 times the energy! This is because in both the case of water and nitrogen, the process requires that the fluid be compressed 100 PSI. However, since the mass of nitrogen is only 0.1% of the mass of water, a much greater amount of energy is required to compress the gas.

Figure 10.3 shows the head produced characteristics of positive displacement and dynamic compressors.

Note that regardless the amount of head required by the process, the flow rate of a positive displacement compressor is not affected. On the other hand, a dynamic compressor's flow rate is significantly affected by changes in the head required by the process. This is because the

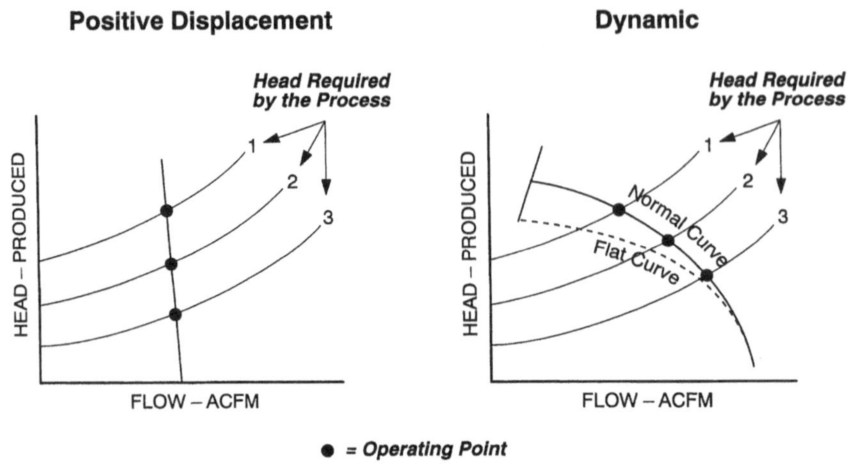

Figure 10.3 Compressor head produced characteristics

characteristic of any dynamic compressor is that it can only produce a greater amount of energy at a lower flow rate. The reason for this characteristic will be explained in a subsequent module. Therefore, any increase in the head required by a process will reduce the flow rate of a dynamic compressor. This is an extremely important fact because reduced flow rate in a dynamic compressor can lead to extreme, long term mechanical damage to the compressor unit. Also note in Figure 10.3 that the flatter the head produced curve a dynamic compressor possesses, the greater the effect of head required upon flow rate.

Head required

Thorough understanding of the concepts of head required by the process and head produced by the compressor is absolutely essential if dynamic compressor operation is to be understood.

It has been my experience that a lack of understanding exists in the area of dynamic compressor performance and often leads to a much greater emphasis upon the compressor's mechanical components (impellers, labyrinths, seals, bearings and shafts). In many cases, the root cause of dynamic compressor mechanical damage is that the head required by the process system exceeded the capability of the dynamic compressor.

Figure 10.4 presents the factors that determine the head (energy) required by any process.

Note that the head required by the process is inversely proportional to the gas density. If the gas density decreases, the head required by the process will increase. Gas density will decrease if gas temperature increases, inlet pressure decreases or molecular weight decreases. If the

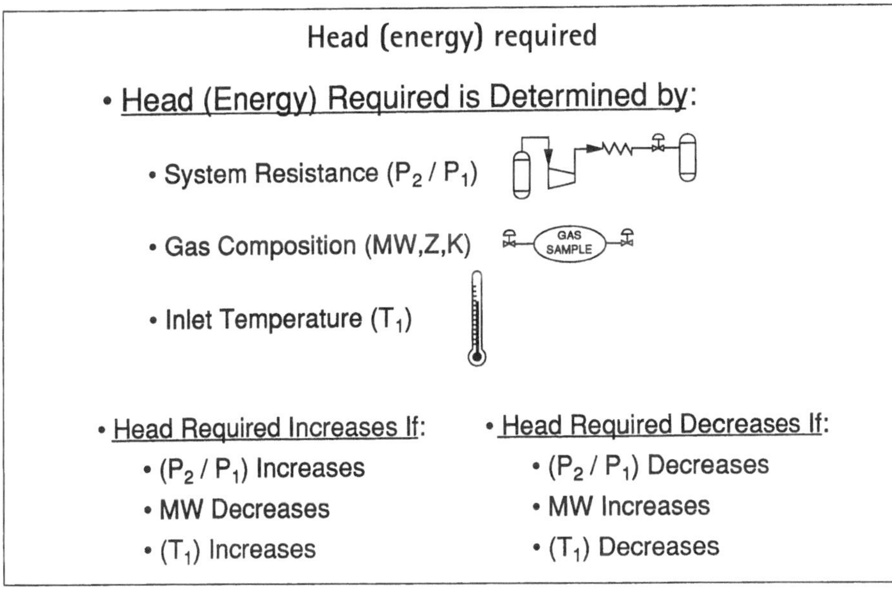

Figure 10.4 Head (energy) required

head required by the process increases, the flow rate of any dynamic compressor will decrease as shown in Figure 10.3. If the gas density increases, the head required by the process will decrease and dynamic compressor flow rate will increase.

Head produced

In Figure 10.5, the factors that determine the head produced by a dynamic compressor are presented.

Simply stated, for a given impeller vane shape, head produced by a dynamic compressor is a function of impeller diameter and impeller speed. Once the impeller is designed, it will produce only one value of head for a given shaft speed and flow rate.

The only factor that will cause a lower value of head to be produced than stated by the compressor performance curve is if the compressor has experienced mechanical damage or if it is fouled.

Figure 10.6 shows how the need for dynamic compressor inspection can be determined.

If for a given flow rate and shaft speed, the head produced falls below the value predicted by greater than 10%, the compressor should be inspected at the first opportunity. Having explained the concept of head required by the process, the method of calculating the head required by a process system needs to be discussed.

Head (energy) produced

- Head (Energy) Produced is Determined by:
  - Compressor Impeller Design

- Head Produced by the Impeller,
  - Increases with Tip Speed
    - Impeller Diameter
    - Compressor Speed (RPM)
  - Increases with Decreasing Flow

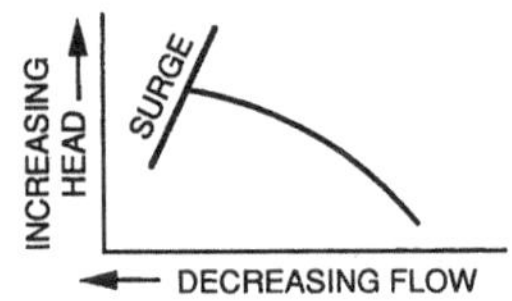

Figure 10.5 Head (energy) produced

The turbo-compressor curve

- The Turbo-Compressor Head (Energy) vs. Flow Curve is *Fixed by Compressor Design*

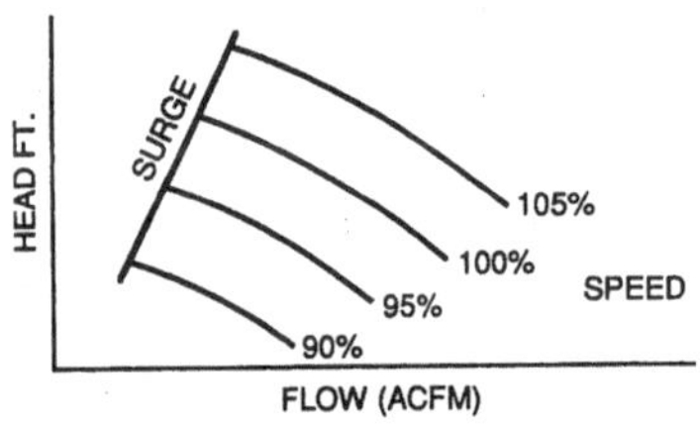

- For a Given Speed and Flow, If an Operating (Test) Point *Does Not Fall on the Curve*,* the Unit Requires Maintenance

* *Within ± 10% (Accounting for Instrument Calibration Error)*

Figure 10.6 The turbo-compressor curve

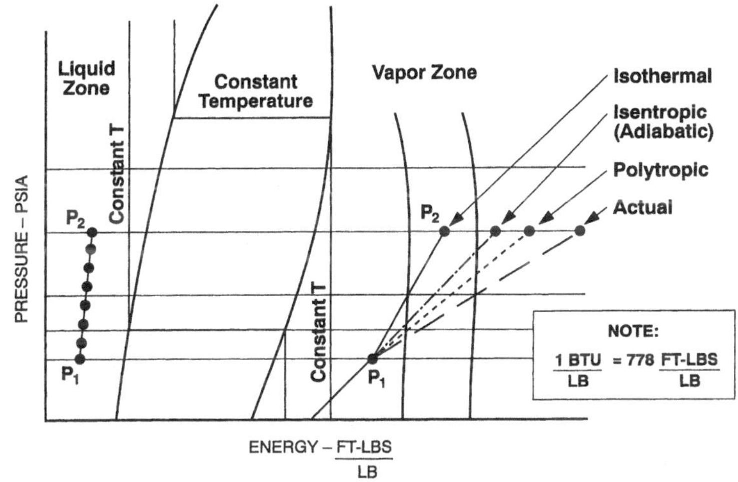

Figure 10.7 Ideal paths of compression

Paths of compression

Figure 10.7 presents a typical Mollier Diagram plotted pressure vs energy.

A Mollier Diagram can be drawn for any pure fluid or fluid mixture. Usually, Mollier Diagrams are prepared only for pure fluids since any change in fluid mixture will require a new Mollier Diagram to be prepared. The Mollier Diagram can be used to determine the head required by the process system for any liquid, saturated vapor or vapor compression process. Observe that for liquid compression, the amount of energy required to increase the pressure from P_1 to P_2 is very small. However, for compression of a vapor, the amount of energy required to compress from P_1 to P_2 is very large as previously explained. Refer back to Figure 10.3 of this module and study the equations that are used to determine the head required for a liquid and a vapor. In addition to many more parameters being required for calculation of head required for a vapor, certain ideal assumptions must be made.

The different types of gas head

A vapor can be ideally compressed by any one of the following reversible thermodynamic paths:

- Isothermal – constant temperature
- Isentropic (adiabatic) – no heat loss

- Polytropic – temperature not constant and heat lost

The actual path that any compressor follows in compressing a vapor from P_1 to P_2 is equal to the reversible path divided by the compressor's corresponding path efficiency. Therefore, we can write the following equation:

$$\textit{Actual Head} = \frac{\textit{Head Isothermal}}{\textit{Eff'y. Isothermal}} = \frac{\textit{Head Isentroic}}{\textit{Eff'y. Isentropic}} = \frac{\textit{Head Polytropic}}{\textit{Eff'y. Polytropic}}$$

When evaluating compressor bids from different vendors quoting different types of reversible heads the above equation proves useful. If the head produced by one vendor divided by the corresponding efficiency is not equal to the corresponding values quoted by the competition . . . Better start asking why!

Figure 10.8 defines the different types of ideal heads and shows the relative difference between their values as compared to polytropic head.

The definition of polytropic head is confusing and difficult to understand if not investigated further.

The ideal gas head equations are described in Figure 10.9.

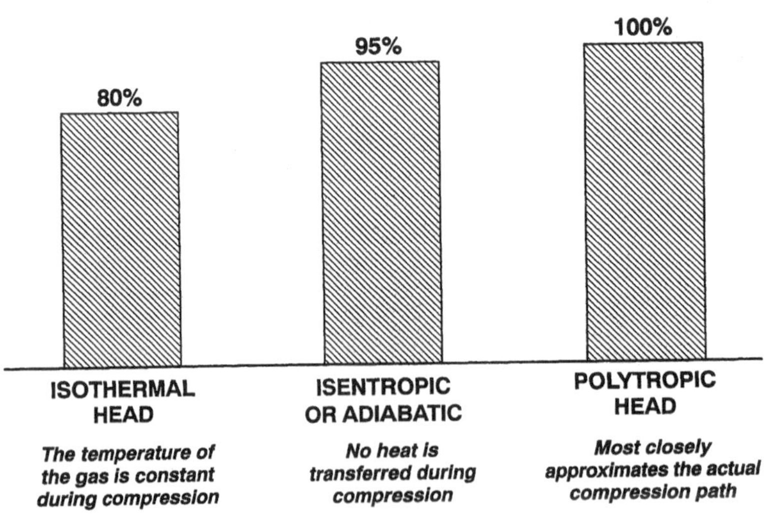

Figure 10.8 Ideal compression heads – relative values

Ideal gas head equations

Isothermal	Isentropic (Adiabatic)	Polytropic
$HD = (\frac{1546}{M.W})(T_1)(Z_{AVG})\left[L_N(\frac{P_2}{P_1})\right]$	$HD = (\frac{1545}{M.W})(T_1)(\frac{K}{K-1})(Z_{AVG})\left[(\frac{P_2}{P_1})^{\frac{(K-1)}{K}} - 1\right]$	$HD = (\frac{1545}{M.W})(T_1)(\frac{n}{n-1})(Z_{AVG})\left[(\frac{P_2}{p1})^{\frac{(n-1)}{n}} - 1\right]$

Where:

$\frac{1545}{M.W}$	=	Gas constant "R"	P_2	=	Discharge Pressure – PSIA
T_1	=	Inlet temp. – °R	P_1	=	Inlet pressure – PSIA
°R	=	460 + °F	K	=	Ratio of specific heats C_p/C_v
Z_{AVG}	=	Average compressibility $\frac{(Z_1 + Z_2)}{2}$	n	=	Polytropic exponent
			$\frac{(n-1)}{n}$	=	$\left(\frac{K-1}{K}\right)\left(\frac{1}{\eta_{POLY}}\right)$
Ln	=	Log to base e	η_{POLY}	=	Polytropic efficiency

Figure 10.9 Ideal gas head equations

Note that the only difference between isentropic and polytropic head is the values;

$$\frac{K-1}{K} \quad \text{and} \quad \frac{n-1}{n}$$

Also note that $\frac{n-1}{n} = \frac{\frac{K-1}{K}}{\eta \text{ poly}}$

Now if η poly = 100%, $\frac{n-1}{n} = \frac{K-1}{K}$ or

Polytropic Head = Isentropic Head

Therefore, I think of $\frac{n-1}{n}$ as a correction

Factor to $\frac{K-1}{K}$ that will most closely approximate the actual

compressor path for a given compressor (refer back to Figure 10.7).

Remember, polytropic head is an ideal reversible compression path. Today, most compressor vendors have adopted polytropic head as their standard for multistage compressors. The main reason is that polytropic head, since it contains an efficiency term, allows individual impeller (stage) heads to be added.

Dynamic compressor curves format

Finally, Figure 10.10 presents the different ways that compressor compression performance can be formatted.

Head vs flow is always preferred because the head produced by a dynamic compressor is not significantly affected by gas density. However, compression ratio or discharge pressure are! That is, compression ratio and discharge pressure curves are invalid if the inlet gas temperature, inlet pressure or molecular weight changes! This may seem confusing at first, but refer back to Figure 10.5 which shows how head is produced by a dynamic compressor. It is a function only of impeller diameter and shaft speed. Gas density influences the head required by the process (refer to Figure 10.4).

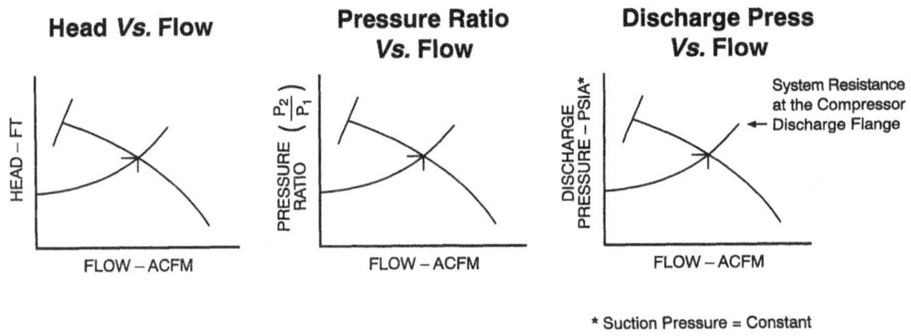

Figure 10.10 Dynamic compressor curves format

11 Performance relationships

- Introduction
- Satisfying the objective
- Gas characteristics
- Compression head
- Impeller types and specific speed
- Efficiency
- Horsepower
- The Fan Laws

Introduction

In this section we will cover the relationships that the COMPRESSOR VENDOR uses to determine the head produced, efficiency, horsepower required and overall design for a particular compressor application. As previously mentioned, the END USER's or the PURCHASER's objective is to deliver a specified amount of a given gas to the process. Therefore, the data that the compressor vendor obtains is required mass flow, inlet pressure, temperature conditions and gas composition. With this data a compressor manufacturer will calculate actual flow, the ideal energy required and the horsepower required to achieve that objective. The calculation for horsepower will require a specific compressor efficiency as well as compressor mechanical losses, that is bearing friction losses, seal losses and disc friction losses. Gas characteristics are defined in this chapter and useful relationships are presented to enable the reader to calculate various compressor requirements. Once the VENDOR obtains the data, the gas head can be calculated. Once the head and required flow are known, the impeller can be selected.

The principle of impeller design is chiefly based on that of specific speed. Specific speed is defined as the ratio of speed times the square root of the actual flow divided by head raised to the three quarters power. It can be shown that increasing values of specific speed will result in increasing impeller efficiencies. Therefore, having been given the required flow and energy (head) the only source of obtaining higher specific speed for the vendor is to increase the compressor speed. This fact is very significant, because while compressors have increased in efficiency over the years, the mechanical requirements have also increased significantly, I.E., higher bore impeller stresses, etc. resulting in potential reliability problems. Therefore, the design of the impeller is a very fine balance between the performance requirements and the mechanical constraints of the components used in the compressor design.

Efficiency is presented as a ratio of ideal energy to actual energy as depicted on a typical Mollier Diagram. In addition, the fan laws are presented showing how increased impeller energy can be obtained via speed change in a compressor application.

Satisfying the objective

The objective of the end user is to deliver a specified amount of a given gas. Refer to Figure 11.1 and note that his objective can best be stated by the relationship:

Gas Flow Produced = Gas Flow Delivered

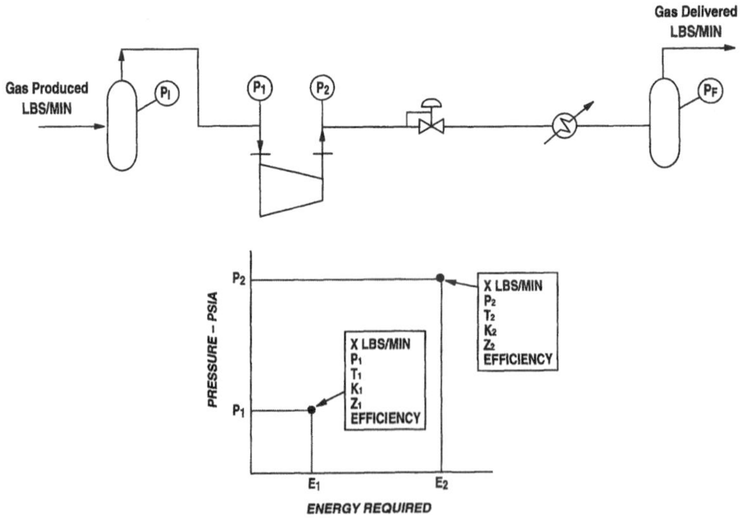

Figure 11.1 The objective: to deliver a specified amount of a given gas

This incidentally is the reason why most process control systems monitor pressure in the process system and install a controller to either modulate flow via a control valve (change the head required by the process) or vary the speed of the compressor (change the head produced by the compressor).

The vendor then, determines the head required by the process based on the parameters given by the contractor and end user on the equipment data sheet. It is very important to note that all possible sources should be used to confirm that the conditions stated on the data sheet are correct and realistic. This fact is especially true for dynamic compressors, since erroneous process conditions will impact the throughput of the compressor.

Gas characteristics

Figure 11.2 presents the relationships used to calculate the design parameters for the compressor. Note that the same relationships are used regardless of the type of compressor (Positive Displacement or Dynamic).

To achieve the client's objective the compressor vendor must calculate the actual flow to the compressor inlet, the actual energy and work required

Actual flow

$$\text{Volume flow rate } (FT^3/MIN) = \text{mass flow rate } \left(\frac{LB's}{Min}\right) + \text{density } \left(\frac{LB}{FT^3}\right)$$

(gas density depends on P_1, T_1, Z, MW)

Energy (ideal)

$$\text{Compression } HEAD_{POLYTROPIC} = \frac{\text{FOOT-LB's}}{\text{LB MASS}}$$

Energy (ideal) to compress and deliver one LB of gas from P_1 to P_2

(Depends on P_1, T_1, Z_{avg}, K_{avg}, MW P_2, efficiency)

Work

$$\text{Power (horsepower)} = \frac{\text{ideal energy } \left(\frac{\text{FT-LB's}}{LB}\right) \text{ mass flow } \left(\frac{LB}{Min}\right)}{33{,}000 \left(\frac{\text{FT-LB's}}{\text{min-HP}}\right) \times \text{efficiency } (\%)}$$

Figure 11.2

The gas characteristics used in the determination of design parameters are defined in Figure 11.3.

Gas characteristics

Compressibility (Z)	-	Accounts for the deviation from an ideal gas
Specific heat (C)	-	The amount of heat to raise one mass of gas one degree
C_P and C_V	-	Specific heat at constant pressure and volume respectively
Specific heat ratio (K)	-	C_P/C_V
MW	-	Molecular weight
Polytropic exponent (n)	-	Used in polytropic head calculations

$$\frac{n-1}{n} = \frac{k-1}{k} \times \frac{1}{\eta \text{ polytropic}}$$

Figure 11.3 Gas characteristics

Figure 11.4 shows useful relationships used in compressor calculations as well as the definitions for constants used.

Useful relationships

Actual flow - FT^3/minute

$$\text{ACFM} = \frac{\text{mass flow} \quad (\text{LBS/MIN})}{\text{Density} \quad (\text{LBS/FT}_3)}$$

$$\text{Density (LBS/FT}^3) = \frac{(\text{P})(144)}{\text{ZRT}}$$

$$\text{ACFM} = \text{SCFM} \times \frac{(14.7)}{\text{P}} \frac{(\text{T})}{520}$$

Energy (Ideal) - FT-LB/LB Mass

Use head equation,
Polytropic is usually used

Efficiency - %

Derived from impeller test results - does not include mechanical losses

Work - horsepower

Brake horsepower = gas horsepower + mech. losses

$$\text{Gas horsepower} = \frac{(\text{HD})(\text{Mass flow})}{(\text{C})\ (\text{eff'y})}$$

where:

$$\text{C} = 33{,}000 = \frac{\text{FT-LBS}}{\text{Min-H.P.}}$$

$$\text{HD} = \frac{\text{HEAD FT-LBS}}{\text{LB}}$$

$$\text{Mass flow} = \frac{\text{LBS}}{\text{Minute}}$$

Eff'y = corresponding efficiency (polytropic, isentropic, etc)

P = pressure - psia

T = temperature - R*

*R = °F + 460

Z = compressibility

R = 1545/mol. wgt

SCFM = standard FT^3/min referenced to 60°F and 14.7 psia

Figure 11.4 Useful relationships

Compression head

The ideal gas head equations are again defined in Figure 11.5. As previously stated, polytropic head is the usual choice among compressor vendors.

Ideal gas head equations

Isothermal	Isentropic (Adiabatic)	Polytropic
$HD = \left(\frac{1546}{M.W}\right)(T_1)(Z_{AVG})\left[L_N\left(\frac{P_2}{P_1}\right)\right]$	$HD = \left(\frac{1545}{M.W}\right)(T_1)\left(\frac{K}{K-1}\right)(Z_{AVG})\left[\left(\frac{P_2}{P_1}\right)^{\frac{(K-1)}{K}}-1\right]$	$HD = \left(\frac{1545}{M.W}\right)(T_1)\left(\frac{n}{n-1}\right)(Z_{AVG})\left[\left(\frac{P_2}{p1}\right)^{\frac{(n-1)}{n}}-1\right]$

Where:

$\frac{1545}{M.W}$	=	Gas constant 'R'	P_2	=	Discharge Pressure – PSIA
T_1	=	Inlet temp. - °R	P_1	=	Inlet pressure – PSIA
°R	=	460 + °F	K	=	Ratio of specific heats C_p/C_v
Z_{AVG}	=	Average compressibility $\frac{(Z_1 + Z_2)}{2}$	n	=	Polytropic exponent
			$\frac{(n-1)}{n}$	=	$\left(\frac{K-1}{K}\right)\left(\frac{1}{\eta_{POLY}}\right)$
Ln	=	Log to base e	η_{POLY}	=	Polytropic efficiency

Figure 11.5 Ideal gas head equations

Impeller types and specific speed

Various types of radial (centrifugal) impellers are shown in Figures 11.6 & 11.7.

Open impellers

Open impellers are shown in Figure 11.6.

The advantages of open impellers is their ability to operate at higher tip speeds and thus produce greater head than closed impellers. Open impellers can produce 15,000 – 25,000 ft-lbs/LB of head per stage. This is because a side plate is not attached to the inlet side of the vanes which results in significantly lower blade stresses. The disadvantages of open impellers are their lower efficiency due to increased shroud (front

Figure 11.6 Compressor impellers

side) leakage and increased number of blade natural frequencies resulting from the cantilevered attachment of the blades to the hub. Most end users restrict the use of open impellers to plant and instrument air applications since the high speeds and intercooling offset the efficiency penalties caused by shroud leakage. Older design multistage centrifugal compressors frequently used open impellers in the first stages since the high flows caused unacceptable side plate stresses in closed impeller design. Modern calculation (finite element) methods and manufacturing methods (attachment techniques – machine welding, brazing, etc.) today make possible the use of enclosed first stage impellers for all multistage compressor applications. Finally, radial bladed impellers (whether open or enclosed) produce an extremely flat (almost horizontal) head curve. This characteristic renders these impellers unstable in process systems that do not contain much system resistance. Therefore, radial impellers are to be avoided in process systems that do not contain much system resistance (plant and instrument air compressors, charge gas compressors and refrigeration applications with side loads.

Enclosed impellers

Enclosed impellers are shown in Figure 11.7.

Note that the first stage impeller in any multistage configuration is always the widest. That is, it has the largest flow passage. As a result, the first stage impeller will usually be the highest stressed impeller. The exception is a refrigeration compressors with side loads (economizers).

Dynamic compressor vendors use specific speed to select impellers based on the data given by the contractors and end user. The vendor is given the total head required by the process and the inlet volume flow. As previously discussed, at the stated inlet flow (rated flow) the head

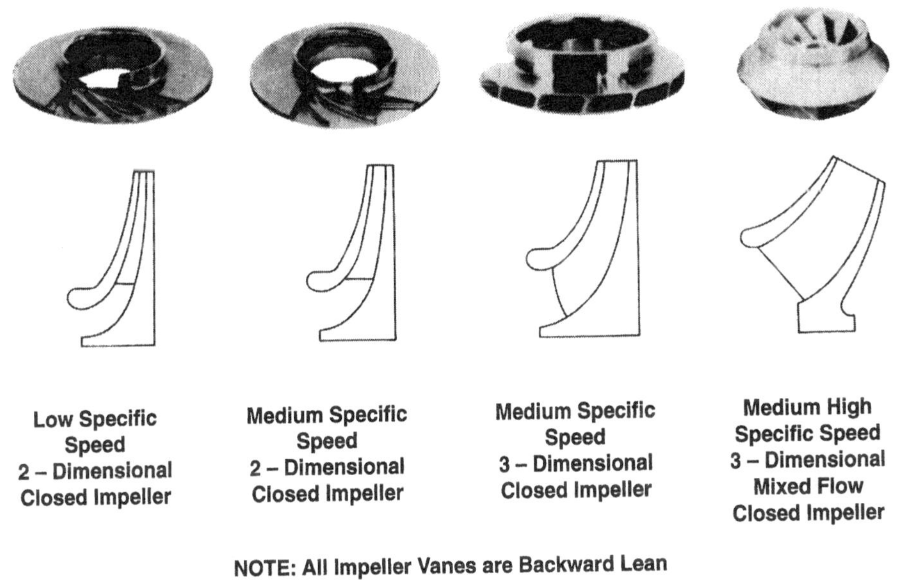

Figure 11.7 Enclosed impellers (Courtesy of IMO Industries, Inc.)

required by the process is in equilibrium with the head produced by compressor. Vendor calculation methods then determine how many compressor impellers are required based on mechanical limitations (stresses) and performance requirements (quoted overall efficiency). Once the head required per stage is determined, the compressor speed is optimized for highest possible overall efficiency using the concept of specific speed as shown in Figure 11.8.

It is a proven fact that the larger the specific speed, the higher the attainable efficiency. As shown, specific speed is a direct function of shaft speed and volume flow and an inverse function of produced head. Since the vendor at this point in the design knows the volume flow and head produced for each impeller, increasing the shaft speed will increase the specific speed and the compressor efficiency.

However, the reader is cautioned that all mechanical design aspects (impeller stress, critical speeds, rotor stability, bearing and seal design) must be confirmed prior to acceptance of impeller selection. Often, too great an emphasis on performance (efficiency) results in decreased compressor reliability. One mechanical design problem can quickly offset any power savings realized by designing a compressor for a higher efficiency.

Referring back to Figure 11.8, calculation of specific speed for the first impeller by the contractor or end user will give an indication of the type of dynamic compressor blading to be used. One other comment,

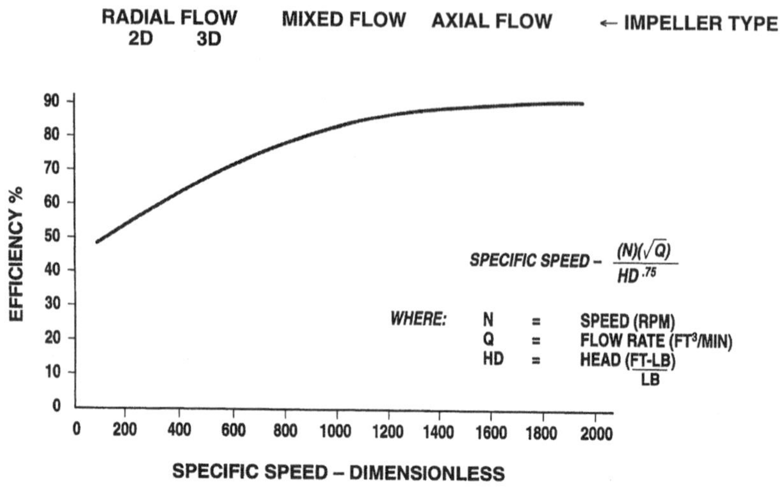

Figure 11.8 Impeller geometry vs. specific speed

Sundstrand Corporation successfully employs an integral high speed gear box design for low flow, high head applications or for low specific speed applications. The use of a speed increasing gear box (for speeds up to 34,000 RPM) enables the specific speed to be increased and therefore resulting in higher efficiency and less complexity than would be obtained with a multistage compressor design approach.

Efficiency

Compressor efficiency, regardless of the type of compressor, can best be understood by referring to a typical Mollier Diagram as depicted in Figure 11.9.

All produced heads shown on performance curves (isothermal, isentropic and polytropic) represent the ideal reversible head produced to compress a given gas from P_1 to P_2. This then is the theoretical compression path of the gas. That is, the energy required to compress a gas if the efficiency is 100%.

However due to friction, sudden expansion etc., the efficiency is less than 100%. Therefore the actual compression path requires more head (energy) to compress the gas from P_1 to P_2. The efficiency then is equal to:

$$\text{Efficiency} = \frac{\Delta E \text{ Ideal } (E_2 \text{ Ideal} - E_1)}{\Delta E \text{ Actual } (E_2 \text{ Actual} - E_1)}$$

$$\textbf{Efficiency} = \frac{\Delta E \text{ Ideal } (E_2 \text{ Ideal} - E_1)}{\Delta E \text{ Actual } (E_2 \text{ Actual} - E_1)}$$

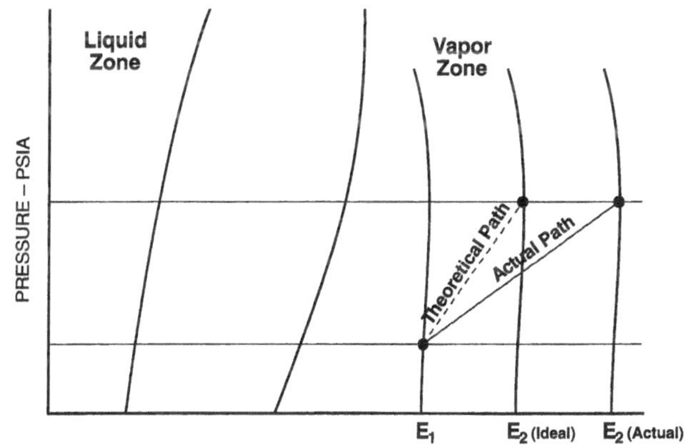

Figure 11.9 Efficiency

Note that ( ) is used to represent any ideal reversible path (isothermal, isentropic, polytropic).

Horsepower

Horsepower is defined as the total actual energy (work) required to compress a given gas from P_1 to P_2 when compressing a given mass flow:

$$\text{G.H.P.} = \frac{\text{Head } (\ ) \frac{\text{ft-lbf}}{\text{lbm}} (\text{Mass Flow - lb/min})}{(33{,}000)\ (\text{Efficiency})(\ )}$$

Note: ( ) must be for the same ideal reversible compression path.

The brake horsepower is the sum of the gas horsepower and the mechanical losses of the compressor.

B.H.P. = G.H.P. + Mechanical losses

The mechanical losses are the total of bearing, seal and windage (disc friction) losses and are provided by the compressor vendor. For estimating purposes, a conservative value of mechanical losses for one centrifugal or axial compressor case would be 150 H.P.

$$\frac{Q_F}{Q_I} = \frac{N_F}{N_I}$$

$$\frac{HD_F}{HD_I} = \left(\frac{N_F}{N_I}\right)^2$$

$$\frac{BHP_F}{BHP_I} = \left(\frac{N_F}{N_I}\right)^3$$

Where:

Q = Flow Rate (FT^3/MIN)

N = Speed (RPM)

HD = Gas Head (FT-LB/LB)

BHP = Horsepower

I = Initial

F = Final

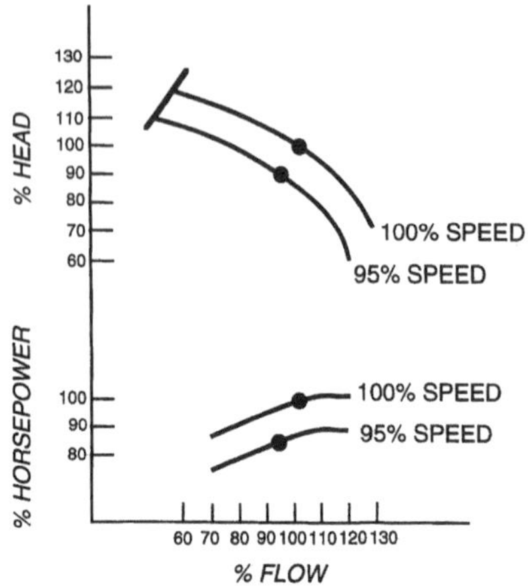

Figure 11.10 The Fan Laws

The Fan Laws

These familiar relationships, sometimes called the affinity laws for pumps were originally derived for a single stage fan which is a low pressure compressor. The Fan Laws are presented in Figure 11.10.

As shown, if speed is changed, the flow, head and horsepower vary by the first, second and third power of speed ratio respectively.

The reader must be cautioned however that the Fan Laws are only an approximation to be used as an estimating tool. Their accuracy significantly decreases with increasing gas molecular weight and increase in the number of compression stages.

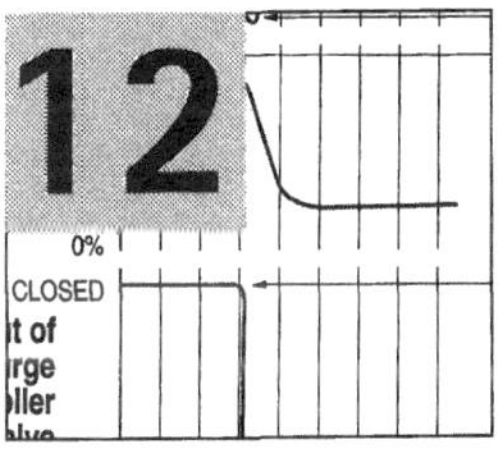

12 Surge (stall) and stonewall

- Introduction
- Surge facts
- The limits of the curve
- What causes surge
- What causes stonewall (choke)

Introduction

In this chapter, we will devote our attention to limits of the dynamic curve, namely surge and stonewall.

The first item that we will discuss will be low flow. This phenomena is known as surge. It should be pointed out immediately that surge is a system phenomena resulting from the inability of the turbo-compressor impeller to produce the amount of required energy that the process system requires.

Some facts concerning surge are that it is a high speed phenomena, flow reversals can occur in less than 150 milliseconds, pressure rapidly fluctuates, noise is generated, temperature increase can be very rapid and mechanical damage can occur.

The intensity of surge varies from application to application and is proportional to the density of the fluid. Higher pressure and higher molecular weight applications can result in greater mechanical damage. Low density applications (hydrogen) can cause surge damage that may not be detectable until the equipment is disassembled.

The cause of surge results from low velocity in the blade or impeller passages. It must be remembered that every dynamic blade or impeller is designed for one and only one operating point. Energy is produced by the action of the vanes and flow between blades or vanes. As the flow through the vanes is reduced, the velocity between the vane passages is also reduced.

As a result, flow changes from laminar (smooth) to turbulent. This causes flow discontinuities. There are many causes of surge. Referring to a cross section of an impeller and a stationary diffuser, we can see and remember from previous discussions that the gas angle of the absolute velocity of the gas exiting an impeller will reduce as the flow decreases. The gas exits the diffuser in a logarithmic spiral fashion. As the exit gas angle becomes less, the path the gas must follow becomes longer and eventually can turn back into the impeller thus causing a backflow. This is one example of a flow discontinuity that limits the amount of energy an impeller can produce. Another example of this phenomena could occur in the impeller or blade passage itself and result in backflow. Regardless of the initial cause, the inability of a rotating blade or impeller to produce the amount of energy required by the system will result in the gas acting on the blade or impeller rather than the blade or impeller acting on the gas. At this point, backflow occurs.

If we examine a typical compressor system and look at the discharge system from the compressor discharge flange to the discharge checkvalve, we can see that at the point of maximum blade or impeller energy production, the pipe and discharge pressure of the compressor can be thought of as an equivalent vessel. As long as the flow into this vessel is equal to the flow out, the pressure in that vessel remains constant. At the onset of surge, the compressor will no longer produce flow into the vessel and stored energy in this vessel now acts on the blade resulting in a flow reversal from the vessel back towards the suction of the compressor. Examining a typical compressor performance curve, we can see that as the head required increases, the operation point will move towards the surge or instability line. When the operating point intersects the surge line, a flow reversal will occur causing the operating point to enter the negative flow region of the performance curve. The equivalent vessel therefore will become evacuated during this reversal. Once the equivalent vessel is evacuated, the system pressure observed by the compressor rapidly reduces thus enabling the compressor to achieve a new operating point in the positive region of the performance curve at high flow. Once the vessel becomes filled, the operating point moves again up the compressor curve to surge and the cycle begins anew. Remember that these cycles can occur as rapidly as six (6) times *per second!*

Naturally, this action produces large amounts of energy which are absorbed by the mechanical components of the compressor and can result in a significant amount of temperature increase inside the compressor. It must be understood that the gas will move back and forth in the compressor upwards of six (6) times per second and not be cooled, therefore temperature increases can be rapid and thermal expansion can cause major damage during surge. This phenomena can occur most easily in axial compressors. Therefore, surge is to be avoided most in axial machines due to their ability to heat up faster than centrifugals and also as a result of the blade configuration which can be more readily excited than centrifugals.

The high flow limit of the dynamic curve is known as choke, or stonewall and is defined as the maximum flow a given impeller can handle. This value will be obtained when the ratio of relative inlet gas velocity to the sonic velocity of the process gas being handled is equal to one. Referring back to the concept of an equivalent orifice, the impeller represents an equivalent orifice. Simply stated, the high flow limit of the curve is attained when the impeller itself acts as a resistance and will not allow additional flow to be passed. We can see by referring to the relationship of the relative mach number, which is the ratio of the inlet relative velocity to the sonic velocity of that gas at suction conditions, that higher molecular weight applications will result in a higher mach number. Therefore, it can be stated that high molecular weight applications result in a shorter range of operation of the compressor curve. We define range of operation as the flow from surge to stonewall.

Typical ranges for surge and carryout are: Surge range can vary between 60% to 80% of design flow and carryout range can vary between 110% to 130% of design flow.

Surge facts

Surge can be a very dangerous phenomena if allowed to continue. Damage caused by surge in any type of dynamic compressor can render that compressor inoperable for periods of time in excess of two (2) months. Considering that most dynamic compressors are not spared, this can cause a significant loss of revenue. Typical daily product revenues for a world class process unit can exceed US$1,000,000 per day! Figure 12.1 shows some of the most common effects of surge.

Surge facts

- Surge is a high speed phenomenon. Flow reversals can occur in less than 150 milliseconds.
- Reversal rate is 30 to 120 cycles/sec.
- Pressure rapidly fluctuates.
- Noise generated.
- Temperature increase (can be rapid).
- Mechanical damage can occur.
- Unit may trip.
- Intensity varies with the application.

Figure 12.1 Surge facts

Of all the effects listed, by far the most damaging is rapid temperature increase for this can cause internal rubs of the compressor at operating speed resulting in impeller breakage, diaphragm breakage, extreme labyrinth seal wear and possible case breakage.

Figure 12.2 shows the results of an actual surge test performed on a 2500 HP solar compressor.

It is the results obtained by a strip chart recorder during a surge cycle. Δp_o is the flow change and Δp_c is the pressure ratio change. Notice how

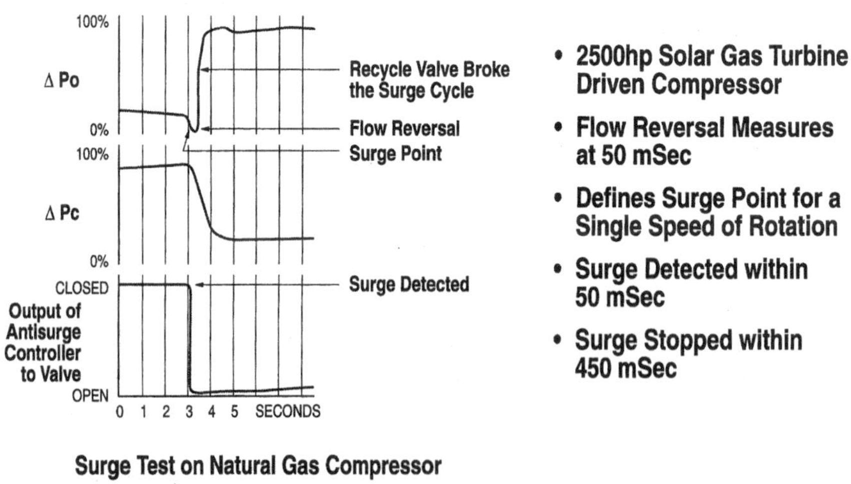

Figure 12.2 A surge test (Courtesy of Compressor Controls Corp.)

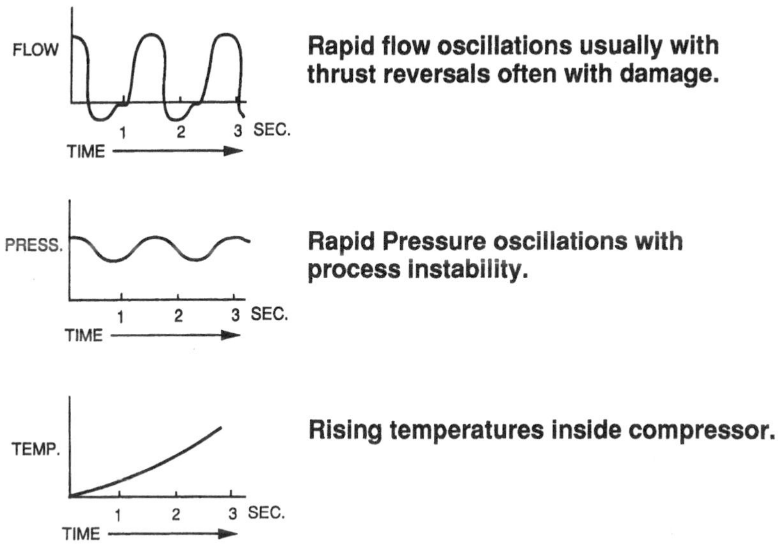

Figure 12.3 The surge phenomena

prior to the surge event that the pressure ratio required (Δp_c) increases while the compressor flow rate (ΔP_o) is decreasing. Observe that once a surge cycle is detected, the output of the controller works to open the anti-surge valve immediately. This fact is important to remember when trying to justify a field surge test to plant operations. A compressor will never be damaged by a surge test if conducted properly. As shown in Figure 12.2, strip chart recorders are used and the surge system is in automatic but the surge control line has been moved to the left of the actual surge location. The surge control line defines the set point of the surge controller and hence the point at which the surge control valve will open.

Figure 12.3 presents the characteristics of flow, pressure and temperature during surge. The degree of parameter change is directly proportional to the density of the gas. The higher the gas density, the greater the effect of the surge.

The limits of the curve

The limits of any type of dynamic compressor curve are the low flow limit (surge) and the high flow limit (choke flow). Surge, as we shall see, is caused by low flow turbulence while choke flow is caused by high velocity friction. Figure 12.4 shows a typical compressor curve and the side view of a compressor stage (top half view only).

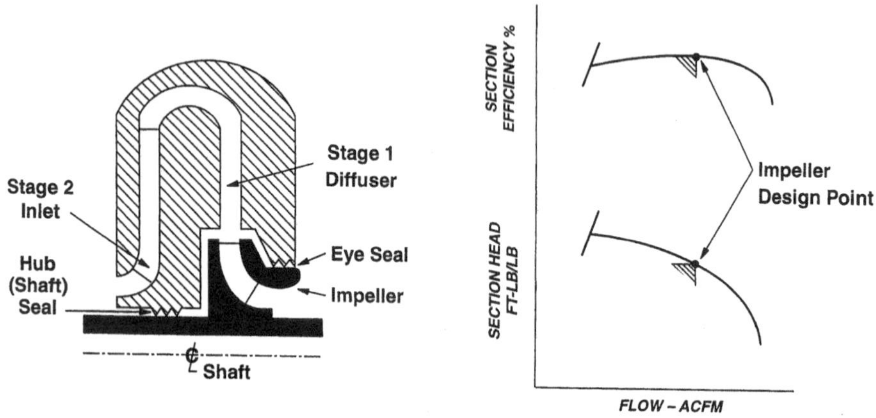

Figure 12.4 The compressor stage and characteristic curve

It can be seen that the limits of any dynamic compressor curve are a consequence of gas velocity in the compressor stage.

What causes surge

Surge is a process system phenomena that is the result of flow separation, caused by low gas velocity, anywhere in a compressor stage (inlet guide vane, impeller suction, impeller mid section, impeller discharge or diffuser). Centrifugal pumps experience the same flow separation phenomena at low flows which sometimes cause the liquid to vaporize resulting in recirculation which can cause cavitation.

As the process system requires more head in any type of dynamic compressor, the flow is reduced to a point that causes flow separation. This event is commonly known as stall. An example of diffuser stall is shown in Figure 12.5.

This diagram depicts the view of a simple impeller with the side plate removed and a vaneless diffuser. As the gas leaves the impeller via vector 'R', the resultant of the gas velocity through the impeller and the impeller tip speed, it passes through the diffuser in a path that approximates a logarithmic spiral. Since the relative gas velocity is the 'y' component of 'R' and will decrease with decreasing flow rate, the gas angle $\propto$ off the blades will decrease. As a result, the path of the gas off the blade will become longer with decreasing flow rate. At surge flow, the velocity off the wheel will be so low that the path of the gas will not leave the diffuser. This will cause a reduction in the head produced by the impeller. At this moment, the compressor is said to be

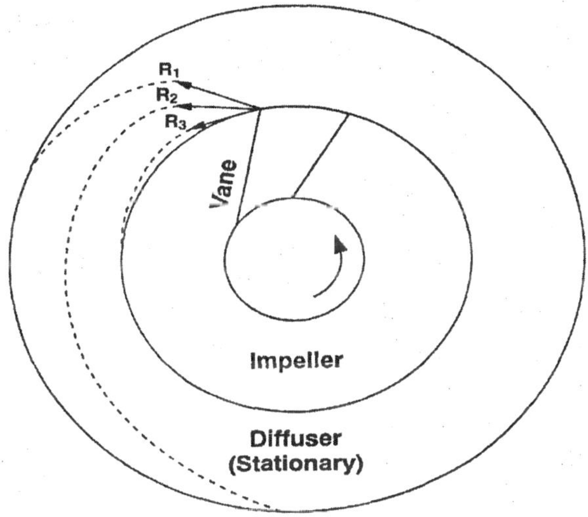

Figure 12.5 Diffuser stall

in a state of flow separation, flow instability or stall. Stall can be initiated by flow separation at any point within a compressor stage (inlet, mid-section, discharge, etc.). Regardless of the location of the flow separation within the compressor stage, a reduction in head produced by the impeller will occur. The dynamic compressor performance curve actually decreases to the left of the surge line since the flow separation (stall) increases losses in the stage and reduces head produced. Please refer to Figure 12.6.

Draw a decreasing head curve to the left of the surge line. On this curve, place an operating point. This point would be the initial point in the surge cycle. A closeup of the process system in Figure 12.6 is shown in Figure 12.7.

Once flow separation occurs, the head produced by any dynamic compressor decreases and the process gas present from the compressor discharge flange to the check valve flows backwards through the compressor. This backflow causes the volume shown in Figure 12.7 to be evacuated resulting in a low discharge pressure. Since the head (energy) required by the process system is a function of discharge pressure, the head required by the process will decrease allowing the compressor to operate in the high flow region of the performance curve. The reversal of dynamic flow caused by flow separation (stall) in the compressor stage and the recovery of flow resulting from reduced discharge pressure is defined as a surge cycle. This surge cycle will continue until either the head produced by the compressor is increased

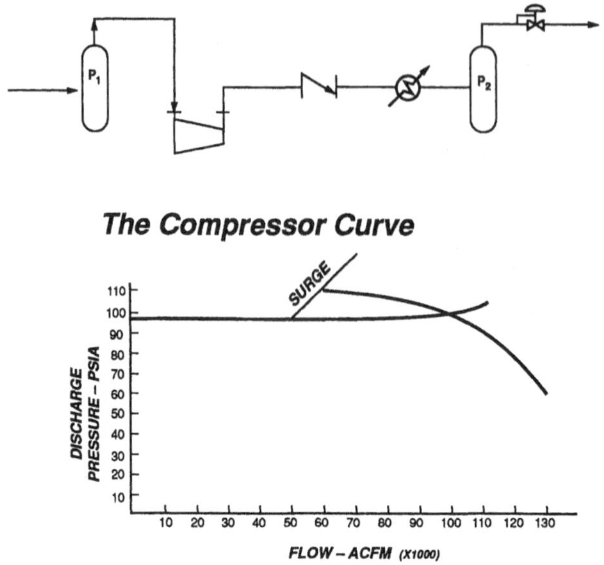

60,000 ft^3/min = 1ft^3/millisecond

Figure 12.6 Surge – The effect of the system

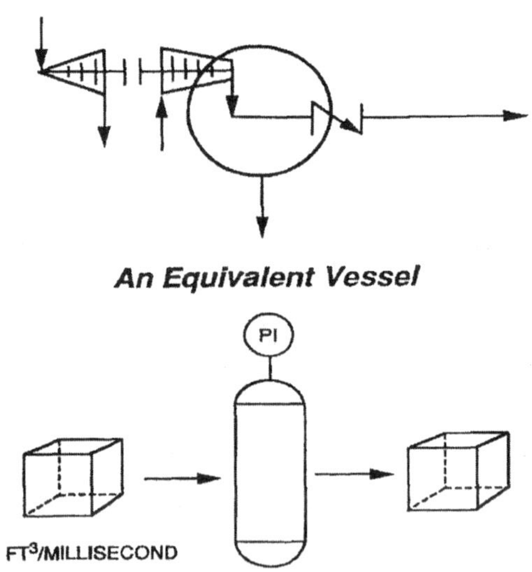

Figure 12.7 Flow in and out of a vessel

or the head required by the process is reduced. The quickest way to eliminate surge is to rapidly reduce the discharge pressure by opening a blow-off or recycle valve in the discharge process system.

Please refer back to the 'Surge Facts' section of this chapter to review the damaging effects of surge.

What causes stonewall (choke)

The definition of stonewall or choke flow is presented in Figure 12.8.

Stonewall

Stonewall or choke flow is defined as the maximum flow a given impeller can handle.

This value will be attained when the ratio of the relative inlet gas velocity to the sonic velocity of the process gas being handled is equal to 1.0.

Figure 12.8 Stonewall

The name 'Stonewall' comes from the fact that the compressor curve suddenly drops off and appears to have come up against a 'Stonewall' (refer to Figure 12.4). The cause of this phenomena is excessive relative gas velocity through the impeller. As the head (energy) required by the process system is reduced, the volume flow through the impeller will increase.

Figure 12.9 shows that any impeller stage is essentially an equivalent orifice with constant flow areas or orifices.

Since volume flow is proportional to areas and gas velocity, an increase in volume flow is a direct increase in gas velocity.

The limit of compressor high volume flow is controlled by the relative mach number which is defined in Figure 12.10.

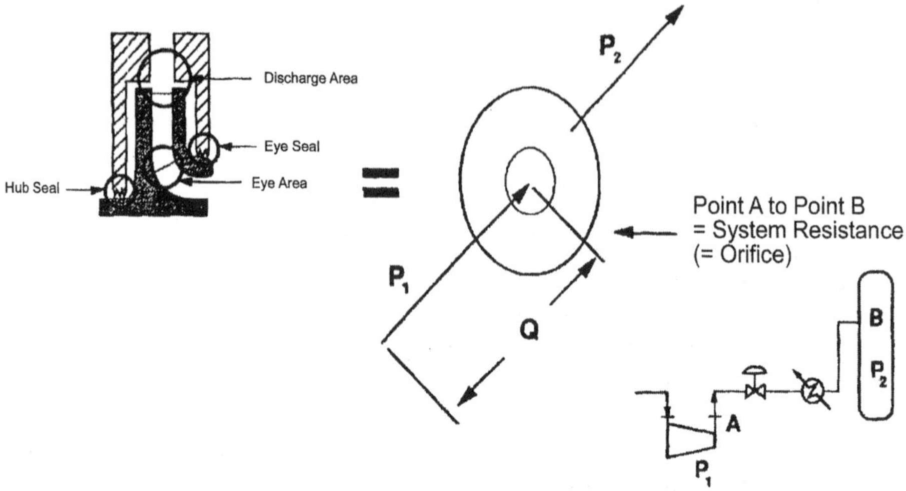

Figure 12.9 Reduce it to an equivalent orifice

Relative Mach Number

$$\text{Relative Mach Number} = \frac{V_{REL\ INLET}}{\sqrt{KgRT_1}}$$

Where::

V_{REL}=	=	Inlet velocity relative to the blade
K	=	Ratio of specific heats (CP/CV)
R	=	1545/Molecular Weight
T_1	=	Inlet temperature (°R)
°R	=	460 + °F
g	=	32 ft/sec^2

Figure 12.10 Relative mach number

When the relative mach number = 1, the maximum possible flow by any dynamic compressor is attained. Relative mach number is the ratio of gas impeller inlet relative velocity to the sonic velocity of the process gas. Figure 12.10 shows that the sonic velocity is directly proportional to K, g and T and inversely proportional to gas molecular weight. Therefore gases with high molecular weight will reach stonewall or choke flow sooner than gases with low molecular weights. This is why the flow range of compressors processing high molecular weight gases is always less than the flow range for compressor processing low molecular weight gases.

Unlike surge, stonewall is not a destructive phenomena. Since horsepower rapidly decreases in stonewall, most dynamic compressor mechanical tests are run in stonewall to reduce shop load horsepower. Also, dynamic compressors can only operate in stonewall if the head (energy) required by the process system is low enough to allow the compressor to operate in this high velocity region of the performance curve. For economic reasons, this is rarely the case since most engineering contractors optimize process pipe design to minimize pipe diameter and therefore increase the head required at high flow rates.

The only case of dynamic compressor damage caused by choke flow that I have experienced involved the axial compressor of a gas turbine. The design of the combustion chambers and turbine section allowed the axial compressor to operate in stonewall under certain operating conditions. The high velocity caused shock waves which excited a natural frequency of a compressor blade row causing blade breakage.

When sizing surge control valves for axial compressors, confirm that the maximum, full open surge valve flow will not excite any axial compressor blade natural frequencies. This is not usually a concern with centrifugal compressor impellers since the impellers are usually of the 'closed type' and therefore rigidly support the blades at both the hub and side plate. In the case of open compressor impellers (blades supported by only the hub), impeller natural frequency excitation should be checked. Open impellers are used in older designs for the first and second stages of large multistage compressors and for centrifugal plant and instrument air compressors.

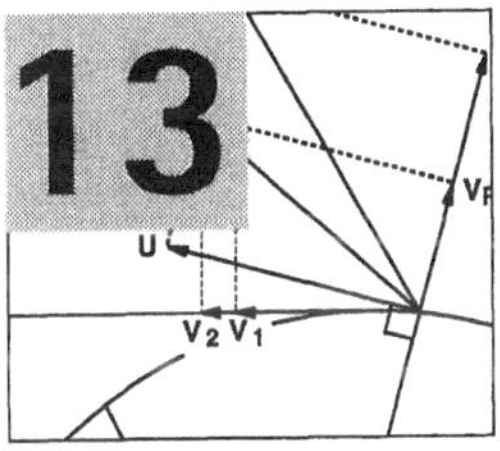

13 The effect of a gas density change

- Introduction
- The factors involved
- The effect on turbo-compressor pressure ratio
- The effect on the compressor head
- The effect on system resistance
- The effect on turbo-compressor flow rate
- The effect on power

Introduction

In this chapter we will discuss the effect of a gas density change on dynamic compressor performance.

The factors that define a given fluid are presented and we can readily observe that there are a significantly larger number of factors necessary to define a vapor since it is a compressible fluid. Referring again to the fluid head equations for a liquid and a vapor, one can see that the pressure ratio developed for a gas is dependent on temperature, molecular weight, the specific heat ratio of the gases and compressibility. The effect of a gas composition or temperature change on turbo-compressor compression ratio is shown and depicted in tabular form. By solving the head equation for pressure ratio, it can readily be seen that pressure ratio will increase with increasing molecular weight and decrease with decreasing molecular weight if the inlet gas temperature is held constant. Conversely, if the molecular weight is held constant, the pressure ratio will increase with decreasing gas inlet temperature and decrease with increasing gas inlet temperature. The effect of a gas composition change on head is assumed to be such that once an impeller is designed, it will remain constant. This is not entirely true for an impeller working on a vapor for the following reasons:

1. Head is generated by impeller tip speed and exit gas velocity relative to the blade.
2. Gas composition and inlet gas temperature changes affect the compression ratio.
3. Volume flow rate varies with pressure, temperature and compressibility.
4. Since the impeller exit area is fixed, a change in volume at the exit will produce a change in absolute gas velocity and gas tangential velocity. Therefore, head does change slightly.

NOTE: For molecular weight changes on the order of 20%, it is common to assume head is constant for a given flow and speed.

The turbo-compressor impeller produced head changes slightly and also the curve shape will change as a result of the changing velocity relative to the blade. A summary table is presented which shows the change of head and the limits of the compressor curve (surge point and choke point) for condition changes of molecular weight and temperature. Also shown is the change of velocity relative to the blade at the exit. One can see that an increasing molecular weight will result in increasing head and a smaller flow range while a decreasing molecular weight will decrease head and produce a greater flow range. On the other hand, increasing temperature will decrease head and increase the flow range. Whereas decreasing temperature will increase head and reduce the flow range. This fact can be explained when we observe the discharge velocity triangle of any compressor stage. Increasing fluid density will result in higher compression through the impeller or blade row. This will result in a reduced exit volume for a given volume of inlet flow. A reduced exit volume flow acting on the same vane area will result in a reduced velocity relative to the blade. A reduced velocity relative to the blade will produce a higher tangential velocity and a correspondingly higher produced head. This same exercise can be presented for a reduction of molecular weight and for temperature changes as well.

It should be mentioned that the effect of molecular weight does have a slight effect on system resistance and this is covered in this section. Also, gas composition will produce a change of the actual volume flow rate since different compression ratios will be produced and as a result, an effect on power will be experienced since both head, mass flow and efficiency will change. Finally, a multistage compressor example is shown to summarize the effect of molecular weight change on multistage compressors.

The factors involved

The parameters necessary to define a given fluid are presented in Figure 13.1. Note that only two parameters are necessary to define a fluid in the liquid state since it is incompressible. On the other hand, three times that number are required to define that fluid in its vapor state since the vapor is compressible.

What factors define a given fluid

Liquid (incompressible)	Gas (compressible)
Specific Gravity (S.G.)	Molecular Weight (M.W.)
Viscosity (υ)	Specific Heat Ratio (K)
	Compressibility (Z)
	Pressure (P-PSIA)
	Temperature (T – °R)
	Viscosity (υ)

Figure 13.1 What factors define a given fluid

Figure 13.2 shows the relationships used to determine the head (energy) required to increase the pressure of a fluid in it's liquid and vapor state. Note how much the density of the fluid influences the amount of energy required to meet a certain process requirement. When one considers that the additional amount of head produced as a centrifugal compressor's flow rate decreases from rated point to surge point is on the order of only 10%, it can be seen that a small change in gas density can result in a significant flow reduction and possibly compressor surge.

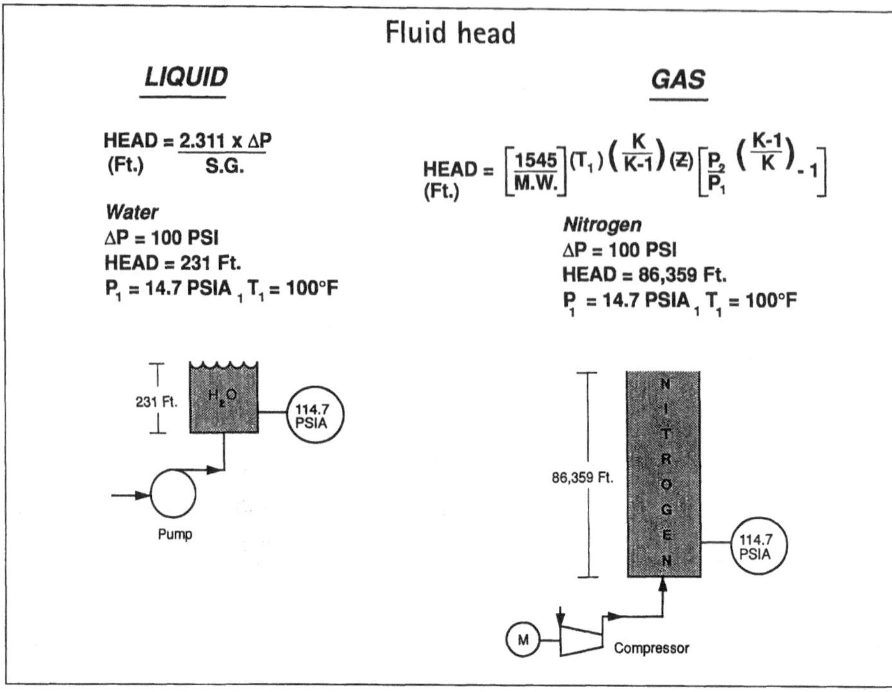

Figure 13.2 Fluid head

The effect on turbo-compressor pressure ratio

The pressure ratio produced by a dynamic compressor is affected by gas density. Figure 13.3 shows that for a given compressor flow and speed the head produced by a dynamic compressor is essentially constant. Therefore, any change in MW, T, K or Z will change the pressure ratio produced. This information is presented in tabular form for changes in molecular weight and inlet gas temperature.

The effect on the compressor head

It is commonly thought that dynamic compressor head produced is always constant for a given flow rate and speed. Figure 13.4 presents this fact for the same compressor operating on different gases (O_2 and N_2).

This statement is not true for a fluid in the vapor state since head in a dynamic compressor is produced by blade velocity and gas velocity. Gas velocity will change will change with gas density since a gas is compressible. These facts are presented in Figure 13.5.

THE EFFECT OF A GAS COMPOSITION AND TEMPERATURE CHANGE ON THE TURBO-COMPRESSOR PRESSURE RATIO

$HEAD_{ISENTROPIC}$ IS RELATED TO PRESSURE RATIO BY:

$$HD_{ISEN} = \left(\frac{1545}{M.W}\right)(T_1)\left(\frac{K}{K-1}\right)(Z)\left[\left(\frac{P_2}{P_1}\right)^{\frac{K-1}{K}}-1\right]$$

ASSUMING HD_{ISEN} IS CONSTANT FOR A GIVEN FLOW,

$$\frac{P_2}{P_1} = \left(1+\frac{(HD_{ISEN})\,(M.W)}{(1545)\,(T_1)\left(\frac{K}{K-1}\right)(Z)}\right)^{\frac{K}{K-1}}$$

THEREFORE THE FOLLOWING TABLE CAN BE DEVELOPED:

EFFECT OF GAS AND T CHANGES ON PRESS. RATIO		
MOLECULAR WGT.	INLET TEMP.	PRESSURE RATIO
INCREASES	CONSTANT	INCREASES
DECREASES	CONSTANT	DECREASES
CONSTANT	INCREASES	DECREASES
CONSTANT	DECREASES	INCREASES

Figure 13.3 The effect of a gas composition and temperature change on the turbo-compressor pressure ratio

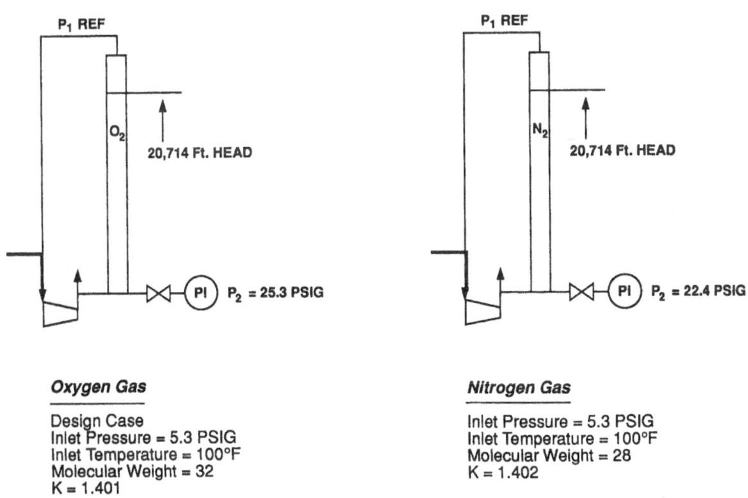

Figure 13.4 The effect of gas composition change on HEAD

The effect of a gas composition and temperature change on turbo-compressor head

The assumption that compressor head remains constant for a given flow with gas composition and temperature changes is not true because:

- Head is generated by impeller tip speed and exit velocity relative to the blade
- Gas composition and temperature changes affect the compression ratio
- Volume flow rate changes with pressure, temperature and compressibility
- Since the impeller exit area is fixed, a change in exit volume rate will produce a change in velocity

Note: For changes on the order of 20%, it is common practice to assume head is constant for a given flow and speed

Figure 13.5 The effect of a gas composition and temperature change on turbo-compressor head

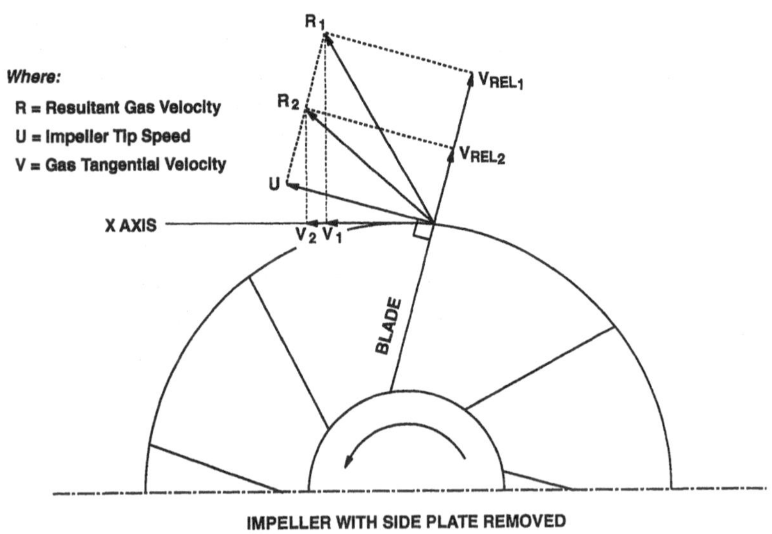

Figure 13.6 Head produced ∝ $(U)(V_T)$

Please refer to Figure 13.6 which shows the relationship between gas velocity (V_{rel}) blade tip speed (U) and tangential gas velocity in a centrifugal compressor.

CONDITION	VEL_{REL} EXIT	HEAD	SURGE POINT	CHOKE POINT
M.W. INCREASES	DECREASES	INCREASES	+ FLOW	- FLOW
M.W. DECREASES	INCREASES	DECREASES	- FLOW	+ FLOW
T_1 INCREASES	INCREASES	DECREASES	- FLOW	+ FLOW
T_1 DECREASES	DECREASES	INCREASES	+ FLOW	- FLOW

ABOVE CHANGES ASSUME A CONSTANT INLET ACTUAL VOLUME FLOW RATE

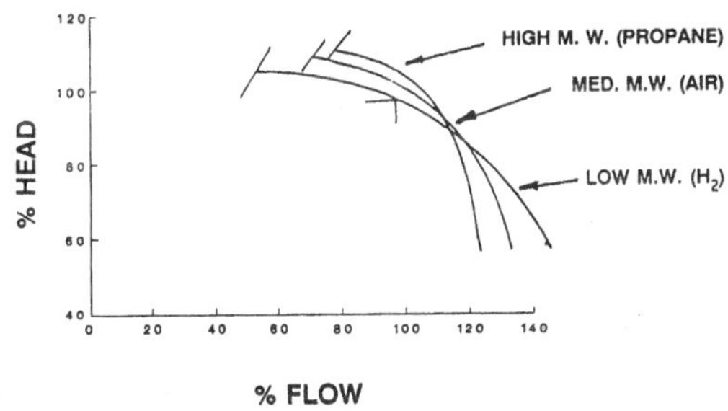

Figure 13.7 Turbo-compressor impeller head change and curve shape summary

Since the head produced by any dynamic impeller is proportional to blade tip speed and gas tangential velocity, reduced gas velocity through the impeller (V_{rel}) will increase the head produced as shown in Figure 13.6. This is the result of increased gas tangential velocity for a given impeller diameter and speed. As shown in Figure 13.5, gas velocity (V_{rel}) will vary with gas density.

Figure 13.7 presents the effect of gas density changes on impeller produced head, surge point and choke point. It can be seen that curve shape is influenced by gas density changes. Therefore, a low density gas will always have a greater flow range than a high density gas.

The effect on system resistance

Figure 13.8 presents the effect of gas density change on the system resistance curve. A slight change in the friction drop in pipes, fittings and vessels results from a change in gas density.

The effect on turbo-compressor flow rate

The effect of gas density changes on actual mass and standard flow rates is shown in Figure 13.9. Note that gas density changes will change the

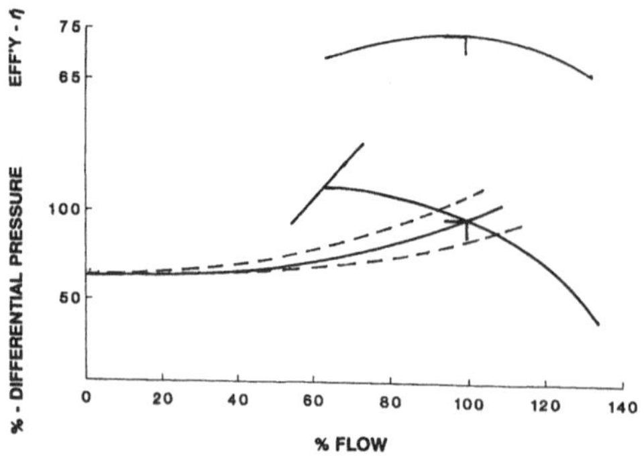

- **FOR A GIVEN DIAMETER AND LENGTH OF PIPE, SYSTEM PRESSURE DROP IS A FUNCTION OF:**

  $\sqrt{(DENSITY)X(PRESSURE)}$

  AND VELOCITY. THEREFORE, A GIVEN SYSTEM RESISTANCE CURVE WILL CHANGE WITH GAS COMPOSITION, PRESSURE, TEMPERATURE AND VELOCITY.

Figure 13.8 The effect of process changes on the system resistance curve

operating point of each compressor stage in a multistage compressor. Depending on the impeller selection, this change could have an adverse affect on the operation of a dynamic compressor causing surge and corresponding high vibration, temperature, flow changes, etc.)

The effect of gas composition on turbo-compressor flow rate

- The actual volume flow rate will vary as a result of the operating point change which is the intersection of the turbo-compressor curve (pressure vs. flow) and the system resistance
- The mass flow rate (lbs/min) will be the product of the new actual volume flow rate (ft^3/min) and the gas density (lb/ft^3) at the new gas conditions (M.W., P, T, Z)
- The standard volume flow rate (SCFM) will be the product of the new actual volume flow rate (ft^3/min) at its pressure temperature corrected for standard conditions (14.7 PSIA and 60°F)

Figure 13.9 The effect of gas composition on turbo-compressor flow rate

The effect on power

As shown in Figure 13.10, dynamic compressor required power increases directly with gas density up to the choke flow or stonewall region of the performance curve. In the choke flow region, the head produced by the compressor approaches zero since the gas velocity is equal to its sonic velocity.

The effect on power

$$\text{Power (BHP)} = \frac{\text{Head}\left(\dfrac{\text{Ft-Lb}}{\text{Lb}}\right) \times \text{Mass flow}\left(\dfrac{\text{Lb}}{\text{Min}}\right)}{33{,}000\left(\dfrac{\text{Ft-Lb}}{\text{Min-H.P.}}\right) \times \eta\ (\%)} + \text{Mech. losses (BHP)}$$

Figure 13.10 The effect on power

Figure 13.11 shows the affect on compressor section performance resulting from a change in the gas molecular weight. As previously discussed molecular weight changes can result in compressor stage mismatching which can cause significant mechanical damage to the compressor train.

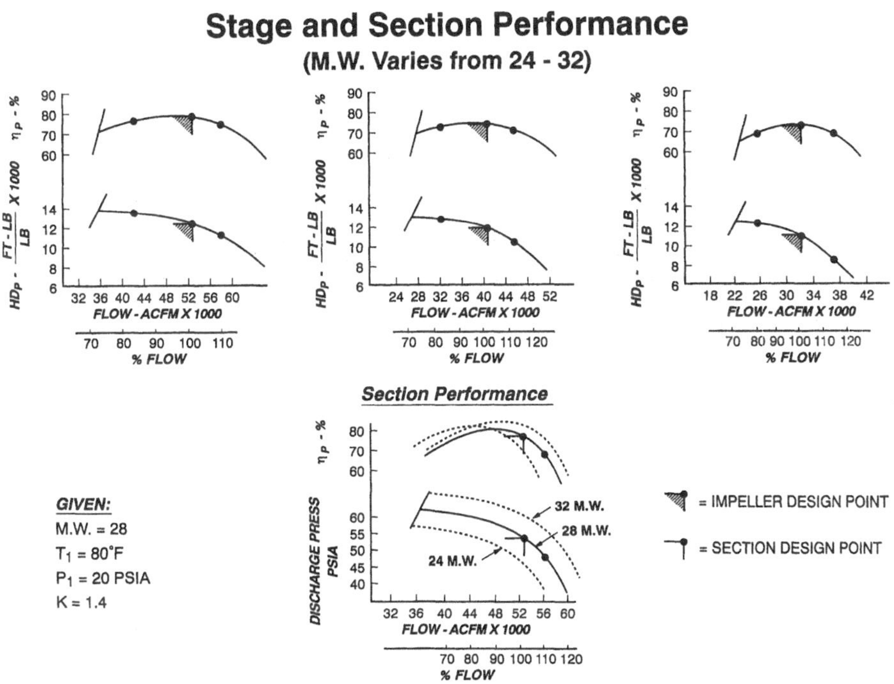

Figure 13.11 Stage and section performance

Turbo-compressor mechanical design overview

- Introduction
- The casing
- The inlet guide vanes
- The rotor
- Diaphragms
- Inter-stage seals
- Shaft end seals
- Journal bearings
- Thrust bearings

Introduction

In this chapter we will define the functions of each major component of a turbo-compressor. That is, what the purpose of each component is or "What it Does". By understanding what each component is supposed to do, you will be in a better position to know if it is performing its duty correctly. We will present each major component starting with the casing, state its function, operating limits and discuss the problems that can occur.

This chapter presents an overview of the mechanical components in a turbo-compressor. Figure 14.1 defines the major components of a typical turbo-compressor.

Function of major components of a turbo-compressor

Gas path components

- Casing
- Guidevanes (stators)
- Rotor (impeller (blades) and shaft)
- Diaphragms (diffuser and return channel)
- Stage seals (impeller and shaft)

Mechanical components

- Shaft end seals
- Journal bearings
- Thrust bearing
- Balance drum

Figure 14.1 Function of major components of a turbo-compressor

Note that these components can be grouped into either gas path or mechanical. That is, their primary function is either to guide the gas or change its velocity or to seal the gas or support the shaft. As we shall see, most components have more than one function.

Figure 14.2A and B shows all of these components in the two major types of Dynamic Compressors – Axial and Centrifugal.

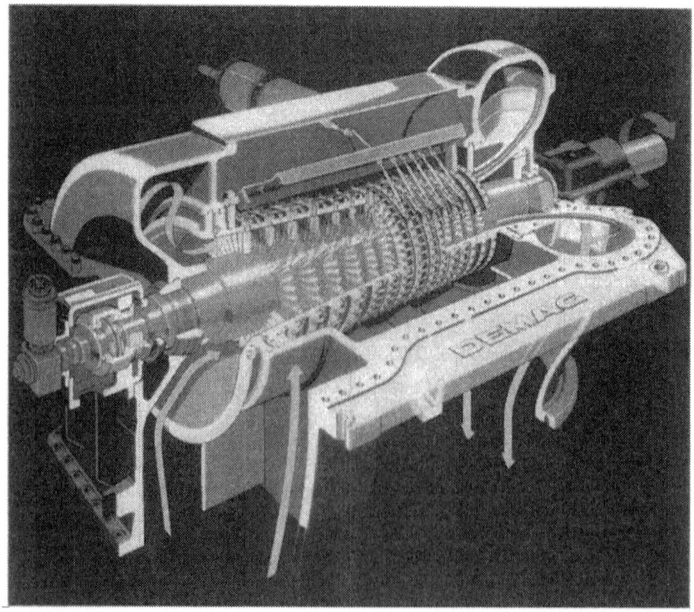

Figure 14.2A Dynamic compressors basic types – Axial (Courtesy of Mannesmann Demag)

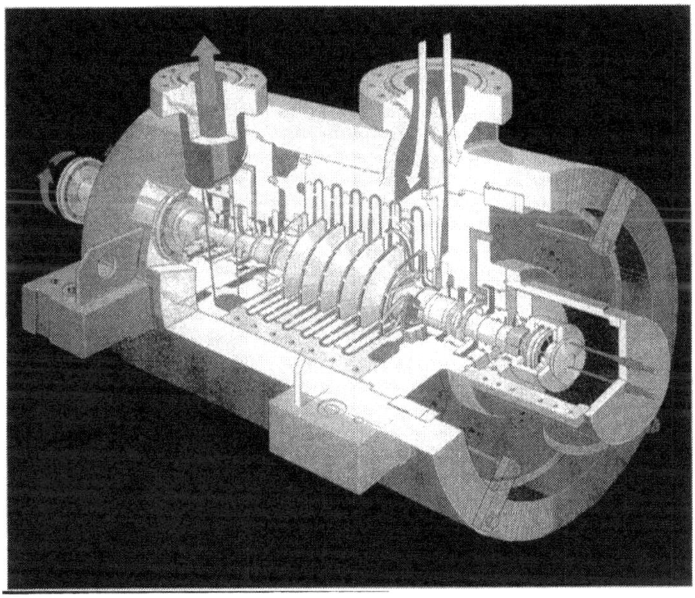

Figure 14.2B Dynamic compressors basic types - Centrifugal (Courtesy of Mannesmann Demag)

The casing

The function of the turbo-compressor casing is defined in Figure 14.3.

Casings can be either horizontally or vertically (barrel type) split. Generally, barrel type casings are used above 600 Psi (40 Bar) pressure or for mixed gases with a high hydrogen content. From a maintenance standpoint horizontally split casings with bottom connections are preferred.

The major problem connected with casing design is distortion of the casing caused by excessive piping or foundation forces. During construction or whenever piping is removed from the casing, dial indicators mounted external to the casing should be positioned on the shaft in the horizontal and vertical planes. Movement of more than 0.002 inch (0.05mm) when piping is bolted up to the casing should require correction to the piping system (pipe flange alignment or spring support modification). The only way for the turbo-compressor shaft to move is if the piping or foundation cause excessive casing strain. A check for excessive casing strain caused by the foundation is commonly known as "soft foot". To check for soft foot, a dial indicator is mounted external to each casing support. If the support moves more than 0.002 inch (0.05mm) when the support foot is unbolted, soft foot exists. Soft foot should be checked with process piping disconnected. The cause of soft foot is that the plane of the support is not parallel to the plane of

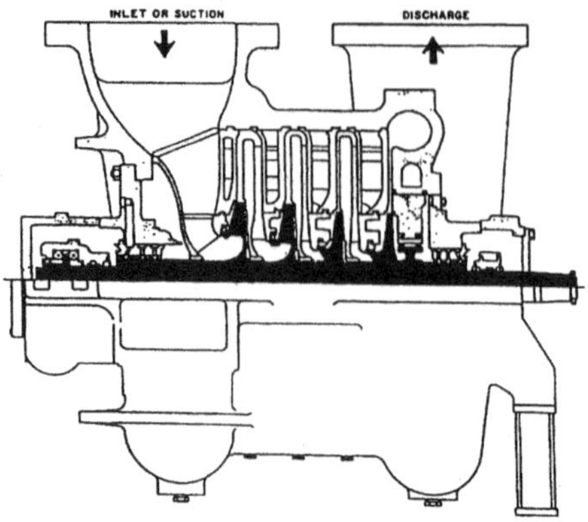

FUNCTION: THE CASING CONTAINS THE PROCESS GAS AND RIGIDLY SUPPORTS ALL STATIONARY INTERNAL PARTS AND THE BEARINGS.

Figure 14.3 The compressor casing

the foundation and/or a gap exists between the foundation and support foot.

The inlet guide vanes

The function of compressor inlet guide vanes is defined in Figure 14.4.

Centrifugal compressor guide vanes are usually fixed. Plant and instrument air compressors can be supplied with adjustable guide vanes. Axial compressors are frequently supplied with adjustable guides vanes.

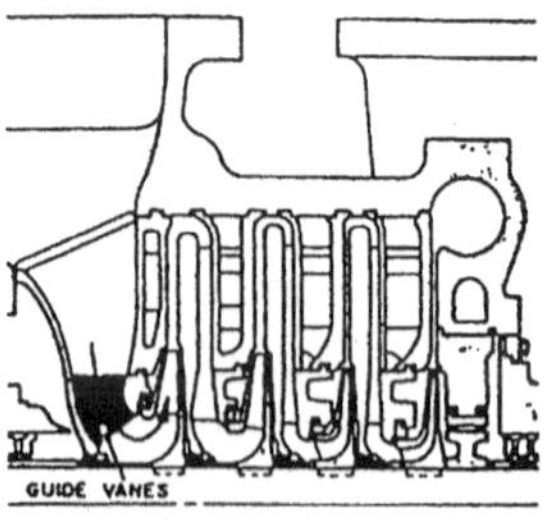

FUNCTION: DIRECTS AND DISTRIBUTES THE GAS TO EACH IMPELLER. CENTRIFUGAL COMPRESSORS GUIDE VANES ARE FIXED. AXIAL COMPRESSOR GUIDE VANES CAN BE ADJUSTABLE WHILE OPERATING.

Figure 14.4 The inlet guide vanes

Adjustable guide vanes alter the dynamic compressor performance curve by changing the velocity and direction of the gas relative to the compressor impeller blades. Impact of the gas velocity vector with the blades (counter rotation) will increase the gas tangential velocity and produce greater energy (head). Leading the gas velocity vector (pre-rotation) will decrease the gas tangential velocity and produce less energy (head).

Guide vanes, stationary or adjustable, positioned at an incorrect angle can alter compressor stage head and efficiency. This problem will be discovered with stationary guide vanes during an initial shop performance test (if required by the End User) and corrected. However, adjustable guide vanes can be improperly set (often known as "scheduled") if the individual guide vanes, bushings or linkages are replaced. Be sure to consult the instruction book and consult the equipment vendor if necessary when replacing any components in an adjustable guide vane assembly.

The rotor

The components and function of a dynamic compressor rotor are presented in Figure 14.5.

Dynamic compressor rotors can contain axial and/or centrifugal impellers. Centrifugal impellers can handle flows from 100 to over 100,000 CFM. Axial impellers (blading rows) are usually used above 100,000 CFM. The head produced by a centrifugal stage can vary from approximately 5,000 to 25,000 $FT/LB_F/LB_M$. An axial stage usually produces approximately 3,000 $FT/LB_F/LB_M$ of head. The most

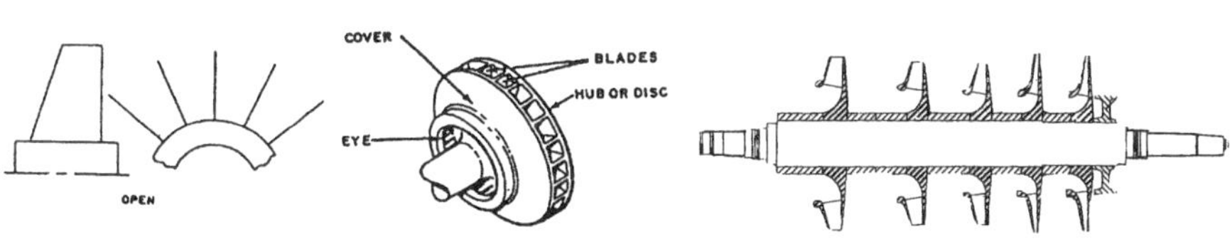

Figure 14.5 The rotor (impeller and shaft)

common problems associated with rotors are; component unbalance, rotor unbalance caused by improper impeller assembly, excessive blade or impeller stress and rotor system natural frequencies. These topics will be discussed in detail later in the book.

Diaphragms

The definition of a diaphragm and its components are presented in Figure 14.6.

Diaphragms can be manufactured of either cast iron or cast steel. Both materials are acceptable. However, in the event of severe axial rubs caused by excessive continuous surge, cast iron diaphragms can crack and will require replacement. Since centrifugal diaphragms are rarely spared, failure will cause long periods of downtime and loss of product revenue.

Diffusers are usually vaneless (parallel walled) but can be vaned or scrolled to increase head produced per stage. Older designs with vaned or scrolled diffusers often encountered problems with impeller side wall breakage caused by fatigue of the impeller side plates resulting form alternating pressures at the impeller discharge. This phenomena was caused by the pressure build up and decay resulting from blade passing interference between the impeller and diffuser vanes. The classic fix is to "scroll" the impeller wheel between each vane. A modified impeller is called a "cabbage cutter wheel" due to its appearance. The idea behind this modification is to remove the effect (side plate cracks) by removing the affected material. This modification has proven successful and does not significantly effect impeller efficiency. Modern "vaned diffuser"

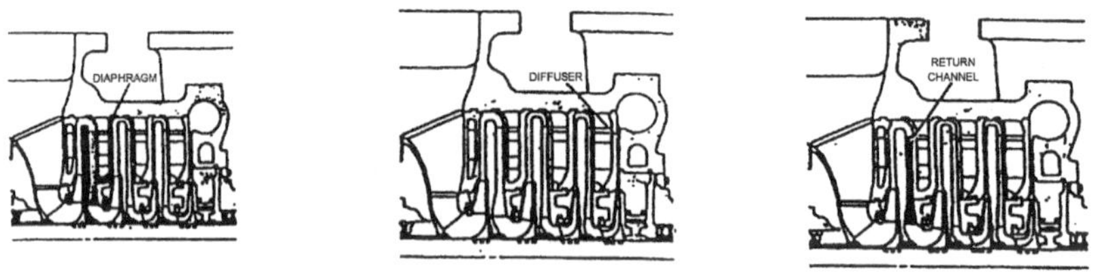

DIAPHRAGM DIFFUSER RETURN CHANNEL

FUNCTION: THE DIAPHRAGM IS THE NON-MOVING PART BETWEEN TWO STAGES AND CONTAINS THE DIFFUSER, RETURN CHANNEL AND STAGE SEALS. THE DIFFUSER CONVERTS VELOCITY INTO PRESSURE. THE RETURN CHANNEL GUIDES THE GAS INTO THE NET STAGE.

Figure 14.6 Diaphragms (diffuser and return channel)

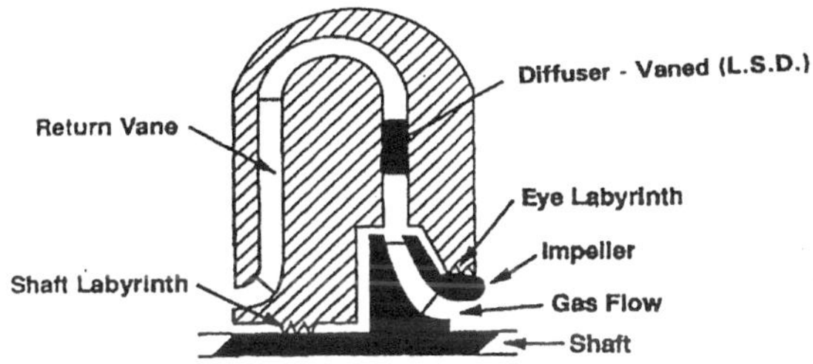

FUNCTION: BOTH IMPELLER AND SHAFT STAGE SEALS PREVENT LEAKAGE FROM RECYCLING FROM THE DISCHARGE OF THE IMPELLER. INCREASED "CLEARANCE" RESULTS IN DECREASED THROUGHPUT.

Figure 14.7 The stage seals (impeller and shaft)

designs like the low solidity diffuser (L.S.D.) move the vane up the diffuser out of the immediate effect of the impeller vanes thus eliminating significant pressure fluctuation. An example of an L.S.D. diffuser is shown in Figure 14.7.

Inter-stage seals

The function of inter-stage seals are defined in Figure 14.7.

Each impeller diaphragm contains one eye seal and one shaft seal. These seals are designed to minimize leakage from the diffuser to the inlet of the impeller (eye seal) and to the inlet of the next stage (shaft seal). Interstage seals are usually manufactured of aluminum to assure that contact with the rotor does not result in an external excitation to cause vibration. However, certain gases and saturated gas mixtures require that alternative materials be used.

Since the early 1970's various abradable material interstage seals have been used. Abradable seals allow for contact with the rotor without causing excitation to the rotor system. This type of interstage sealing system had it's origin in the aircraft engine industry. The advantage of an abradable seal is increased efficiency. Increases in stage efficiency as high as 3% have been obtained. Frequently used abradable seal materials are fluorisint (mica impregnated teflon) and honeycomb arrangements of hastelloy. The temperature limits of these materials are approximately 350°F and 1100°F respectively. Usually, the labyrinth (knife edges) are positioned on the rotors whenever abradable seals are used. It has proven cost effective not to attempt to machine knife edges in the abradable material. One note of caution regarding fluorisint labyrinths

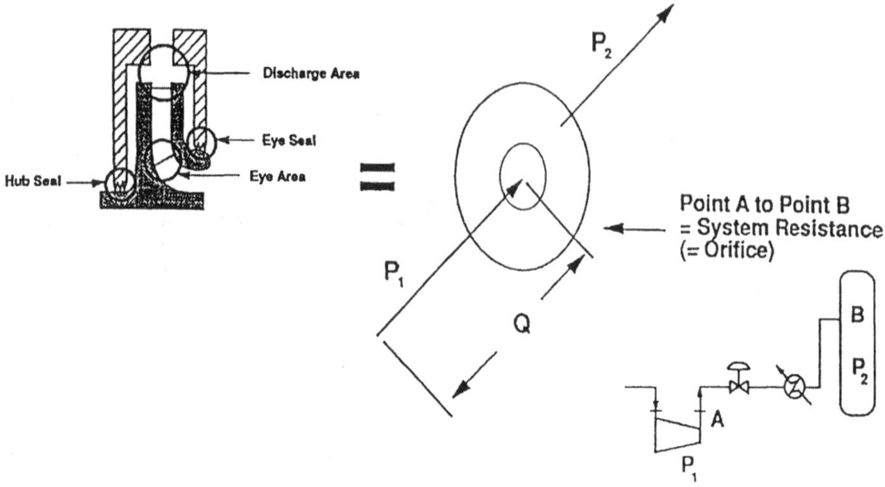

Figure 14.8 Reduce it to an equivalent orifice

... they should be pressure balanced. Pressure balancing of each interstage labyrinth assures that the high pressure side of the seal will not deflect or force the material into the rotor surface. The lack of pressure balance has resulted in premature wear of labyrinth seals. Pressure balance is easily achieved by assuring the pressure behind the seals is less than or equal to the pressure between the labyrinth surface and rotor surface.

Deterioration of labyrinth seals in a centrifugal compressor is one of the major causes of reduced performance. The eye seal is effected to a much greater extent than the shaft seal since the ΔP across this seal is greater. The causes of labyrinth seal wear are: rotor vibration, excessive moisture and fouling. If centrifugal compressor head and efficiency fall off greater that 10%, the compressor should be inspected at the next opportunity if the compressor is known not to be fouled.

Refer to Figure 14.8 and observe how this conclusion can be made by thinking of the interstage labyrinth as equivalent orifices.

Shaft end seals

Compressor shaft end seals perform the function of containing the process fluid by minimizing external leakage and directing the leakage to a safe location. The types of shaft end seals used in compressor applications are shown opposite.

Seal type	Typical applications	Limitations Speed	Pressure	Seal system type
Labyrinth	■ Inert Gases ■ Hydrocarbon	None	None	■ Eductor
Mechanical Contact (oil)	■ Hydrocarbon	Appx.* 12,000 RPM	None	■ ΔP (Oil to Gas) Control
Liquid film (Bushing)	■ Hydrocarbon	None	None	■ Level (Oil to Gas) Control
Dry Gas	■ All Applications (since 1990)	Appx.* 15,000 RPM	Appx.* 1000/ seal	■ ΔP(Buffer Gas)

*Consult vendor's experience list

With the exception of labyrinth seals used for inert gas service, all compressor seals are configured as double seals. That is, the sealing fluid is supplied at a pressure greater than the process fluid and therefore requires a seal between the process and the sealing fluid and between atmosphere and the sealing fluid. Figure 14.9 shows a liquid film bushing seal used in hydrocarbon service.

Follow the path of the sealant and observe in Figure 14.9 that this seal is in a double seal configuration. The inner bushing seal will leak a small amount (typically less than 5 gallons/day) into the seal oil drainers. The outer seal, frequently called the atmospheric bushing directs oil flow back to seal oil reservoir. Typical values vary from 1–10 gallons/ minute. The seal shown in Figure 14.9 also has a "flow-thru" feature. This feature provides additional cooling to the seal and is an advantage in applications that start up with low process gas pressure but normally operate at high process gas pressures.

Figure 14.10 presents the seal oil system for the seal shown in Figure 14.9.

Trace the seal oil supply from the reservoir to the seal. Note the seal oil, which is maintained at approximately 3 psi above reference gas, is distributed into three different paths: the gas side bushing, the atmospheric bushing and the flow through oil. Compressor seals and sealing systems will be covered in separate modules later in the course. Before leaving this subject, all seal systems are designed to automatically maintain a constant seal fluid to process fluid differential across the inner seal. **However**, the differential pressure across the outer seal varies directly with the process gas pressure. This fact presents a

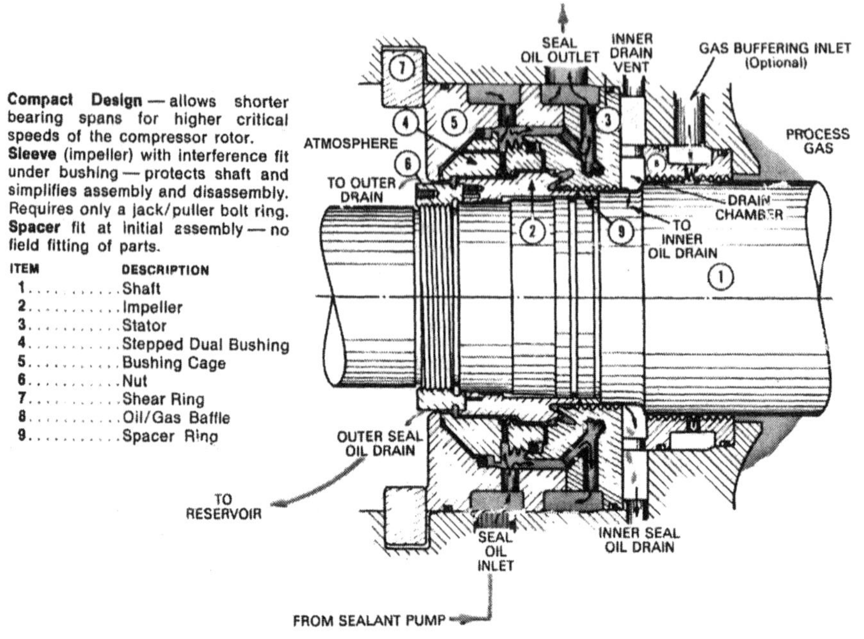

Figure 14.9 Shaft end oil seals (Courtesy of AC Compressor Company)

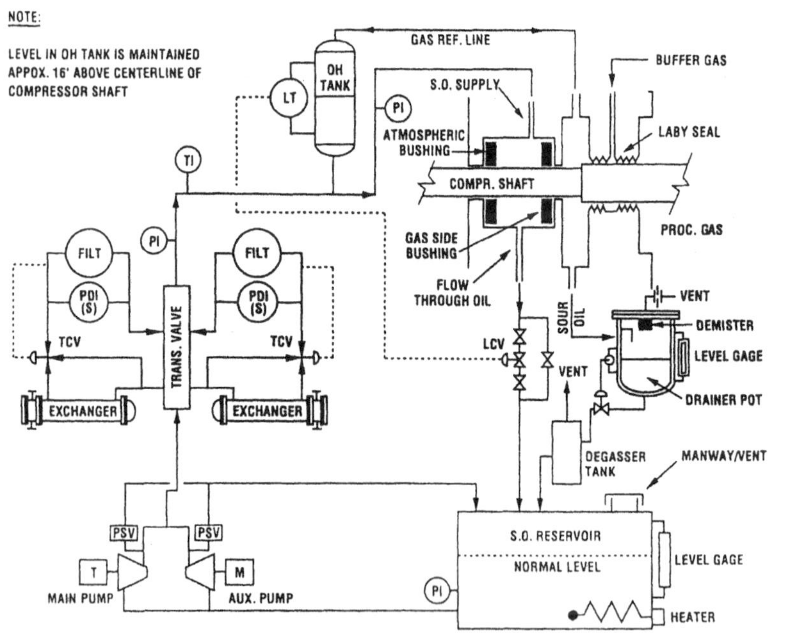

FUNCTION: THE SEAL OIL SYSTEM CONTINUOUSLY SUPPLIES CLEAN, COOL, OIL TO THE SEALS AT THE PROPER DIFFERENTIAL (OIL TO GAS) PRESSURE AND FLOW RATE. IT REMOVES THE HEAT OF FRICTION FROM THE SEALS.

Figure 14.10 The seal oil system (Courtesy of M. E. Crane, Consultant)

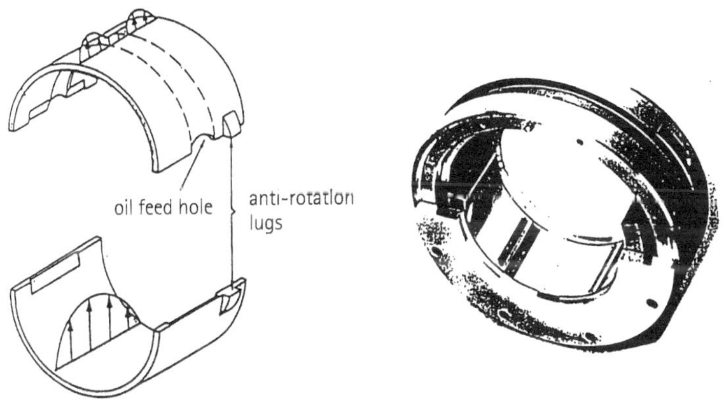

FUNCTION: JOURNAL BEARINGS REGARDLESS OF TYPE CONTINUOUSLY SUPPORT THE ROTOR WITH AN OIL FILM OF LESS THAN ONE THOUSANDTH OF AN INCH.

Figure 14.11 Journal bearings

problem during start-up with low process gas pressure since the seal fluid supply pressure is low and the temperature in the outer seal can become excessive causing seal damage.

Journal bearings

The function of a journal bearing is defined in Figure 14.11.

Most compressors use hydrodynamic bearings as shown. A hydrodynamic bearing supports the rotor with a thin film oil wedge. Small compressors, fans or blowers sometimes use anti-friction (ball or roller) bearings. Regardless of the type of bearing, the basic principle in bearing design is as follows:

$$P = \frac{F}{A}$$

Where P is the pressure in psi on the bearing oil film or component (ball or roller) that results from all the rotor forces (static and dynamic) acting on the bearing area. If the forces are too large or the bearing area is too small, the pressure load will become excessive and the bearing will fail.

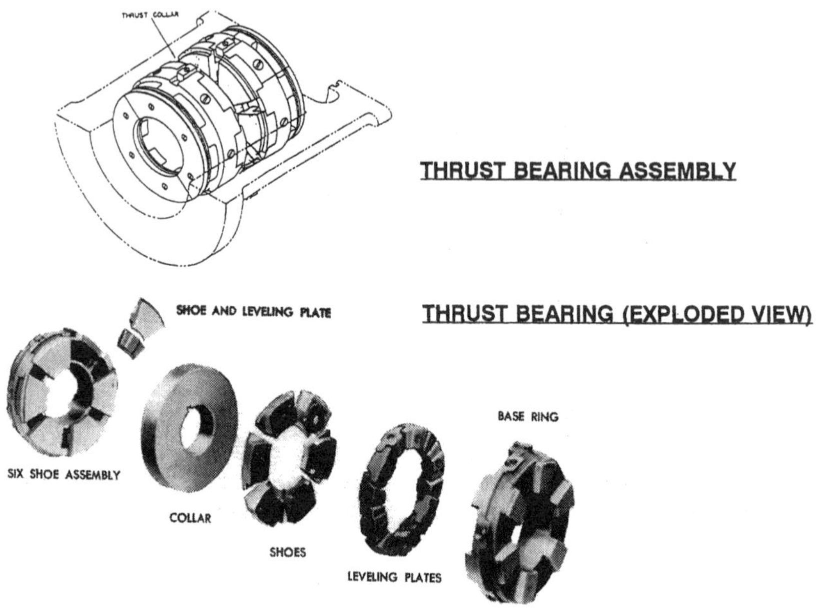

FUNCTION: THE THRUST BEARING POSITIONS THE ROTOR AXIALLY IN THE CASING AND TAKES THE LOAD GENERATED BY THE BLADES AND BALANCE DRUM*

* NOTE: ONLY INSTALLED ON REACTION TURBINES

Figure 14.12 Thrust bearing (Courtesy of Kingsbury, Inc)

Thrust bearings

Figure 14.12 shows a typical hydrodynamic double acting, self equalizing thrust bearing.

The same principle applies in the design of thrust bearing. That is,

$$P = \frac{F}{A}$$

The thrust bearing supports all compressor axial loads. For any hydrodynamic bearing, journal or thrust, the approximate pressure that would cause a bearing failure is 500 psi. Most industry specifications limit the load pressure to 50% of this value or approximately 250 psi.

It should be noted that there are exceptions to this rule. Gearboxes and compressors with integral gearboxes usually have design loads in the 300–400 psi range. This is because major source of bearing load in a gear box is torque. The designer must consider the entire torque range. That is, start up conditions (low torque) and maximum torque conditions. As a result, during start-up the load pressure P will be relatively small on the bearings. This is the reason that many gears are unstable (vibrate) at low loads. The only way to correct this is to

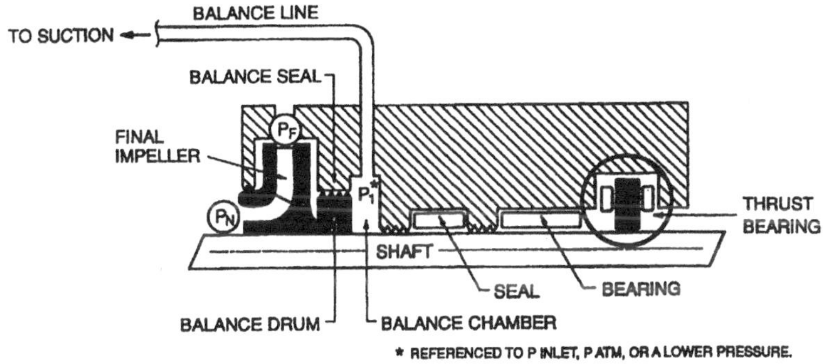

Total Impeller Thrust (LB) = Σ Individual Impeller Thrust
Balance Drum Thrust (LB) = (P_F - P_1) x (Balance Drum Area)
Thrust Bearing Load (LB) = Total Impeller Thrust - Balance Drum Thrust

FUNCTION: The Balance Drum and Seal Reduce Thrust (Force) on the Thrust Bearing by Maintaining Design Clearance Between the Drum and Seal

Figure 14.13 Balance drum and seal

increase the bearing load pressure at low loads at the expense of having a larger value of load pressure at maximum loads.

In many thrust bearing applications, it is not possible to provide sufficient bearing area to maintain the load pressure at 250 psi. In these cases, a thrust balancing system is required to reduce the thrust load. Please refer to Figure 14.13.

The balance system reduces thrust by providing a thrust force which counteracts the total impeller thrust force thus reducing the thrust bearing load P (P.S.I.) This is accomplished by designing the face dimensions of the balance drum for sufficient area (A) to provide the desired force (F) for the differential pressure (P_2–P_1) across the balance drum. Therefore the balance force is calculated by:

$$F_{BALANCE} = (\Delta P)\,(A_{BALANCE\ DRUM})$$

The balance line, which usually runs from the balance chamber to the compressor suction, maintains the balance chamber pressure at P_1 + Balance Line ΔP. An effective method of balance drum condition monitoring is to monitor balance drum ΔP. An increasing ΔP, at the same operating point, will indicate increased balance drum clearance. Often, thrust bearing failures are caused by balance seal wear. Monitoring balance line ΔP will indicate the root cause of a thrust bearing failure in this case. Other important facts concerning thrust balance are:

- Thrust can be balanced to any value (0 thrust, towards suction or towards discharge)

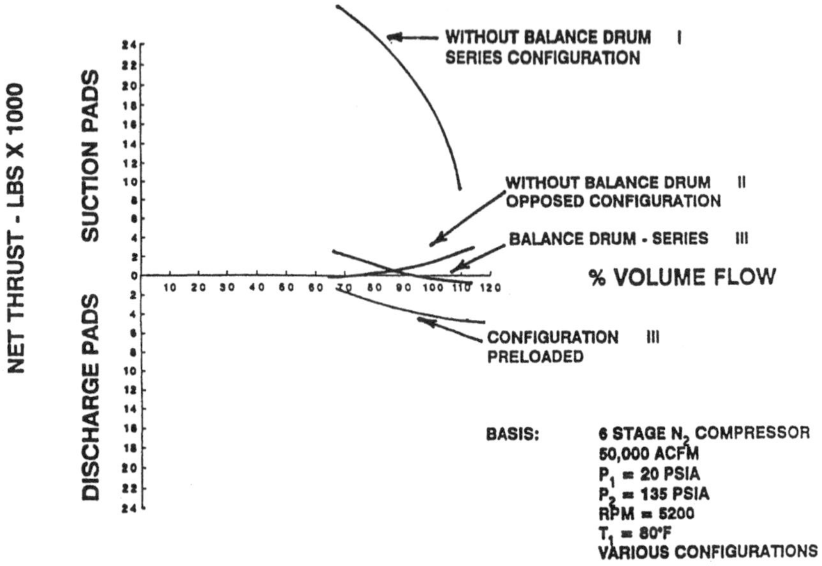

Figure 14.14 Net thrust load vs. % flow

- Active thrust direction is the direction the rotor thrusts at rated conditions
- Thrust position is a function of performance and will change with flow rate, speed and gas density.

Figure 14.14 shows how thrust load can change with balance system design and compressor performance.

Note that a compressor thrust vs flow curve without a balance drum (I) actually follows the shape of a performance curve. A single stage compressor without a balance drum will have its highest operating thrust at surge. An example would be a circulator compressor application. Curves II and III show various other thrust balancing arrangements. Note that the thrust balance can be designed in many different ways! Every custom designed compressor has a unique thrust vs flow curve based on the design of the thrust balancing system.

Figure 14.15 shows a typical lubrication system and its function.

Such a system would continuously provide lubrication to the journal and thrust bearings present in a compressor train.

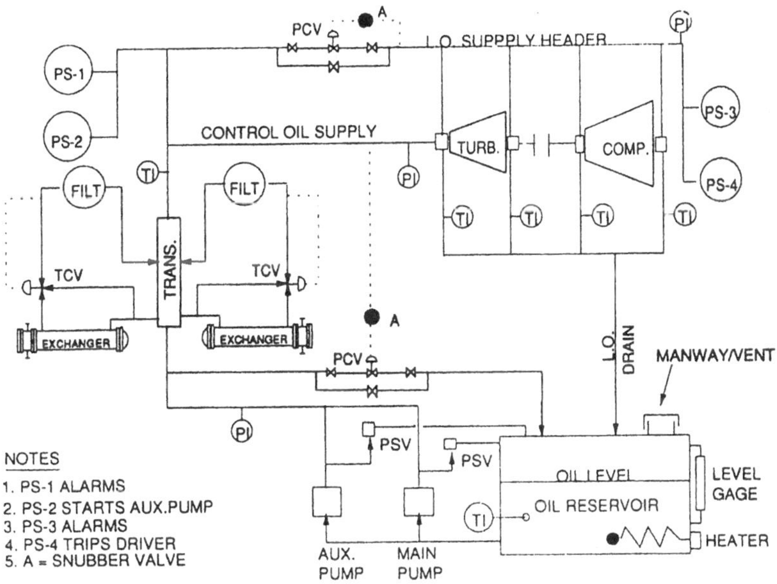

FUNCTION: THE LUBE OIL SYSTEM CONTINUOUSLY SUPPLIES CLEAN, COOL, OIL TO ALL BEARINGS AT THE PROPER PRESSURE AND FLOW RATE. IT REMOVES THE HEAT OF FRICTION FROM THE BEARINGS.

Figure 14.15 The lube oil system (Courtesy of M. E. Crane, Consultant)

15 Radial bearing design

- Introduction
- Anti-friction bearings
- Hydrodynamic bearings
- Hydrodynamic bearing types
- Condition monitoring
- Vibration instabilities

Introduction

Radial bearings fall into two major categories:

- Anti-friction
- Hydrodynamic

Anti-friction bearings rely on rolling elements to carry the load of the equipment and reduce the power losses or friction.

Hydrodynamic bearings rely on a liquid film, usually lubricating oil, to carry the load of the equipment and minimize friction. In general, anti-friction bearings are used for equipment of low horsepower. Hydrodynamic bearings are generally used for all rotating equipment above approximately 500 horsepower. During this section we will concentrate on the subject of hydrodynamic bearings since they are the principle bearing type used in turbo-compressor operation.

A bearing in the radial position is responsible for mainly supporting the weight of the shaft and any dynamic loads that are present in the system. It can be generally stated that the dynamic loads are in the order of 20% of the static loads on the bearing journal. In the case of gears however, the radial load component is made up principally of the meshing force of the gear teeth and the load will vary from zero to maximum torque. One must be careful in this design to assure that the various load angles occur in zones of the bearing that can support these loads.

We will examine loads on hydrodynamic bearings and present an example of the bearing sizing based on static and dynamic loads of a rotor system. We will see that there are specific oil film pressure limits which dictate the bearing dimensions (length and diameter).

The types of hydrodynamic bearings will be reviewed; plain, stationary anti-whirl types and tilt pads.

We will conclude this section by discussing the condition monitoring requirements for radial bearings and briefly discuss shaft vibration and vibration troubleshooting (diagnostics).

Anti-friction bearings

Anti-friction radial bearings support the rotor using rolling elements to reduce friction losses. They are used in low horsepower applications (below 500 H.P.) and in aero-derivative gas turbines. Examples of roller and ball type anti-friction bearings are shown in Figure 15.1.

As previously mentioned, all bearings are designed on the basis of sufficient bearing area to support all the forces acting on the bearing.

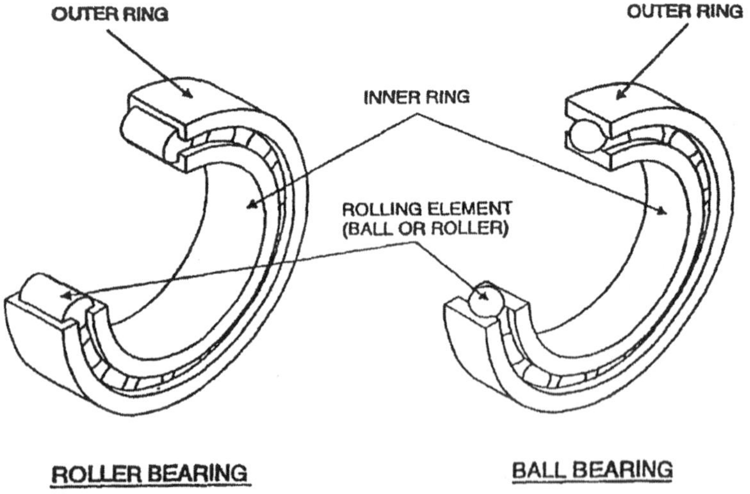

Figure 15.1 Anti-friction bearings

That is:

$$P = \frac{F}{A}$$

Where: P = Pressure on the bearing elements – P.S.I.
F = The total of all static and dynamic forces acting on the bearing
A = Contact area

For anti-friction bearing applications, the pressure, P is the point contact or 'Hertzian' stress on the bearing elements and rings or 'races'. For an anti-friction bearing to be properly designed, its D-N number and bearing life must be determined. Figure 15.2 presents the definition of D-N number and its uses.

D-N Number

- Is a measure of the rotational speed of the anti-friction bearing elements
- D-N number = bearing bore (millimeters) x speed (RPM)
- Approximate lubrication ranges

Lubrication type	D-N Range
Sealed	below 100,000
Regreaseable	100,000–300,000
Oil lube (unpressurized)	300,000–500,000
Oil lube (pressurized)	above 500,000

Figure 15.2 D-N number

Each type of anti-friction bearing has a maximum operating D-N number. If this value is exceeded, rapid bearing failure can occur. In addition, D-N numbers are typically used to determine the type of lubrication required for bearings. A common practice in the turbo-machinery industry has been to use hydrodynamic bearings when the D-N number exceeds approximately 500,000.

The exception to this rule is aircraft gas turbines which can have D-N numbers in excess of 2,000,000. In these applications, hydrodynamic bearings are not used since the size and weight of the required lubrication system would be prohibitive.

Anti-friction bearings possess a finite life which is usually specified as 'B-10' or 'L-10' life as defined in Figure 15.3.

'B' OR 'L' – 10 Life

'B' or 'L' – 10 Life is defined as the life in hours that 9 out of 10 randomly selected bearings would exceed in a specific application.

'B' or 'L' – 10 Life =

$$\frac{16700}{N}\left[\frac{C}{F}\right]^3$$

Where: N = RPM
C = Load in LBS that will result in a bearing element life of 1,000,000 revolutions.
F= Actual load in LBS

Figure 15.3 'B' or 'L' – 10 life

An important fact to note is that the life of any anti-friction bearing is inversely proportional to cube of the bearing loads. As a result, a small change in the journal bearing loads can significantly reduce the bearing life. When anti-friction bearings suddenly start failing where they did not in the past, investigate all possible sources of increased bearing loads (piping forces, foundation forces, misalignment, unbalance, etc.). Anti-friction bearings are usually designed for a minimum life of 25,000 hours continuous operation.

Hydrodynamic bearings

Hydrodynamic bearings support the rotor using a liquid wedge formed by the motion of the shaft (see Figure 15.4).

Oil enters the bearing at supply pressure values of typically 15–20 psig. The shaft acts like a pump which increases the support pressure to form a wedge. The pressure of the support liquid (usually mineral oil) is determined by the area of the bearing by the relationship:

$$P = \frac{F}{A}$$

Where: P = Wedge support pressure (P.S.I.)
F = Total bearing loads (static and dynamic)
A = Projected bearing area ($A_{PROJECTED}$)

$A_{PROJECTED} = L \times d$

Where: L = Bearing axial length
d = Bearing diameter

As an example, a 4" diameter bearing with an axial length of 2" (L/d = .5) would have $A_{PROJECTED}$ = 8 in^2.

If the total of static and dynamic forces acting on the bearing are 1600 lbs. force, the pressure of the support wedge is:

$$P = \frac{1600\ Lb_{FORCE}}{8\ in^2}$$
$$= 200\ P.S.I.$$

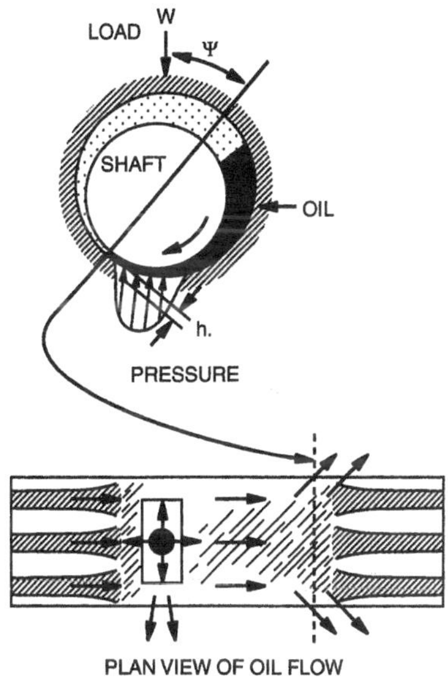

Figure 15.4 Hydrodynamic Lubrication (Courtesy of Bently Nevada Corp.)

The maximum desired design wedge pressure for oil is approximately 500 psi. However, it has been common practice to limit hydrodynamic bearing loads to approximately 250 psi in compressor applications. Figure 15.5 is a side view of a simple hydrodynamic bearing showing the dynamic load forces.

The primary force is the load which acts in the vertical direction for horizontal bearings. However, the fluid tangential force can become

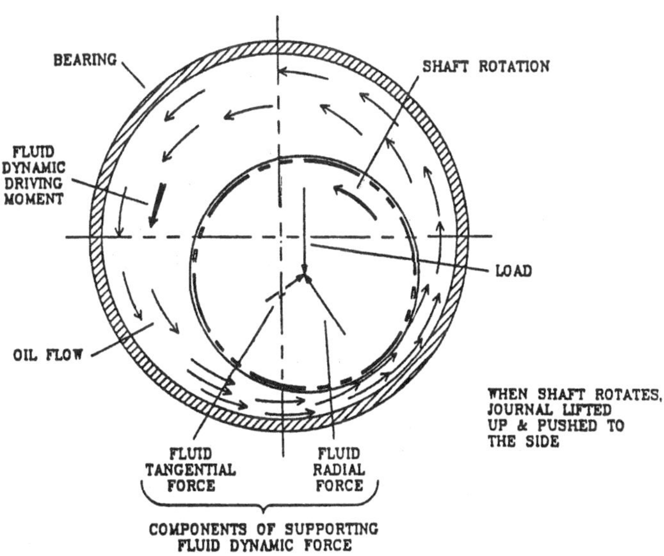

Figure 15.5 Shaft/bearing dynamics (Courtesy of Bently Nevada Corp.)

large at high shaft speeds. The bearing load vector then is the resultant of the load force and fluid tangential force. The fluid radial force opposes the load vector and thus supports the shaft. It has been demonstrated that the average velocity of the oil flow is approximately 47–52% of the shaft velocity. The fluid tangential force is proportional to the journal oil flow velocity. If the fluid tangential force exceeds the load force, the shaft will become unstable and will be moved around the bearing shell. This phenomena is known as oil whirl.

Hydrodynamic bearing types

Regardless of the type of hydrodynamic bearing, all bearing surfaces are lined with a soft, surface material made of a composition of tin and lead. This material is known as Babbitt. Its melting temperature is above 400°F, but under load will begin to deform at approximately 320°F. Typical thickness of Babbitt over steel is 0.060 (1.5mm). Bearing embedded temperature probes are a most effective means of measuring bearing load point temperature and are inserted just below the Babbitt surface. RTD's or thermocouples can be used. There are many modifications available to increase the load effectiveness of hydrodynamic bearings. Among the methods available are:

- Copper backed Babbitt or 'Trimetal' – to aid in heat removal
- Back pad cooling – used on tilt pad bearings to remove heat
- Direct cooling – directing cool oil to maximum load points

A typical straight sleeve hydrodynamic journal bearing is shown in Figure 15.6.

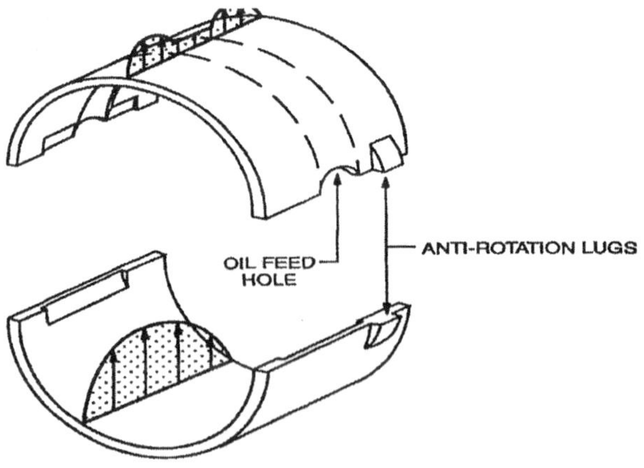

Figure 15.6 Straight sleeve bearing liner (Courtesy of Elliott Co.)

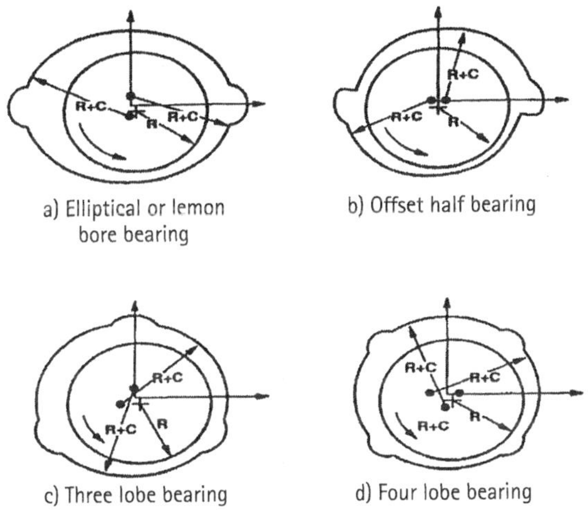

Figure 15.7 Prevention of rotor instabilities

Straight sleeve bearings are used for low shaft speeds (less than 5,000 RPM) or for older turbo-compressor designs. Frequently, they are modified to incorporate a pressure dam, in the direction of rotation. The pressure dam must be positioned in the top half of the bearing to increase the load vector (see Figure 15.5). This action assures that the tangential force vector will be small relative to the load vector thus preventing shaft instability. It should be noted that incorrectly assembling the pressure dam in the lower half of the bearing would render this type of bearing unstable. When shaft speed is high, other alternatives to prevent rotor instabilities are noted in Figure 15.7.

Shown are examples of anti-whirl bearings. The most common types of these bearings are the 3 and 4 lobe design. Elliptical and offset bearing designs do prevent instabilities but tend to increase shaft vibration if the load vector passes through the major axis of the bearing. These types of bearings may have to be rotated in the bearing brackets to prevent this occupance.

The most common hydrodynamic bearing for higher speed applications is the tilt pad journal bearing shown in Figure 15.8.

A tilting pad bearing offers the advantage of increased contact area since the individual pads conform to the shaft orbit. In addition, this type is also a highly effective anti-whirl bearing since the spaces between the pads prevent oil whirl.

Most end users specify tilt pad radial and thrust bearings for turbo-compressor applications.

Figure 15.9 shows the mechanical frictional losses and oil flow requirements for a tilt pad journal bearing as a function of shaft speed.

Figure 15.8 Tilting pad journal bearing assembly (Courtesy of Kingsbury, Inc.)

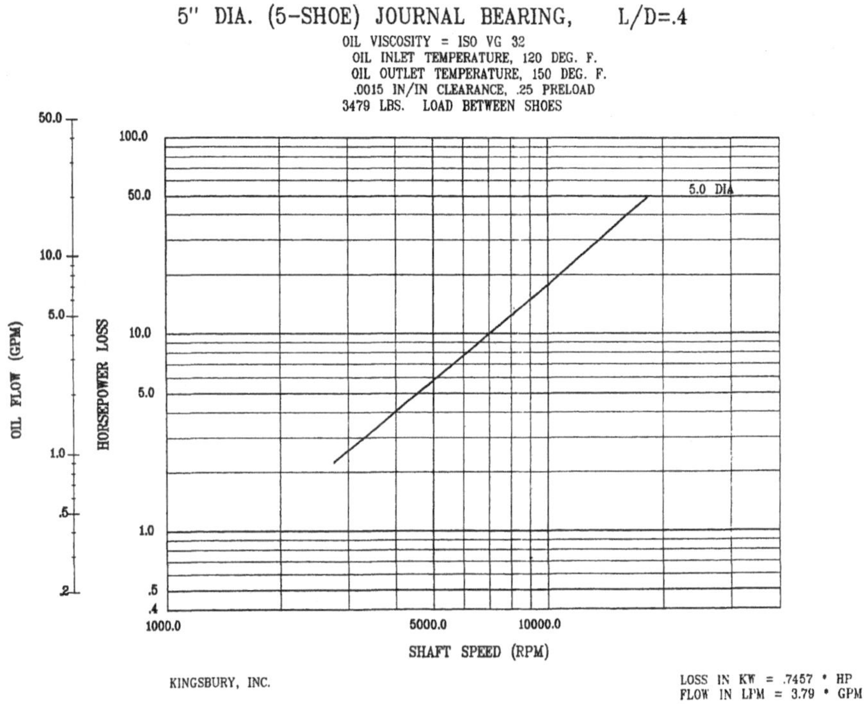

Figure 15.9 Typical journal bearing selection curve (Courtesy of Kingsbury, Inc.)

Note that the basis for horsepower loss and oil flow is an oil temperature rise of 30°F. This is the normal design ΔT for all hydrodynamic bearings. Also given in this figure is the data necessary to calculate bearing pressure at the load point.

As an exercise calculate the following for this bearing:

- Projected Area

$A_{PROJECTED}$ = 5" x 2"

= 10 square inches

- Pressure

= 3479 Lb force ÷ 10 square inches

= 347.9 psi on the oil film at load point

Condition monitoring

In order to determine the condition of any journal bearing, all the parameters that determine its condition must be monitored.

Figure 15.10 presents the eight parameters that determine the condition of a hydrodynamic journal bearing along with typical limits. Attendees are advised to consult the manufacturers instruction book for vendor recommended limits.

Parameter		Limits
1	Radial vibration (peak to peak)	2.5 mils (60 microns)
2	Bearing pad temperature	220°F (108°C)
3	Radial shaft position (except for gearboxes where greater values are normal from unloaded to loaded operation)	>30° change and/or 30% position change
4	Lube oil supply temperature	140°F (60°C)
5	Lube oil drain temperature	190°F (90°C)
6	Lube oil viscosity	Off spec 50%
7	Lube oil flash point	Below 200°F (100°C)
8	Lube oil particle size	Greater than 25 microns

Condition monitoring parameters and their alarm limits according to component:

1. Journal bearing (hydrodynamic)

Figure 15.10 The eight parameters that determine the condition of a hydrodynamic journal bearing along with typical limits

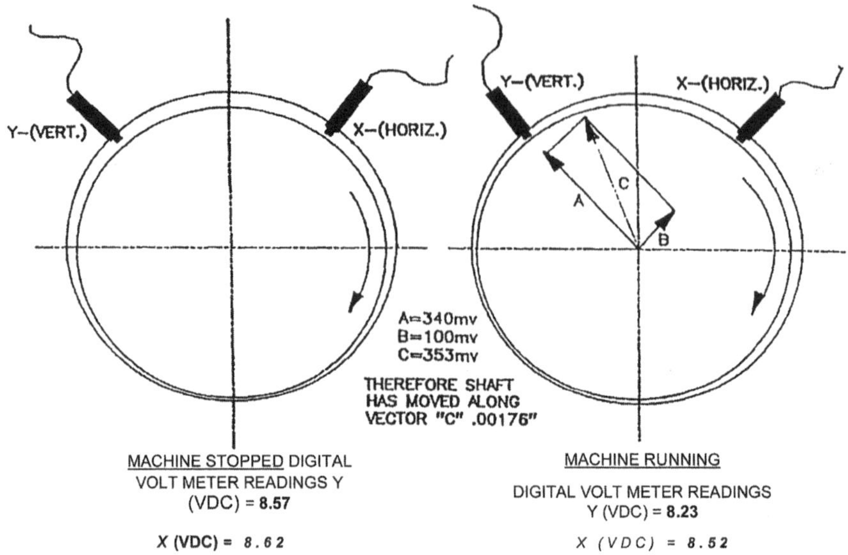

Figure 15.11 Shaft movement analysis (relative to bearing bore) (Courtesy of M.E. Crane Consultant)

One important parameter noted in Figure 15.10 that is frequently overlooked is shaft position. Change of shaft position can only occur if the forces acting on a bearing change or if the bearing surface wears. Figure 15.11 shows how shaft position is determined using standard shaft proximity probes.

Regardless of the parameters that are condition monitored, relative change of condition determines if and when action is required. Therefore, effective condition monitoring requires the following action for each monitored condition.

- Establish baseline condition
- Record condition trend
- Establish condition limit

Figure 15.12 presents these facts for a typical hydrodynamic journal bearing.

Based on the information shown in this trend, the bearing should be inspected at the next scheduled shutdown. A change in parameters during month 6 has resulted in increased shaft position, vibration and bearing pad temperature.

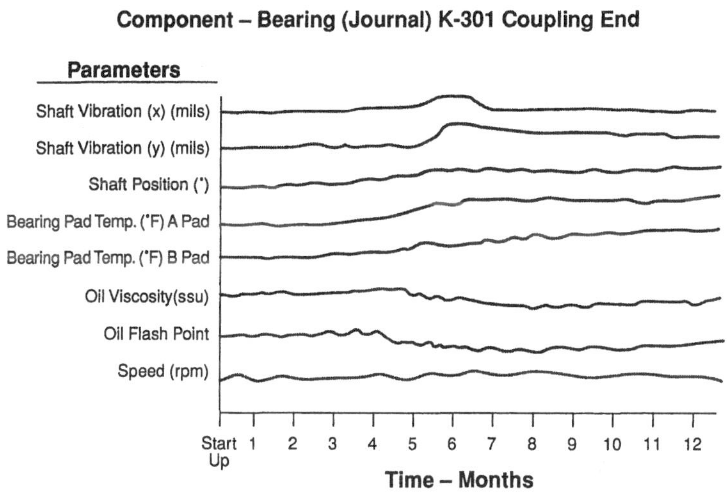

Figure 15.12 Trending data for a typical hydrodynamic journal bearing

Vibration instabilities

Vibration is an important condition associated with journal bearings because it can provide a wealth of diagnostic information valuable in determining the root cause of a problem.

Figure 15.13 presents important information concerning vibration.

Vibration

- Vibration is the result of a system being acted on by an excitation.
- This excitation produces a dynamic force by the relationship:

$$F_{DYNAMIC} = Ma$$

Where: M = Mass (Weight/g)
g = Acceleration due to gravity (386 IN/SEC^2)
a = Acceleration of mass M (IN/SEC^2)

- Vibration can be:
  - Lateral ________ ↕ ________
  - Axial →________________ ←
  - Torsional ________________

Figure 15.13 Vibration

CATEGORY	EXAMPLES	EXCITATION TYPE	VIBRATION TYPE
FORCED VIBRATIONS 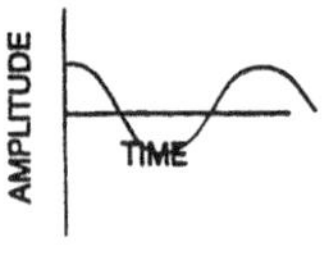	UNBALANCE MISALIGNMENT PULSATION	CONSTANT CONSTANT PERIODIC	LATERAL LATERAL AND AXIAL TORSIONAL
TRANSIENT VIBRATIONS 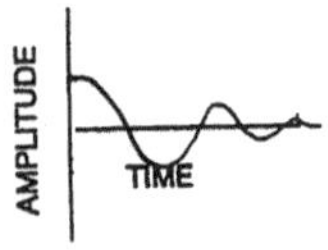	SYNCHRONOUS MOTOR START-UP IMPULSE (SHOCK FORCE)	RANDOM RANDOM	TORSIONAL TORSIONAL, AXIAL, RADIAL
SELF EXCITED	INTERNAL RUB OIL WHIRL GAS WHIRL	RANDOM CONSTANT CONSTANT	LATERAL LATERAL LATERAL

Figure 15.14 Excitation forces with examples

Figure 15.14 defines excitation forces with examples that can cause rotor (shaft) vibration.

Turbo-compressors generally monitor shaft vibration relative to the bearing bracket using a non-contact or 'proximity probe' system as shown in Figure 15.15. The probe generates a D.C. eddy current which continuously measures the change in gap between the probe tip and the shaft. The result is that the peak to peak unfiltered (overall) shaft vibration is read in mils or thousandth of an inch. The D.C. signal is normally calibrated for 200 milli volts per mil. Probe gaps (distance between probe and shaft) are typically 0.040 mils or 8 volts D.C. to assure the calibration curve is in the linear range. It is important to remember that this system measures shaft vibration relative to the bearing bracket and assumes the bearing bracket is fixed. Some systems incorporate an additional bearing bracket vibration monitor and thus record vibration relative to the earth or 'seismic vibration'.

As previously discussed, vibration limits are usually defined by:

$$\text{Vibration(mils p-p)} = \sqrt{\frac{12000}{RPM}}$$

This value represents the allowable shop acceptance level. A.P.I. recommends alarm and trip shaft vibration levels be set as follows:

$$V_{ALARM} = \sqrt{\frac{24000}{RPM}}$$

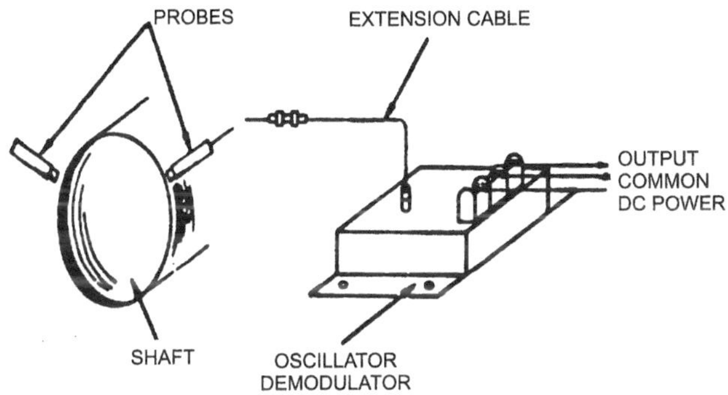

Figure 15.15 Non contact displacement measuring system

$$V_{TRIP} = \sqrt{\frac{36000}{RPM}}$$

In the writers' opinion, shaft vibration alarm and trip levels should be based on the following parameters as a minimum and should be discussed with the machinery vendor prior to establishing levels:

- Application (critical or general purpose)
- Potential loss of revenue
- Application characteristics (prone to fouling, liquid, unbalance, etc.)
- Bearing clearance
- Speed
- Rotor actual response (Bode Plot)
- Rotor mode shapes (at critical and operating speeds)

Figure 15.16 presents a vibration severity chart with recommended action.

A schematic of a shaft vibration and shaft displacement monitor are shown in Figure 15.17.

As mentioned above, vibration is measured unfiltered or presents 'overall vibration'. Figure 15.8 shows a vibration signal in the unfiltered and filtered conditions. All vibration diagnostic work (troubleshooting) relies heavily on filtered vibration to supply valuable information to determine the root cause of the vibration.

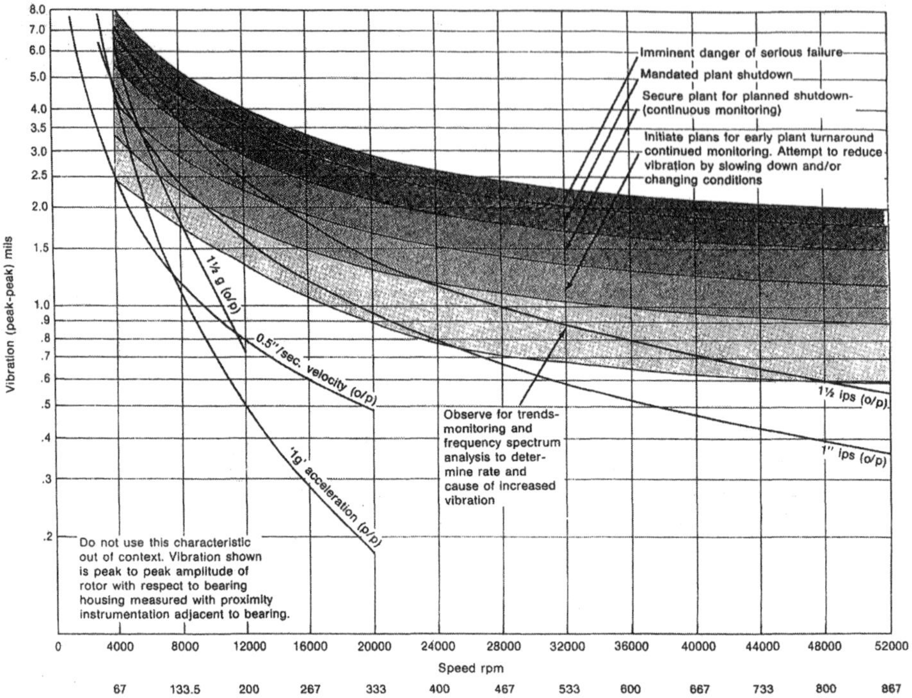

Figure 15.16 Vibration severity chart (Courtesy of Dresser-Rand and C. J. Jackson P.E.)

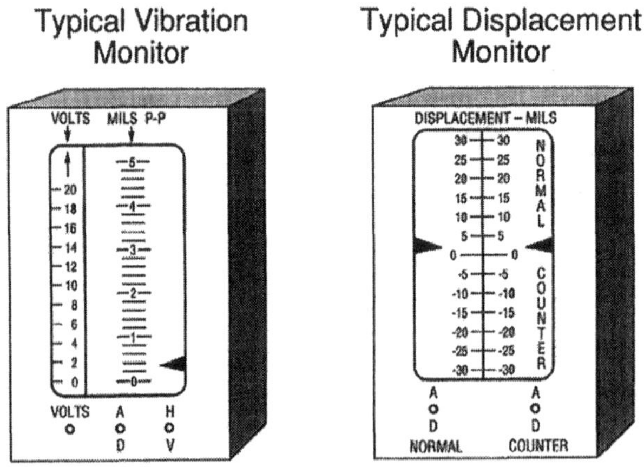

Figure 15.17 Shaft vibration and displacement

ANY VIBRATION SIGNAL IS MADE UP OF ONE OR MORE FREQUENCIES. A TYPICAL UNFILTERED (OVERALL) AND FILTERED SIGNAL ARE SHOWN BELOW:

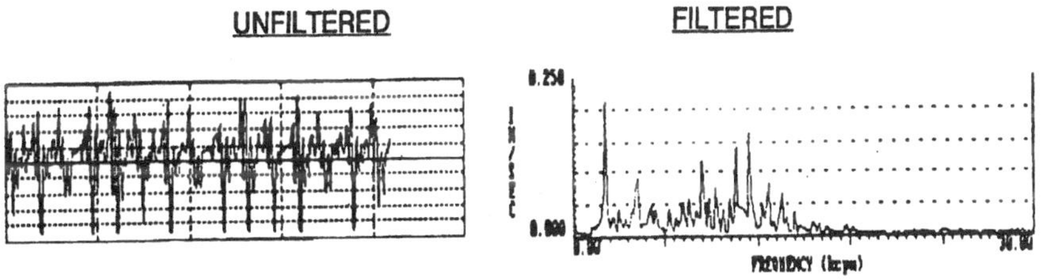

AN ANALOGY TO A FILTERED SIGNAL IS A RADIO. IN A GIVEN LOCALITY, MANY STATIONS ARE TRANSMITTING SIMULTANEOUSLY. ANY GIVEN STATION IS OBTAINED BY ADJUSTING THE TUNER TO THE CORRECT TRANSMISSION FREQUENCY.

Figure 15.18 Vibration frequency

Figure 15.19 presents an example of a radio tuner as an analogy to a filtered vibration signal.

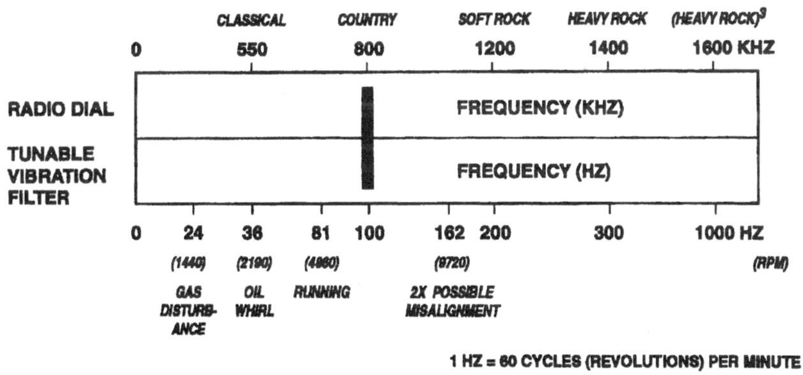

By Adjusting the Tuner (Filter) to a Selected Station (Frequency) the Desired Program (Vibration Frequency Signal) can be Obtained if it is "On the Air" (Present)

Figure 15.19 Radio tuner/vibration filter analogy

By observing the predominant filtered frequencies in any overall (unfiltered) vibration signal, valuable information can be gained to add in the troubleshooting procedure and thus define the root cause of the problem.

Rotor axial (thrust) forces

- Introduction
- The hydrodynamic thrust bearing
- Impeller thrust forces
- Rotor thrust balance
- Thrust condition monitoring

Introduction

In every rotating machine utilizing reaction type blading, a significant thrust is developed across the rotor by the action of the impellers or blades. Also in the case of equipment incorporating higher than atmospheric suction pressure, a thrust force is exerted in the axial direction as a result of the pressure differential between the pressure in the case and atmospheric pressure.

In this section we will cover a specific rotor thrust example and calculate thrust balance for a specific case. We will see the necessity in some applications of employing an axial force balance device known as a balance drum. In many instances, the absence of this device will result in excessive axial (thrust) bearing loadings. For the case of a machine with a balance device, the maintenance of the clearances on this device are of utmost importance. In many older designs the clearances are maintained by a fixed close clearance bushing made out of babbitt which has a melting temperature of approximately 350°F, depending on the pressure differential across the balance drum. If the temperature in this region should exceed this value, the effectiveness of the balance drum would suddenly be lost and catastrophic failures can occur inside the machine. Understanding the function of this device and the potential high axial forces involved in its absence is a very important aspect of condition monitoring of turbo-compressors.

We will also examine various machine configurations including natural balanced (opposed) thrust and see how thrust values change even in the case of a balanced machine as a function of machine flow rate.

Finally, we will examine thrust system condition monitoring and discuss some of the confusion that results with monitoring these machines.

The hydrodynamic thrust bearing

A typical hydrodynamic double acting thrust bearing is pictured in Figure 16.1.

The thrust bearing assembly consists of a thrust collar mounted on the rotor and two sets of thrust pads (usually identical in capacity) supported by a base ring (Michell Type).

The Kingsbury type includes a set of leveling plates between each set of pads and the base ring. This design is shown in Figure 16.2.

Both the Michell and Kingsbury types are used. Figure 16.3 provides a view of the leveling plates providing the self-equalizing feature in the Kingsbury design. The self-equalizing feature allows the thrust pads to lie in a plane parallel to the thrust collar.

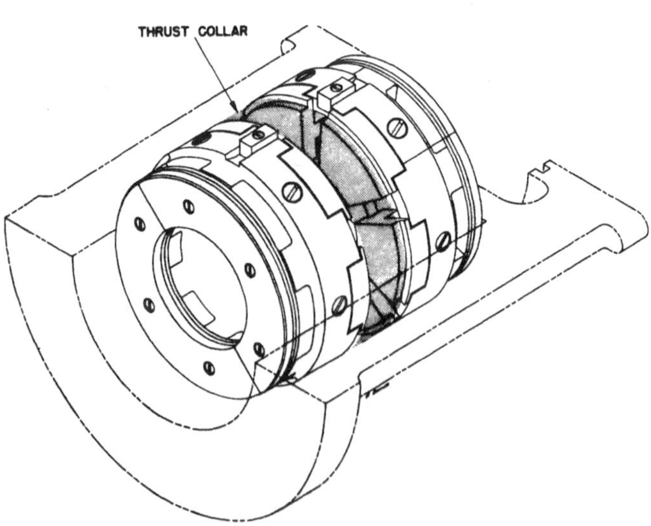

Figure 16.1 Double acting self-equalizing thrust bearing assembly (thrust collar removed) (Courtesy of Elliott Company)

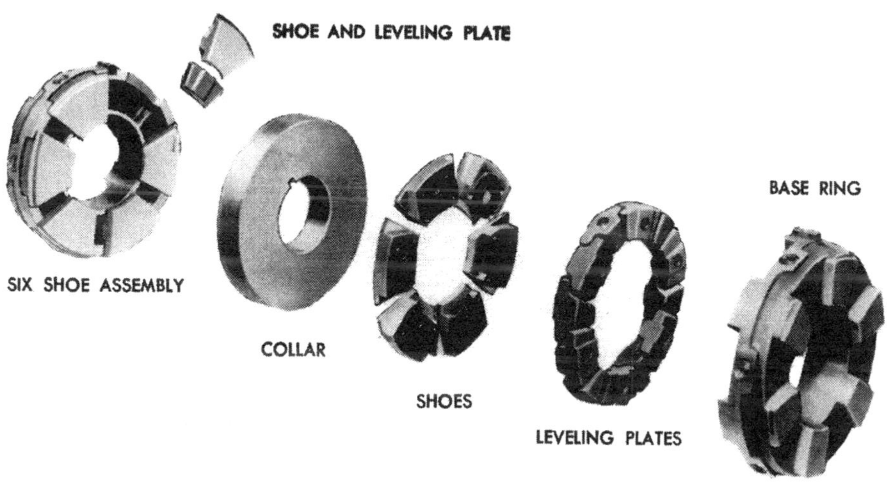

Figure 16.2 Small Kingsbury six-shoe, two direction thrust bearing. Left-hand group assembled, except for one shoe and 'upper' leveling plate. Right-hand group disassembled (Courtesy of Kingsbury, Inc.)

Regardless of the design features, the functions of all thrust bearings are:

- To continuously support all axial loads
- To maintain the axial position of the rotor

The first function is accomplished by designing the thrust bearing to provide sufficient thrust area to absorb all thrust loads without exceeding the support film (oil) pressure limit (approximately 500 psi).

Figure 16.4 shows what occurs when the support film pressure limit is exceeded.

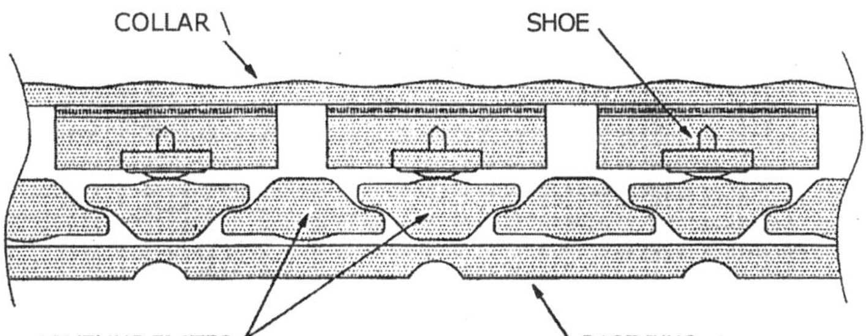

Figure 16.3 Self-equalizing tilt-pad thrust bearing (View – looking down on assembly) (Courtesy of Kingsbury, Inc.)

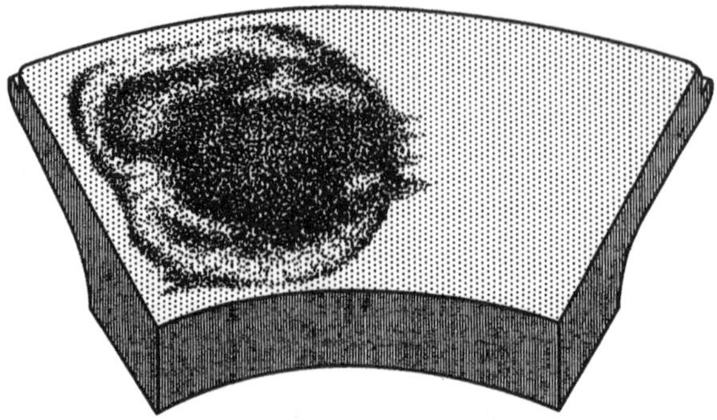

Figure 16.4 Evidence of overload on a tilt-pad self-equalizing thrust bearing pad (Courtesy of Kingsbury Corp.)

The oil film breaks down, thus allowing contact between the steel thrust collar and soft thrust bearing pad overlay (Babbitt). Once this thin layer (1/16") is worn away, steel to steel contact occurs resulting in significant turbo-compressor damage.

Thrust pad temperature sensors, located directly behind the babbitt at the pad maximum load point protect the compressor by tripping the unit before steel to steel contact can occur.

Figure 16.5 presents different Kingsbury bearing size rated capacities as a function of speed.

Figure 16.6 shows how thrust pad temperature and thrust load are related for a given thrust bearing size and shaft speed. Note that the

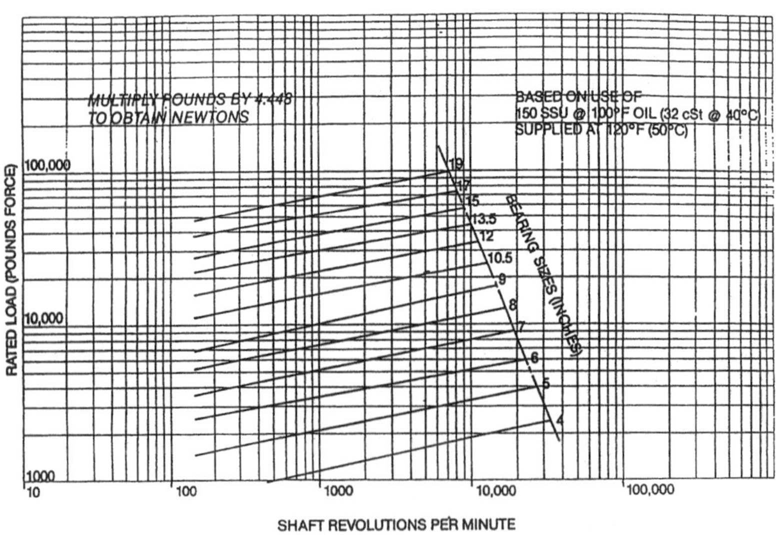

Figure 16.5 Thrust bearing rated load vs. speed (Courtesy of Kingsbury Corp.)

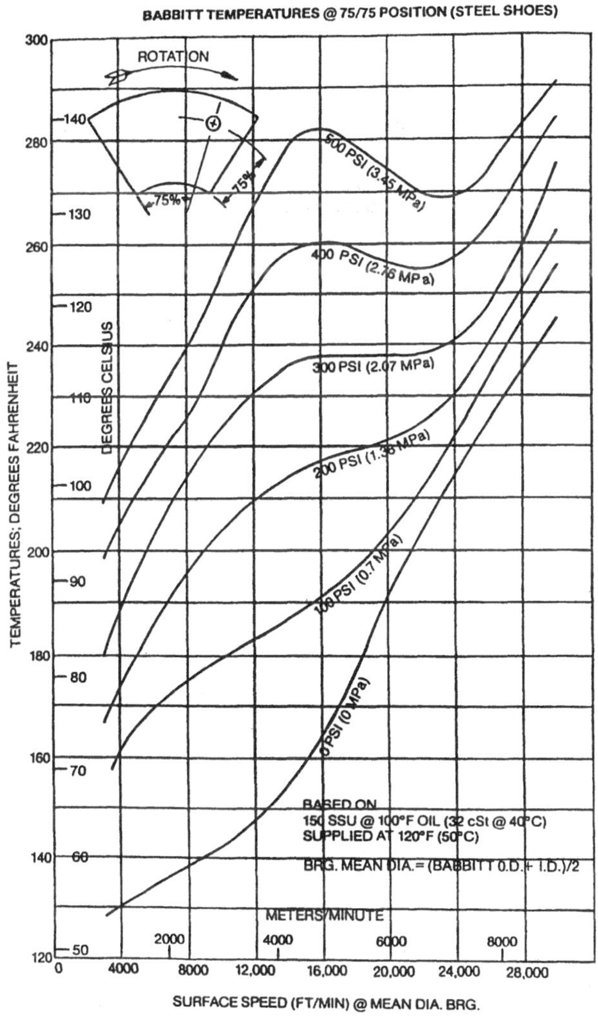

Figure 16.6 The relationship between thrust pad temperature and thrust load (Courtesy of Kingsbury, Inc.)

greater the thrust load (P.S.I.), the smaller the oil film and the greater the effect of oil viscosity on oil flow and heat removal. Based on a maximum load of 500 psi, it can be seen from Figure 16.6 that a turbo-compressor thrust bearing pad temperature trip setting should be between 260° and 270°F.

Other than to support the rotor in an axial direction, the other function of the thrust bearing is to continuously maintain the axial position of the rotor. This is accomplished by locating stainless steel shims between the thrust bearing assembly and compressor axial bearing support plates. The most common thrust assembly clearance with the thrust shims installed is 0.011 – 0.014. These values vary with thrust bearing size. The vendor instruction book must be consulted to determine the proper clearance.

The following procedure is used to assure that the rotor is properly positioned in the axial direction.

1. With thrust shims removed, record total end float by pushing rotor axially in both directions (typically .250"–.500").
2. Position rotor as stated in instruction book.
3. Install minimum number of stainless steel thrust shims to limit end float to specified value.*

*An excessive number of thrust shims act as a spring resulting in a greater than specified axial clearance during full thrust load conditions.

Proper running position of the rotor is critical to obtaining optimum efficiency and preventing axial rubs during transient and upset conditions (start-up, surge, etc.)

Impeller thrust forces

Every reaction type compressor blade set or impeller produces an axial force towards the suction of the blade or impeller. Refer to Figure 16.7.

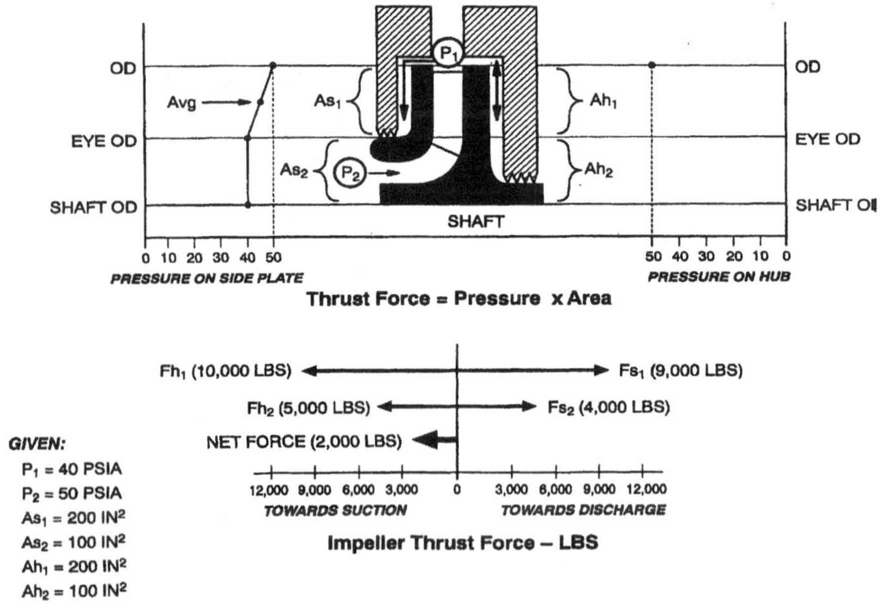

Figure 16.7 Impeller thrust force

In this example, the net force towards the compressor suction is 2,000 psi for the set of conditions noted. Note that the pressure behind the impeller is essentially constant (50 psi), but the pressure on the front side of impeller varies (from 50 psi to 40 psi) because of the pressure drop across the eye labyrinth. Every impeller in a multistage compressor will produce a specific value of axial force towards it's suction at a specific flow rate, speed and gas composition. A change in any or all of these parameters will produce a corresponding change in impeller thrust.

Rotor thrust balance

Figure 16.8 shows how a balance drum or opposed impeller design reduces thrust force. The total impeller force is the sum of the forces from the individual impellers. If the suction side of the impellers is opposed, as noted in Figure 16.8, the thrust force will be significantly reduced and can approach 0. If the suction side of all impellers are the same (in series), the total impeller thrust force can be very high and may exceed the thrust bearing rating. If this is the case, a balance drum must be mounted on the rotor as shown in Figure 16.8. The balance drum face area is varied such that the opposing force generated by the

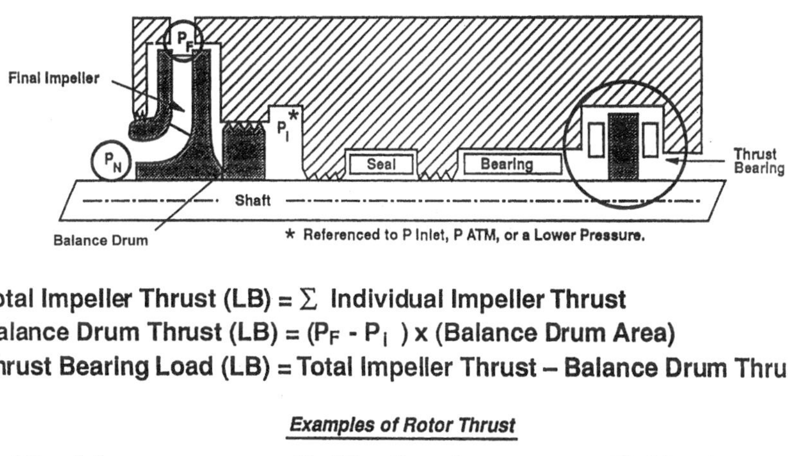

Total Impeller Thrust (LB) = Σ Individual Impeller Thrust
Balance Drum Thrust (LB) = (P_F - P_I) x (Balance Drum Area)
Thrust Bearing Load (LB) = Total Impeller Thrust – Balance Drum Thrust

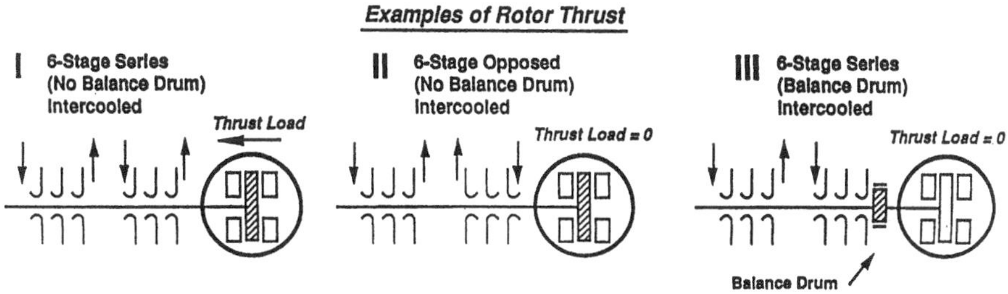

Figure 16.8 Rotor thrust force

balance drum reduces the thrust bearing load to an acceptable value. The opposing thrust force results from the differential between compressor discharge pressure (P_F) and compressor suction pressure (P_1) since the area behind the balance drum is usually referenced to the suction of the compressor. This is accomplished by a pipe that connects this chamber to the compressor suction. This line is typically called the 'balance line'.

It is very important to note that a balance drum is used only where the thrust bearing does not have sufficient capacity to absorb the total compressor axial load. And the effectiveness of the balance drum depends directly on the balance drum seal. Fail the seal, (open clearance significantly) and thrust bearing failure can result.

A common misunderstanding associated with balance drum systems is that a balance drum always reduces the rotor thrust to zero. Refer to Figure 16.9 and observe that this statement may or may not be true depending on the thrust balance system design. And even if it is, the thrust is zero only at one set of operating conditions.

Figure 16.9 shows a rotor system designed four (4) different ways. Note how the thrust **always** changes with the flow rate regardless of the design. Another misconception regarding thrust balance systems is the normal or 'active' direction of thrust. In many cases, the active thrust is assumed to always be towards the suction of the compressor.

NET THRUST LOAD VS. % FLOW

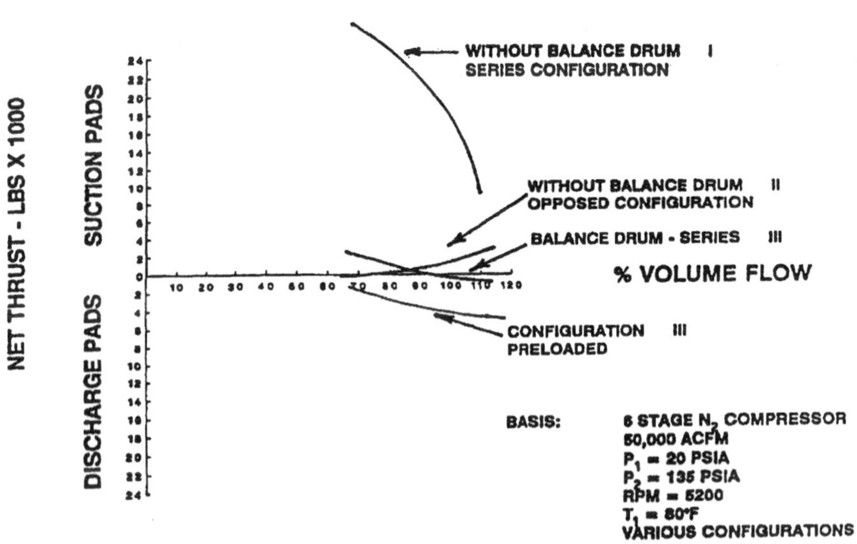

Figure 16.9 Rotor system designed four different ways

Observing Figure 16.9, it is obvious that the 'active' direction can change when the turbo-compressor has a balance drum or is an opposed design. It is recommended that the use of active thrust be avoided where possible and that axial displacement monitors be labeled to allow determination of the thrust direction at all times.

Please refer to Figure 16.10 which shows a typical thrust displacement monitor.

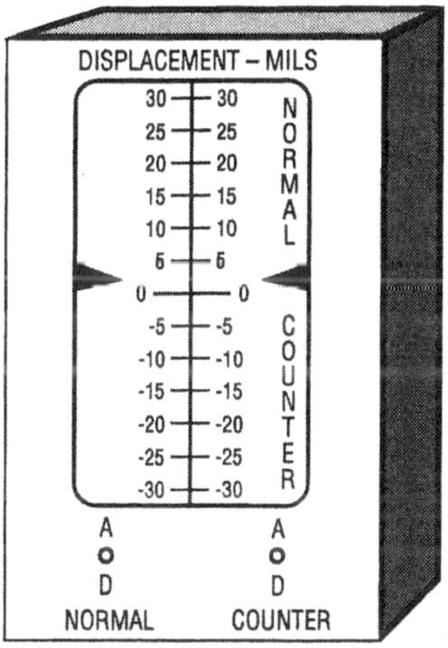

Figure 16.10 Typical axial thrust monitor

These monitors detect thrust position by targeting the shaft end, thrust collar or other collar on the rotor. Usually two or three probes (multiple voting arrangement) are provided to eliminate unnecessary compressor trips. The output of the probes is noted on the monitor as either + (normal) or – (counter). However, this information gives no direct indication of the axial direction of the thrust collar. The following procedure is recommended:

1. With compressor shutdown, push rotor towards the suction and note direction of displacement indicator.
2. Label indicator to show direction towards suction of compressor.

Knowing the actual direction of the thrust can be very useful during troubleshooting exercises in determining the root cause of thrust position changes.

Thrust condition monitoring

Failure of a thrust bearing can cause long term and possibly catastrophic damage to a turbo-compressor. Condition monitoring and trending of critical thrust bearing parameters will optimize turbo-compressor reliability.

The critical thrust bearing condition monitoring parameters are:

- Rotor position
- Thrust pad temperature
- Balance line ΔP

Rotor position is the most common thrust bearing condition parameter and provides useful information regarding the direction of thrust. It also provides an indication of thrust load but does not confirm that thrust load is high. Refer to Figure 16.11.

All axial displacement monitors have pre-set (adjustable) values for alarm and trip in both thrust directions. Typically, the established procedure is to record the thrust clearance (shims installed) during shutdown and set the alarm and trip settings as follows:

$$\text{Alarm} = \frac{\text{Clearance}}{2} + 10 \text{ mils (each direction)}$$

$$\text{Trip} = \text{Alarm Setting} + 5 \text{ mils (each direction)}$$

The above procedure assumes the rotor is in the mid or zero position of the thrust clearance. An alternative method is to hand push the rotor to the assumed active position and add appropriate values for alarm and trip.

The writer personally recommends the first method since an active direction of thrust does not have to be assumed.

As noted, axial displacement monitors only indicate the quantity of thrust load. False indication of alarm or even trip settings can come from:

- Compression of thrust bearing components
- Thermal expansion of probe adaptors or bearing brackets
- Loose probes

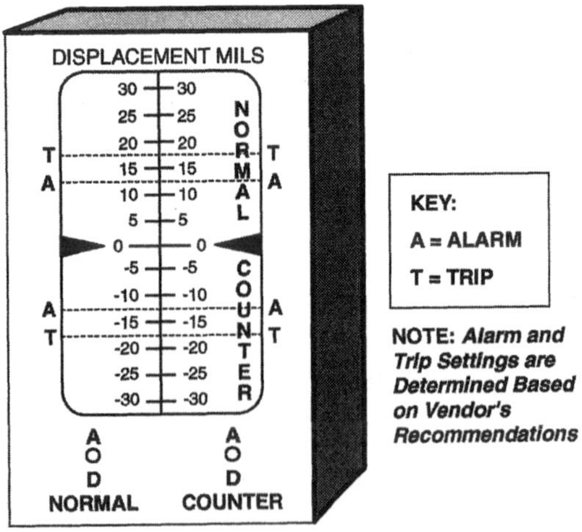

Figure 11 Typical axial displacement monitor

It is strongly recommended that any alarm or trip displacement value be confirmed by thrust pad temperature if possible prior to taking action. Please refer back to Figure 16.6 of this chapter and note that the thrust pad temperature in the case of thrust pad overload is approximately 250°F. If an axial displacement alarm or trip signal is activated **observe** the corresponding thrust pad temperature. If it is below 220°F, take the following action:

- Observe thrust pads. If no evidence of high load is observed (pad and back of pad) confirm calibration of thrust monitor and change settings if necessary.

The last condition monitoring parameter for the thrust system is balance line pressure drop. An increase of balance line ΔP will indicate increased balance drum seal leakage and will result in higher thrust bearing load. Noting the baseline ΔP of the balance line and trending this parameter will provide valuable information as to the root cause of a thrust bearing failure. In many field case histories, the end user made many thrust bearing replacements until an excessive balance drum clearance was discovered as the root cause of the thrust bearing failure. It is a good practice to always check the balance line ΔP after reported machine surge. Surging will cause high internal gas temperatures which can damage the balance drum seal.

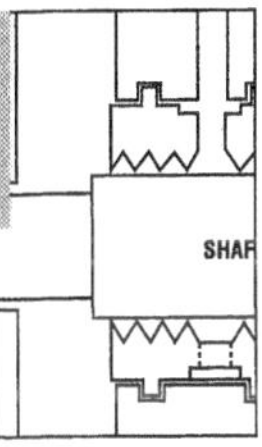

17 Compressor seal system overview and types

- Introduction
- The supply system
- The seal housing system
- Seal supply systems
- Seal supply system summary
- Seal liquid leakage system

Introduction

There are numerous types of fluid seal systems since the types of seals utilized, sealing fluids and sealing pressures vary widely. Regardless of the type of seal used, the function of a critical equipment seal system is as follows: '*To continuously supply clean fluid to each specified seal interface point at the required differential pressure, temperature and flow rate*'. A typical seal system for a centrifugal compressor is shown in Figure 17.1.

The system shown is for use with clearance bushing seals. Let's examine this figure by proceeding through the system from the seal oil reservoir through the compressor shaft seal and back through the reservoir. As previously discussed, the concept of sub-systems can be useful here. The seal oil system shown can be divided into four major sub-systems:

A The supply system

B The seal housing system

C The atmospheric drain system

D The seal leakage system

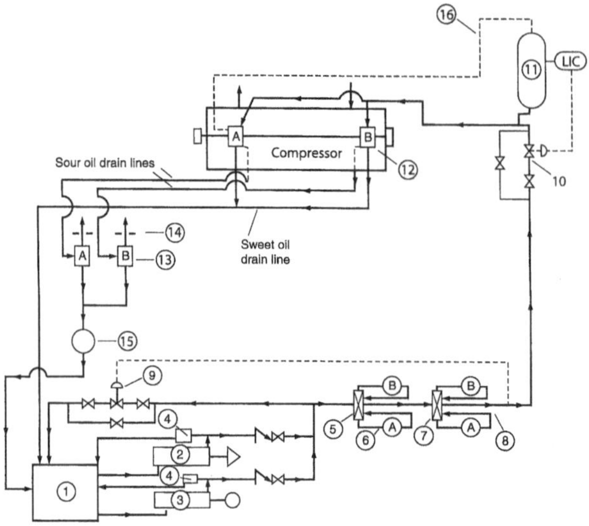

1. Oil reservoir
2. Main screw pump - turbine driven
3. Aux screw pump - motor driven
4. Relief valves
5. Transfer valve
6. Oil coolers (A & B)
7. Transfer valve
8. Oil filters (A & B)
9. Back pressure control valve
10. Level control valve
11. Overhead seal oil tank (typically 15 ft above compressor centerline)
12. ISO sleeve seals (also called liquid bushing seals)
13. Seal oil drainers (A & B)
14. Drainer orifice vents (A & B)
15. Degassing tank
16. Reference gas line to seal oil tank

Note: component condition instrumentation and autostarts not shown

Figure 17.1 API 614 lube/seal oil system for ISO-sleeve seals (Courtesy of Elliott Co.)

A The supply system

This system consists of the reservoir, pumping units, the exchangers, transfer valves, temperature control valves, and filters. The purpose of this sub-system is to continuously supply clean, cool sealing fluid to the seal interfaces at the correct differential pressure.

B The seal housing system

This system is comprised of two different seals. A gas side bushing, and an atmospheric bushing. The purpose of the seal housing system is to positively contain the fluid in the compressor and not allow leakage to the atmosphere. The seal fluid is introduced between both seal interfaces, thus constituting a double seal arrangement. Refer to Figure 17.2 for a closer examination of the seal.

The purpose of the gas side bushing seal is to constantly contain the reference fluid and minimize sour oil leakage. This bushing can be conceived as an equivalent orifice. This concept is similar to bearings previously discussed with the exception that the referenced downstream pressure of the gas side bushing can change. In order to assure a constant flow across this 'orifice,' the differential pressure must be maintained constant. Therefore, every compressor seal system is designed to maintain a constant differential against the gas side seal. The means of obtaining this objective will be discussed as we proceed.

The other seal in the system is the atmospheric bushing whose purpose is to minimize the flow of seal liquid to an amount that will remove

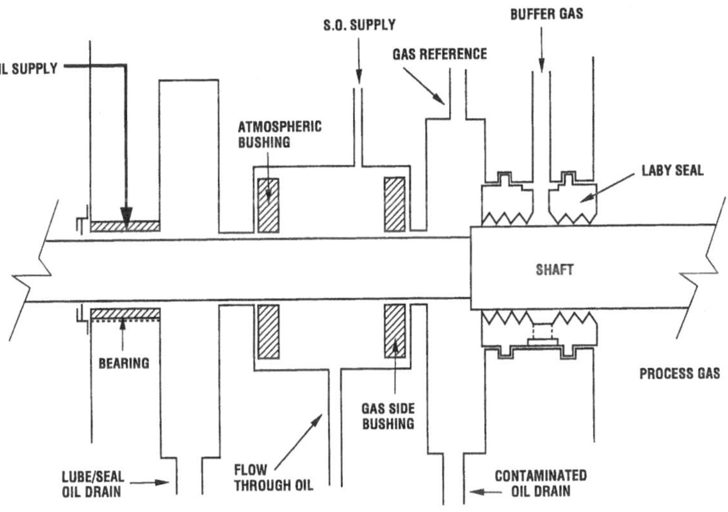

Figure 17.2 Bushing seal schematic (Courtesy of M.E. Crane Consultant)

frictional heat from the seal. This bushing can be conceptualized as a bearing, since the downstream pressure is usually atmospheric pressure. In systems that directly feed into a bearing the atmospheric bushing downstream pressure will be constant (approximately 20 psi). However, the upstream supply pressure will vary with the pressure required by the sealing media in the compressor.

As an example, if a seal system is designed to maintain a constant differential of 5 lbs. per square inch between the compressor process gas and the seal oil supply to the gas side bushing, the supply pressure with 0 PSIG process gas pressure, would be 5 psi to both the gas side bushing and atmospheric bushing. Therefore, gas side bushing and the atmospheric bushing differential would both be equal to 5 psi. If the process gas pressure were increased to 20 psi, the seal oil system would maintain a differential of 5 psi across the gas side seal, and the supply pressure to the gas side bushing and atmospheric bushing would be 25 psi. In this case, the differential across the gas side bushing would remain constant at 5 psi, but the atmospheric bushing differential pressure would increase from 5 to 25 psi. As a result, a primary concern in any seal liquid system is the assurance that the atmospheric bushing receives proper fluid flow under all conditions. After the seal fluid exits the seal chamber, it essentially returns through two additional subsystems.

C The atmospheric draining system

The flow from the atmospheric bushing, if it does not directly enter the bearing system, will return to the seal oil reservoir. In addition, flow from any downstream control valve will also return through the atmospheric drain system to the seal oil reservoir. Both these streams should be gas free since they should not come in contact with the process gas.

D The seal leakage system

The fluid that enters the gas side bushing is controlled to a minimum amount such that it can be either discarded or properly returned to the reservoir after it is degassed. Typically, this amount is limited to less than 20 gallons per day per seal. Since this liquid is in contact with the high speed shaft it is atomized and combines with sealing gas to enter the leakage system. This system consists of:

- An automatic drainer
- A vent system
- Degassing tank (if furnished)

The function of each component is as follows:

The drainer

The drainer contains the oil-gas mixture from the gas side seal. The liquid level under pressure in the drainer, is controlled by an internal float or external level control valve to drain oil back to the reservoir or the de-gassing tank, as required.

The vent system

The function of the seal oil drainer vent system is to assure that all gas side seal oil leakage is directed to the drainer. This is accomplished by referencing the drainer vent to a lower pressure than the pressure present at the gas side seal in the compressor. The drainer vent can be routed back to the compressor suction, suction vessel or a lower pressure source.

The degassing tank

This vessel is usually a heated tank, with ample residence time (72 hours or greater) to sufficiently de-gas all seal oil such that it will be returned to the reservoir and meet the seal oil specification. (Viscosity, flashpoint, dissolved gasses, etc.) These items will be discussed in detail later.

We will now proceed to discuss each of the major sub-systems in detail. Defining the function of each such that the total operation of a seal system can be simplified.

The supply system

Referring to definition of a seal oil system, it can be seen that the function is identical to that of a lube oil system, with one exception. The exception is that the seal fluid must be delivered to the seals at the specified differential pressure. Let's examine this requirement further.

Refer again to Figure 17.2 which shows an equivalent orifice diagram for a typical compressor shaft seal. Notice, that the atmospheric bushing downstream pressure is constant (atmospheric pressure). However, the gas side bushing pressure is referenced to the compressor process pressure. This pressure can and will vary during operation. If it were always constant, the requirement for differential pressure control would not be present in a seal system and would be identical to that of a lubricating system. Another way of visualizing the systems is to understand that the lube system utilizes differential pressure control as well, but the reference pressure (atmospheric pressure) is constant and consequently all control valves need only control lube oil pressure. However seal systems require some means of constant differential pressure control (reference gas pressure to seal oil supply pressure). This objective can be accomplished in many different ways. Referring back to Figure 17.1 it can be seen that the supply system function is identical to that of a lube oil system with the exception that the liquid is referenced to a pressure that can vary and must be controlled to maintain a constant differential between the referenced pressure and the seal system supply pressure. The sizing of the seal oil system components is also identical to that of the lube oil system components. Refer back and observe the heat load and flow required of each seal is determined in a similar way to that of the bearings. Seals are tested at various speeds and a necessary flow is determined to remove the heat of friction under various conditions. The seal oil flow requirements and corresponding heat loads, are then tabulated and pumps exchangers, filters, and control valves are sized accordingly.

The seal oil reservoir is sized exactly the same way as lube oil reservoir in our previous example. The only major difference between the component sizing of a seal and a bearing is that the seal flows across the atmospheric bushing will change with differential pressures. As previously explained, any liquid compressor seal incorporates a double seal arrangement. The gas side seal differential is held constant by system design. The atmospheric side seal differential varies with varying seal reference (process) pressure. Therefore, the total flow to the seals will vary with process pressure and must be specified for maximum and minimum values when sizing seal system components. Remembering the concept of an equivalent orifice, a compressor at atmospheric conditions will require significantly less seal oil flow than it will at high pressure (200 psi) conditions. This is true since the differential across the atmospheric seal and liquid flow will increase from a low value to a significantly higher value, while the gas side bushing differential and liquid flow will remain constant provided seal clearances remain constant.

Many seal system problems have been related to insufficient seal oil flow through the atmospheric bushing at low suction pressure conditions. Close attention to the atmospheric drain cavity temperature is recommended during any off design (low suction pressure condition) operation.

The seal housing system

Regardless of the type, the purpose of any seal is to contain the fluid in the prescribed vessel (pump, compressor, turbine, etc.). Types and designs of seals vary widely. Figure 17.3 shows a typical mechanical seal used for a pump.

Since the contained fluid is a liquid, this seal utilizes that fluid to remove the frictional heat of the seal and vaporize the liquid, thus attaining a perceived perfect seal. A small amount of vaporized liquid constantly exits the pump across the seal face. It is a fact that all seals leak. This is the major reason that many pump applications today are required to utilize seal-less pumps to prevent emission of toxic vapors. The following is a discussion of major types of seal combinations used in centrifugal compressor seal applications.

Gas seals

A typical gas seal is shown in Figure 17.4.

Gas seals have recently drawn attention since their supply systems appear to be much simpler than those of a traditional liquid seal system.

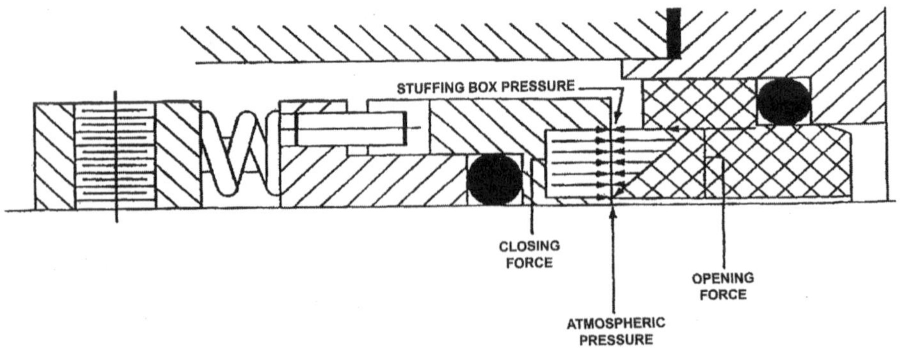

Figure 17.3 Typical pump single mechanical seal

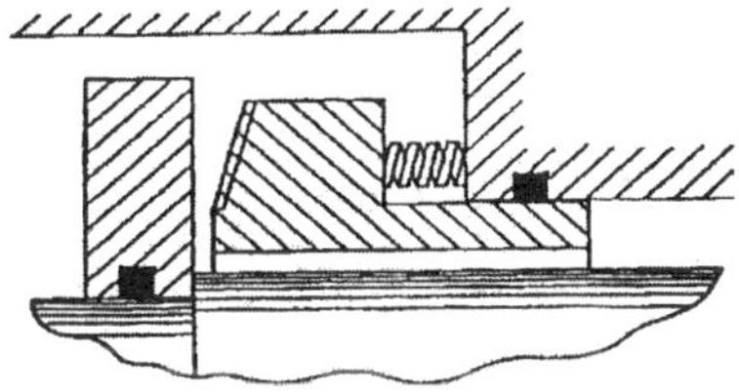

Typical Design For Curved Face — Spiral Groove Non-contact Seal;
Curvature May Alternately Be On Rotor

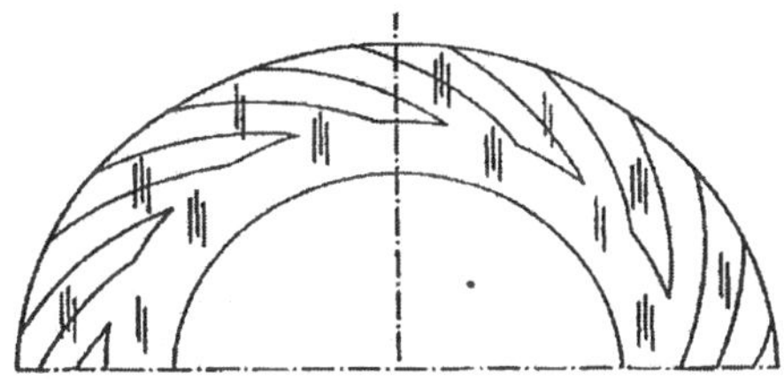

Typical Spiral Groove Pattern On Face Of Seal
Typical Non-contact Gas Seal

Figure 17.4 Typical gas seal (Courtesy of John Crane Co.)

Since gas seals utilize the sealed gas or a clean buffer gas, a liquid seal-system incorporating pumps, a reservoir and other components, is not required. However, one must remember that the sealing fluid still must be supplied at the proper flow rate, temperature and cleanliness. As a result, a highly efficient, reliable source of filtration, cooling, and supply must be furnished. If the system relies upon inert buffer gas for continued operation, the supply source of the buffer gas must be as reliable as the critical equipment itself. Gas seal configurations vary and will be discussed in detail in the next section. They can take the form of single, tandem (series), or multiple seal systems. The principle of operation is to maintain a fixed minimum clearance between the rotating and non-rotating face of the seal. The seal employed is essentially a contact seal with some type of lifting device to maintain a fixed minimum clearance between the rotating faces. It is essential that the gas between these surfaces be clean since any debris will quickly clog areas and reduce the effectiveness of the lifting devices, consequently resulting in rapid damage to the seal faces.

Liquid seals

Traditionally, the type of seal used in compressor service has been a liquid seal. Since the media that we are sealing against is a gas, a liquid must be introduced that will remove the frictional heat of the seal and assure proper sealing. Therefore, all compressor liquid seals take the form of a double seal. That is, they are comprised of two seals with the

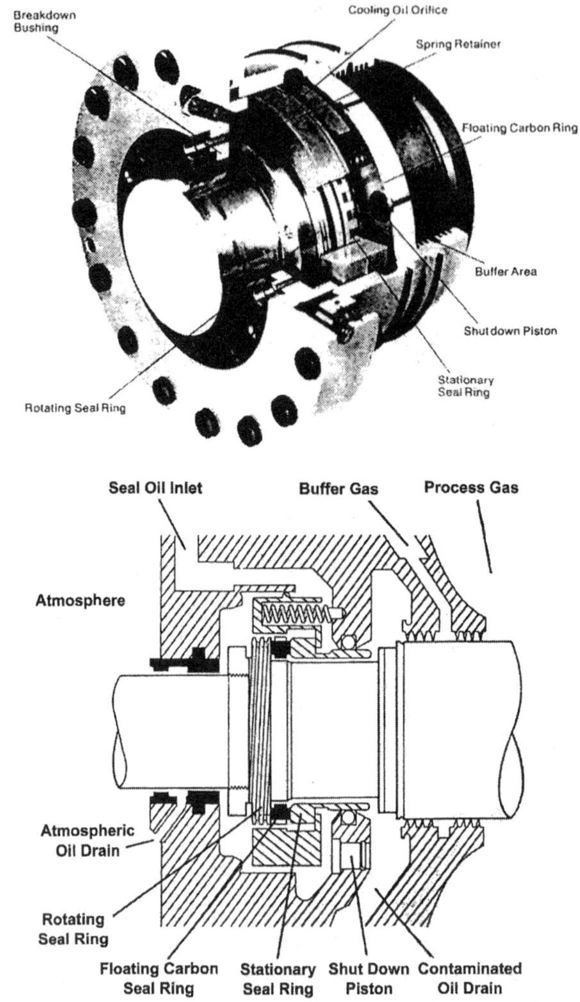

Figure 17.5 ISO carbon seal (Courtesy of Elliott Co.)

sealing liquid introduced between the sealing faces. Refer to Figure 17.5. To assure proper lubrication of both the gas side (inboard) and atmospheric side (outboard) seals, the equivalent 'orifices' of each seal must be properly designed such that the differential pressure present provides sufficient flow through the seal to remove the heat of friction at the maximum operating speed. The type of gas side seal used in Figure 17.5 is a contact seal similar to that used in most pump applications. This seal provides a minimum of leakage (five to ten gallons per day per seal) and provides reliable operation. (Continuous operation for 3+ years.) As will be discussed below, the specific types of seals used in the double seal (liquid) configuration can vary.

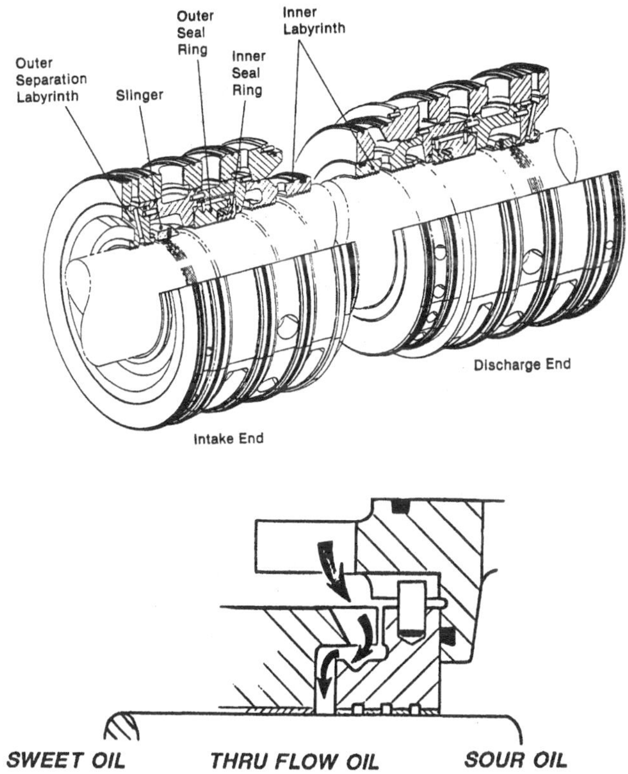

Figure 17.6 Bushing seal – top: oil film seal: bottom: seal oil flow (Courtesy of Dresser-Rand)

Liquid bushing seals

A liquid bushing seal can be used for either a gas side or an atmospheric side seal application. Most seals utilize a liquid bushing seal for an atmospheric bushing application. A typical bushing seal is shown in Figure 17.6.

The principle of a bushing seal is that of an orifice. That is, a minimum clearance between the shaft and the bushing surface to minimize leakage. The bushing seal is designed such that the clearance is sufficient to remove all the frictional heat at the maximum power loss condition of that bushing with the available fluid differential across the bushing. It is important to realize that while acting as a seal, the bushing must not act as a bearing. That is, it must have degrees of freedom (float) to assure that it does not support the load of the rotor. Since its configuration is similar to a bearing, if not allowed freedom of movement, it can act as an equipment bearing and result in a significant change to the dynamic characteristics of equipment with potential to cause damage to the critical equipment. In order to achieve the objectives of a bushing seal, clearances are on the order of 0.0005" diametrical clearance per inch of shaft diameter.

Liquid bushing seals are also used for gas side seals, however, their leakage rate will be significantly larger than that of a contact seal since they are essentially an orifice. When used as a gas side bushing, therefore, the system must be designed to minimize the differential across the bushing. As a result, the differential control system utilized must be accurate enough to maintain the specified oil/gas differential under all operating conditions. The typical design differential across a gas side bushing seal is on the order of five to ten psid. The accurate control of this differential is usually maintained by a level control system.

Referring back to Figure 17.6, one can see that functioning of the bushing seal totally depends on maintaining a liquid interface between the seal and shaft surface. Failure to achieve this results in leakage of gas outward through the seal. It must be fully understood that all bushing seals must continuously maintain this liquid interface to assure proper sealing. All systems incorporating gas side bushing seals must have the seal system in operation whenever pressurized gas is present inside the compressor case. If a liquid interface is not maintained, gas will migrate across the atmospheric bushing seal and proceed through the system returning back to the supply system. There have been cases in such system designs where failures to operate the seal system when the compressor is pressurized have resulted in effectively turning the gas side bushing into a filter for the entire process gas system! This resulted in the supply side of the seal oil system being filled with extensive debris that required lengthy flushing and system cleaning operations prior to putting the unit back into service. Remember, any system incorporating a gas side bushing seal must be designed such that the entrance of process gas into the supply system is prohibited at all time. This can be accomplished by either:

- Continuous buffer gas supply
- A check valve installed as close as possible to seals in the seal oil supply header
- Rapid venting and isolation of the compressor case on seal system failure

In the second and third cases above, supply seal oil piping must be thoroughly checked for debris prior to re-start of the compressor. It is our experience that many bushing seal system problems have resulted from improper attention to the above facts.

Contact seals

Figure 17.7 shows a typical compressor contact seal. As mentioned, these seals are similar in design to pump seals. In order to remove the

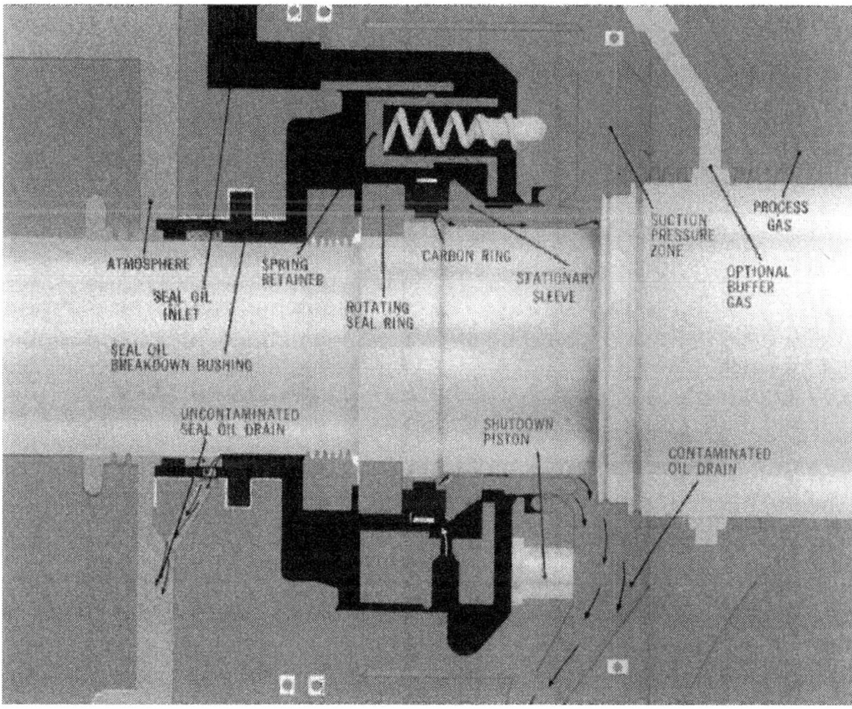

Figure 17.7 Compressor contact seal (Courtesy of Elliott Co.)

heat of friction for this type of seal, a sufficient differential pressure above the reference gas must be maintained. Typical differentials for contact seals vary between 35 and 50 lbs. per square inch differential pressure. Leakage rates with a properly installed seal can be maintained between five to ten gallons per day per seal.

A limitation in the use of contact seals are shaft speed, since the contact seal operates on a surface perpendicular to the axis of rotation, the rubbing speed of the seal surface is critical. As a result, contact seals are speed limited. Typical maximum speeds are approximately 12,000 revolutions per minute. Above those speeds, bushing seals are used, since the sealing surface is maintained at a lower diameter and correspondingly lower rubbing speed. The maximum limit of differential pressure across contact seals is controlled by the materials of construction and is approximately 200 lbs. per square inch differential. As a result, contact seals are usually used for gas side seal applications. They are very seldom utilized for atmospheric seal applications since they are differential pressure limited.

Since the differential pressure required across the seal face is relatively high as compared to a bushing seal, contact seals utilize differential pressure control as opposed to level control for most bushing seals. This fact will be discussed in the next section.

Restricted bushing seals

The last type of seal to be discussed is a restricted bushing type seal. This type of seal is shown in Figure 17.8.

This particular type of restricted bushing seal utilizes a small pumping ring in the opposite direction of bushing liquid flow to compensate for the relatively large leakage experienced with bushing seals by introducing an opposing pumping flow in the opposite direction. Seals of this type can be designed for practically zero flow leakages. However, it must be pointed out that in variable speed applications, the pumping capability of the trapped seal ring must be calculated for both minimum and maximum speeds. Failure to do so can result in the actual pumping of gas from the compressor into the sealing system. It is recommended that such seals be designed to leak a small amount at maximum operating speed. Any retrofits of equipment employing this type of seal should be investigated when higher operating speeds are anticipated. A restricted bushing seal is used exclusively for gas side service.

In summary, the basic types of liquid seals used for compressor applications can be either: open bushing types, contact types, or restricted bushing types. Contact types are used primarily on the gas side. Liquid bushing types are used on either the gas or atmospheric side. Restricted bushing types are used exclusively on the gas side.

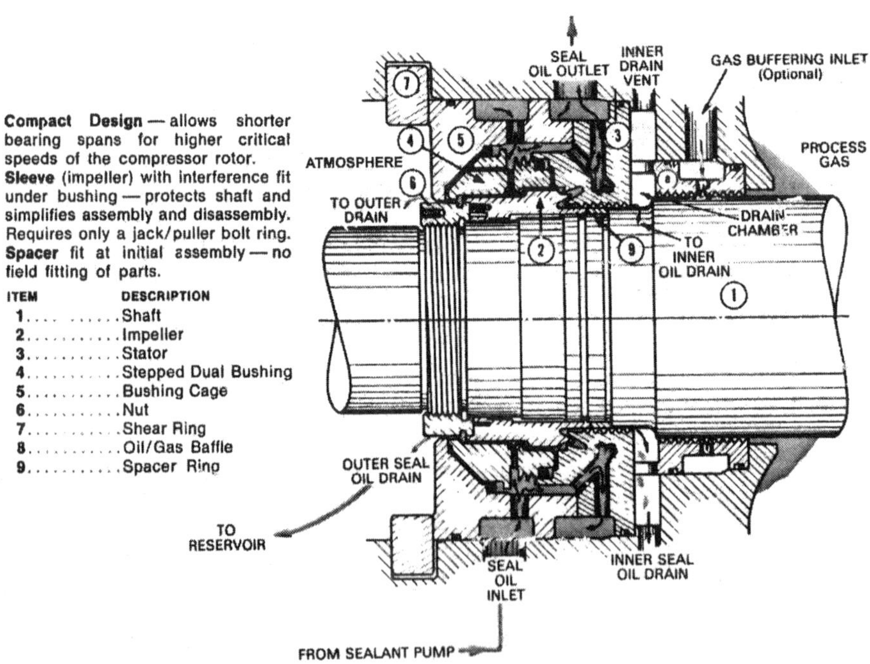

Figure 17.8 Turbo-compressor 'trapped bushing seal' (Courtesy of A.C. Compressor Corp.)

We will now investigate various seal system designs using various seal combinations employing the types of seals that have been discussed in this section.

Seal supply systems

As can be seen from the previous discussion, the type of seal system will depend on the type of seal utilized. We will now examine five different types of seal systems, each utilizing a different type of main compressor shaft seal system. As we proceed through each type, the function of each system will become clear.

Example 1: Contact type gas side seal – bushing type atmospheric side seal with cooling flow

This system incorporates a contact seal on the gas side and a bushing seal on the atmospheric side of each end of the compressor. The inlet pressure of the seal fluid on each end is referenced to the suction pressure of the compressor. It should be noted that some applications

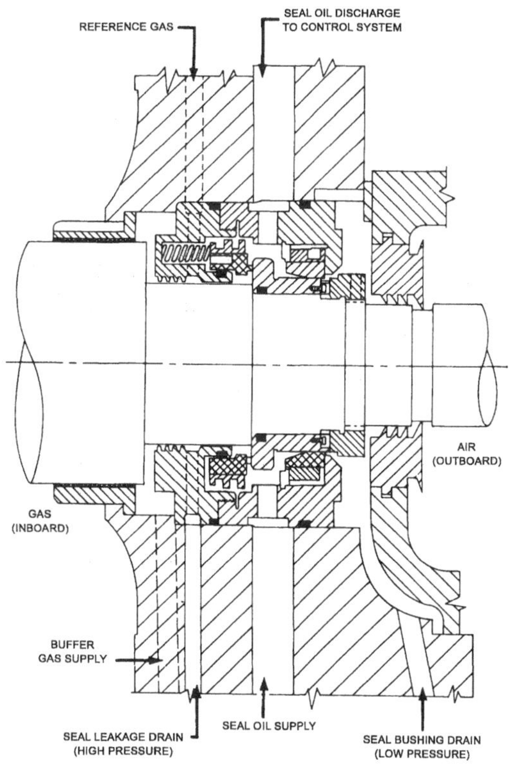

Figure 17.9 Compressor shaft seal (Courtesy of IMO Industries)

employ different reference pressures on each end of the compressor. The reference pressure should be taken off the balance drum end, or high pressure end of the compressor, to assure that the oil to gas differential pressure is always at a minimum acceptable value. Therefore, the low pressure end may experience a slightly higher oil to gas differential then the reference end of the seal. Refer to Figure 17.9.

Proceeding through the seal, the seal oil supply, which is referenced to the gas reference pressure, enters the seal chamber. The differential across the gas side contact seal is maintained by a differential pressure control valve located downstream of the seal. Seal oil flows in three separate directions:

- Through the seal chamber (cooling flow)
- Through the gas side contact seal (10–20 gallons/day)
- Through the atmospheric seal

Let's examine the variants of flows across the equivalent orifice of each portion of this configuration.

The gas side contact seal will experience a constant flow, that for purposes of discussion, can be assumed to be zero gallons per minute (since the maximum flow rate will usually be on the order of ten gallons per day).

The atmospheric side bushing seal flow will vary based upon the referenced gas pressure. At low suction pressure conditions, this flow will be significantly less than it will be under high pressure conditions. The seal system design must consider the maximum reference pressure to be experienced in the compressor case to assure that sufficient seal oil flow is available at maximum pressure conditions.

The seal chamber through flow in this seal design is used to remove any excess frictional heat of the seals and is regulated by the downstream control valve. As an example, let us assume the following values were calculated for this specific seal application.

1. *Gas Side Seal Flow* = 0 GPM
2. *Atmospheric Side Seal Flow*
   Reference Pressure = 0 PSIG
   Seal Flow = 5 GPM
   Reference Pressure = 200 PSIG
   Seal flow = 12 GPM
3. *Flow Through Flow*
   Minimum = 3 GPM (occurring at high ATM bushing flow = 12 GPM)
   Maximum = 12 GPM (occurring at low ATM bushing flow = 3 GPM)
4. *Seal Oil Supply Flow* in both cases = 15 GPM

As shown in the previous example, the required seal oil supply at maximum operating speed required to remove frictional heat is 15 gallons a minute. At start-up, low suction pressure conditions, the control valve must open to allow an additional ten gallons a minute flow through to the seal chamber. At maximum operating pressure, however, the valve only passes a flow of three gallons a minute since 12 gallons a minute exit through the atmospheric bushing. This type of system is less sensitive to low suction pressure operation since flow through oil will remove frictional heat around the atmospheric bushing.

Example 2: Contact type gas side seal – bushing type atmospheric seal with orificed through flow

The only difference between this type of system and the previous system is that the back pressure is maintained constant by a permanently installed through flow orifice. As a result, the differential pressure control valve is installed on the inlet side of the system. The process gas reference is still the same as before, that is, to the highest pressure side of the compressor. Figure 17.10 shows this type of system.

Let us examine the previous example case for this system and observe the differences.

1. *Gas Side Seal Flow = 0 GPM*
2. *Atmospheric Side Seal Flow*
   Reference Pressure = 0 PSIG
   Seal Flow = 5 GPM
   Reference Pressure = 200 PSIG
   Seal Flow = 12 GPM
3. *Flow Through Flow (Orifice)*
   Minimum = 0.5 GPM
   Maximum = 3 GPM

As can be seen, this system is more susceptible to high temperature atmospheric bushing conditions at low suction pressures and must be observed during such operation to assure integrity of the atmospheric bushing. In this system, the control valve will sense supply oil pressure to the seal chamber and control a constant set differential, approximately 35 psid, between the reference gas pressure and the supply pressure. If continued low pressure operation is anticipated with such a system, consideration should be given to a means of changing the minimum flow and maximum flow orifice for various operation points. Externally piped bypass orifices could be arranged such that a bypass line with a large orifice for minimum suction pressure conditions could be installed and opened during this operation. It is important to note, however, that the entire supply system must be designed for this flow condition and control valve must be sized properly to assure

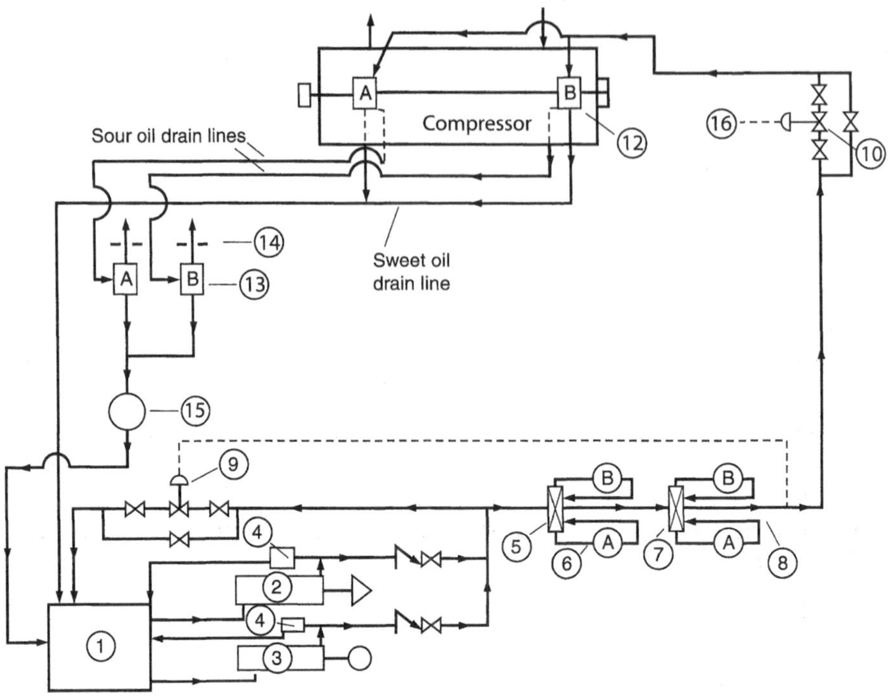

1. Oil reservoir
2. Main screw pump - turbine driven
3. Aux screw pump - motor driven
4. Relief valves
5. Transfer valve
6. Oil coolers (A & B)
7. Transfer valve
8. Oil filters (A & B)
9. Back pressure control valve
10. Differential pressure control valve
11. Overhead seal oil tank (typically 15 ft above compressor centerline)
12. ISO sleeve seals (also called liquid bushing seals)
13. Seal oil drainers (A & B)
14. Drainer orifice vents (A & B)
15. Degassing tank
16. Reference gas line

Note: component condition instrumention and autostarts not shown

Figure 17.10 API 614 Lube-seal oil system for ISO-carbon seals (Courtesy of Elliott Company)

proper flow at this condition. In addition, the low pressure bypass line must be completely closed during normal high pressure operation.

Example 3: Bushing gas side seal – bushing atmospheric side seal with no flow through provision

Figure 17.1 shows this type of seal system. In this type of system, the differential control valve becomes a level control valve sensing differential from the level in an overhead tank and is positioned upstream of the unit. Both bushings can be easily conceived as equivalent orifices. The gas side bushing flow will remain constant regardless of differential. The atmospheric bushing flow will vary according to seal chamber to atmospheric pressure differential.

Therefore, the atmospheric bushing must be designed to pass a minimum flow at minimum pressure conditions that will remove frictional heat and thus prevent overheating and damage to the seal. Since a gas side bushing seal is utilized, a minimum differential across this orifice must be continuously maintained.

Utilizing the concept of head, the control of differential pressure across the inner seal is maintained by a column of liquid.

As an example, if the required gas side seal differential of oil to gas is 5 psid, by the liquid head equation:

$$\text{Head} = \frac{2.311 \times 5 \text{ psid}}{.85} = 13.6 \text{ ft.}$$

Therefore maintaining a liquid level of 13.6 ft. above the seal while referencing process gas pressure will assure a continuous 5 psid gas side bushing differential. In this configuration, the control valve which senses its signal from the level transmitter, will be sized to continuously supply the required flow to maintain a constant level in the overhead tack. As an example, consider the following system changes from start up to normal operation.

Seal System Flow

Item	Start-up condition	Normal operation
Compressor suction pressure	0 PSIG	200 PSIG
Overhead tank reference pressure	0 PSIG	200 PSIG
Gas side seal bushing flow	0 GPM	0 GPM
Atmospheric side seal bushing flow	5 GPM	12 GPM
Seal pump flow	20 GPM	20 GPM
Bypass valve flow	15 GPM	8 GPM

In this example a change from the start-up to operating condition will increase gas reference pressure on the liquid level in the overhead tank and would tend to push the level downward. Any movement of the level in the tank will result in an increasing signal to the level control valve to open, thus increasing the pressure (assuming a positive displacement pump) to the overhead tank and reestablishing the pre-set level.

In the above example, at 200 psi reference pressure, the bypass valve would close considerably. To increase the pressure supply of the seal oil from 5 psi to 205 psi, the difference of bypass flow through the valve (7 gpm) is equal to the increased flow through the atmospheric bushing at this higher differential pressure condition. Utilizing the concept of equivalent orifices, it can be seen that the additional differential pressure across the atmospheric bushing orifice is compensated for by reducing the effective orifice area of the bypass control valve. This is accomplished by sensing the level in the head tank and maintaining it at a constant value by opening the seal oil supply valve.

As in the case of the orificed through flow example above, this configuration is susceptible to high atmospheric bushing temperatures at low suction pressures and must be monitored during this condition. Repeated high temperatures during low suction pressure conditions should give consideration to re-sizing of atmospheric bushing clearances during the next available turnaround. The original equipment manufacturer should be consulted to assure correct bushing sizing and supply system capability.

Example 4: Gas side bushing seal – atmospheric side bushing seal with through flow design

Refer to Figure 17.11. The only difference between this system and the previous system is that a through flow option is added to allow sufficient flow through the system during changing pressure conditions. The bypass valve in the previous system is replaced in this system by a level control valve referenced from a head tank level transmitter, and is installed downstream of the seal chamber. This system functions in exactly the same way as the system in Example 1. The only difference being that a level control valve in this example replaces the differential control valve in the previous example. Both valves have the same function, that is, to control the differential in the seal chamber between the seal oil supply and the referenced gas pressure. A level control valve is utilized in this example, however, since a bushing seal requires a significantly lower differential between the seal oil supply pressure and the gas reference pressure.

Consider the following example. Assume that a differential control valve would be used as opposed to a level control valve for the system in Figure 17.11. For the start-up case, the differential control valve would have to maintain a differential of 5 psi over the reference gas. When the reference gas pressure were 0 psi, the oil upstream pressure to the valve would be approximately 5 psi. For the operating case, maintaining the same 5 psi differential, the upstream pressure across the valve would be approximately 205 lbs instead of 5 psi. Consequently, the valve position would change significantly, but still would have to control the differential accurately to maintain 5 psi. Reduction of this pressure in any amount below 5 psi could result in instantaneous bushing failure. However, if a level control valve were installed, the accuracy of the valve would be measured in inches of oil instead of psi. Any level control system could control the level within two inches, which would be only a .06 psid variation in pressure differential!

This example shows that the accurate means of controlling differential pressure for systems requiring control of small differential values, is to use level instead of differential control. This system would be designed such that the combination of the atmospheric flow and the through flow through the seal would be equal to the flow from the pump.

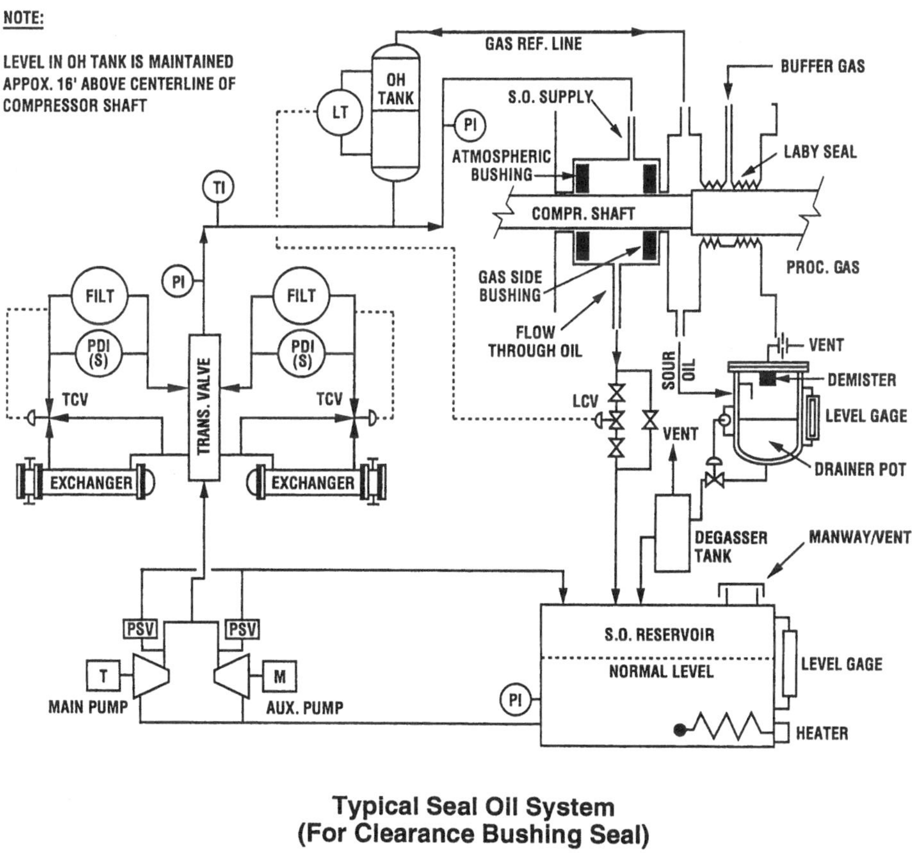

Figure 17.11 Typical seal oil system (Courtesy of M.E. Crane Consultant)

Example 5: Trapped bushing gas side seal: atmospheric side bushing seal with flowthrough design

This system would follow exactly the same design as the system described in Example 4. The only difference would be in the amount of flow registered in the seal oil drainer. A trapped bushing system is designed to minimize seal oil drainer pot leakage. Typical values can be less than five gallons/day.

Seal supply system summary

All of the above examples have dealt with a system incorporating one seal assembly. It must be understood that most systems utilize two or more seal system assemblies. Typical multi-stage compressors contain two seal assemblies per compressor body and many applications contain upwards of three compressor bodies in series, or six seal assemblies. Usually each compressor body is maintained at the suction pressure to that body, therefore three discreet seal pressure levels would be required and three differential pressure systems would be utilized. The concepts discussed in this section follow through regardless of the amount of seals in the system. Sometimes, the entire train, that is, all the seals referenced to the same pressure. In this case, one differential seal system could be used across all seals.

In conclusion, remembering the concept of an orifice will help in understanding the operation of these systems. Remember, the gas side bushing is essentially zero flow, the atmospheric side bushing flow varies with changing differential across the seal and any seal chamber through flow will change either as a result of differential across a fixed orifice or the repositioning of the control valve.

Seal liquid leakage system

This seal system sub-system's function is to collect all of the leakage from the gas side seal and return it to the seal reservoir at specified seal fluid conditions. Depending upon the gas condition in the case, this objective may or may not be possible. If the gas being compressed has a tendency to change the specification of the seal oil to off specification conditions, one of two possibilities remain:

- *Introduce a clean buffer gas* between the seal to assure proper oil conditions
- *Dispose of the seal oil leakage*

In most cases, the first alternative is utilized. Once the seal oil is in the drainer, a combination of oil and gas are present. A vent may be installed in the drainer pot to remove some of the gas, or a degassing tank can be incorporated.

This concludes the overview section of seal oil systems. As can be seen from the above discussion, it is evident that the design of a seal oil system follows closely to that of a lube system. The major difference is that the downstream reference pressure of the components (seal) varies, whereas in the case of a lube system it does not. In addition, the collection of the expensive seal oil is required in most cases and a downstream collector, or drainer system, must be utilized. Other than these two exceptions, the design of the seal system is very similar to that of a lube system and the same concepts apply in both cases.

18 Reciprocating compressors major component functions

- Introduction
- Frame and running gear
- Cylinder distance piece
- Piston rod packing
- Cylinder and liner
- Reciprocating compressor cylinder valves
- Piston assembly
- Pulsation dampeners
- Cylinder and packing lubricators
- Cooling system

Introduction

In this chapter we will define the functions of each major component of a reciprocating compressor. That is, what the purpose of each component is or "What It Does". By understanding what each component is supposed to do, you will be in a better position to know if it is performing its duty correctly. We will present each major component starting with the crankcase, state its function, operating limits and what to look for. After presenting each component's general information, we will present specific information concerning site compressors.

Frame and running gear

Figure 18.1 presents a picture of a seven throw crank shaft arrangement along with a sectional view of two throws. The crankcase supports the crank shaft bearings, provides a sump for the bearing and crosshead lube oil and provides support for the crosshead assembly.

Typical crank case condition monitoring and safety devices are:

- Relief device – To prevent crank case breakage in the event of explosion (caused by entrance of process gas into the crank case).
- Breather vent – To allow removal of entrained air from the lube oil.
- Crank case oil level gauge – Allows continuous monitoring of crank case lube oil level.
- Crank case oil temperature gauge – Allows continuous monitoring of crank case lube oil temperature.
- Crank case vibration detector (optional) – Provides information concerning crank case vibration useful in detecting dynamic changes in running gear.
- Crank case low oil level switch (optional) – Provides alarm signal on low crank case oil level.
- Main lube oil pump – shaft driven (optional) – Directly connected to crank shaft and usually discharges oil directly to crank shaft

FUNCTIONS:
- TRANSMITS POWER
- CONTAINS LUBE OIL
- CONVERTS ROTARY TO RECIPROCATING MOTION

Figure 18.1 Frame and running gear (Crank case and crosshead) (Courtesy of Dresser Rand)

bearings, connecting rod bearing, crosshead shoes and crosshead pin bushing via precision bore in crank shaft and connecting rod bearing.

- Main lube oil pump discharge pressure gauge (when supplied) – Allows continuous monitoring of main lube oil pump discharge pressure.

An important reliability consideration is to assure that the crank case is securely mounted and level. This requires proper grouting and maintaining a crack free (continuous) crank case base support. Since the dynamic forces on the crank case and crosshead mounting feet can be very large, it is usually common to use an epoxy grout. Epoxy grouts provide high bond strengths and are oil resistant. All reciprocating baseplates should be continuously checked for any evidence of grout foundation cracks (discontinuities) and repaired at the first opportunity.

Figures 18.2 and 18.3 show plan, elevation and side views of a two throw balanced opposed crank case assembly.

The crosshead assembly shown has the function of continuously assuring vibration free reciprocating motion of the piston and piston rod. The crosshead pads (or shoes) and supports are usually made from

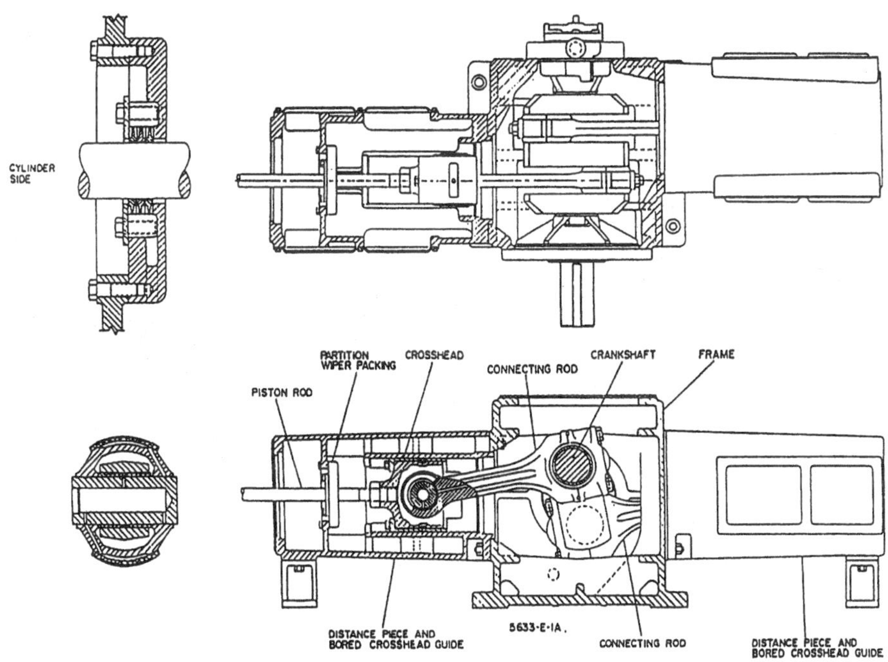

Figure 18.2 HDS off-gas running gear (Courtesy of Dresser Rand)

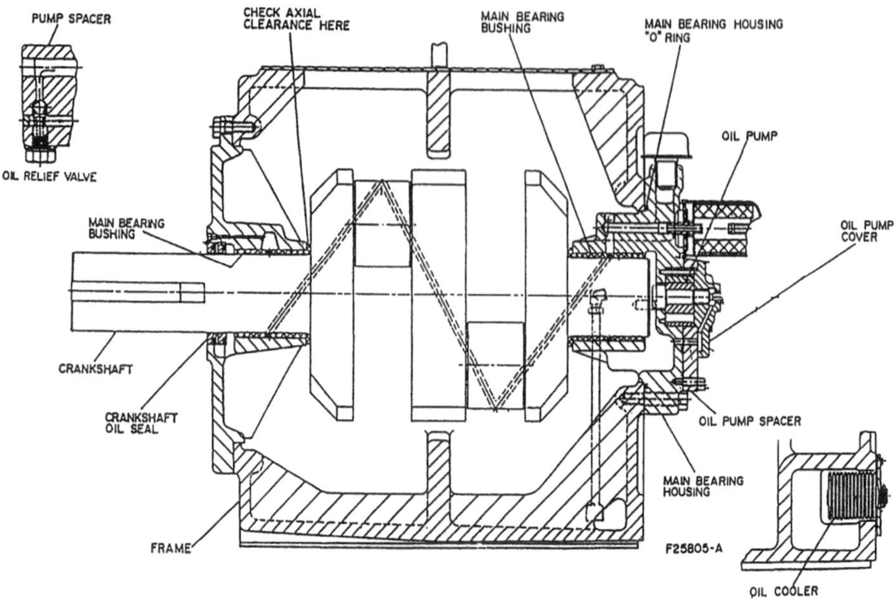

Figure 18.3 HDS off-gas running gear (Courtesy of Dresser Rand)

Babbitt or aluminum (smaller size units). Crosshead assembly lubrication is supplied via a pressure drilled hole (rifle drilled) in the connecting rod which in turn lubricates the crosshead pin bushing and crosshead shoes (see Figure 18.3).

Cylinder distance piece

Figure 18.4 presents the functions of the cylinder distance piece.

The proper operation of the distance piece baffles and seals is essential to maintaining reciprocating compressor safety and reliability in process gas applications. In most refinery process gas applications, a double compartment distance piece is used to assure contamination of the crank case or cylinders does not occur. Usually, the cylinder end compartment contains a partial N_2 atmosphere since the packing rings is usually N_2 purged.

Reliability considerations concerning this assembly are assuring proper packing, partition packing and wiper ring clearances.

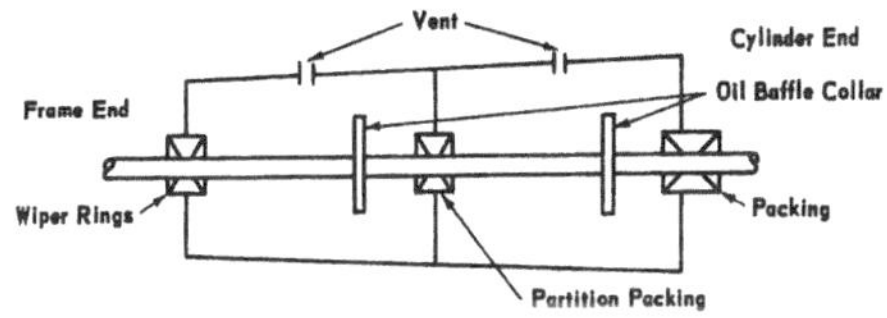

DOUBLE COMPARTMENT DISTANCE PIECE

DOUBLE COMPARTMENT SCHEMATIC

FUNCTIONS:
- PREVENTS CONTAMINATION OF PROCESS GAS
- PREVENTS CONTAMINATION OF CRANKCASE OIL

Figure 18.4 Cylinder distance piece

Piston rod packing

Figure 18.5 depicts a sectional and exterior view of a cartridge packing assembly.

The number of packing rings and type of arrangement is varied according to the cylinder maximum operating pressures. It is important to note that the packing does not provide an absolute seal, but only minimizes the leakage from the cylinder. Shown in the upper portion of the section drawing in Figure 18.5 is the vent port which carries the leakage gas either to a safe vent location (atmosphere, flare or fuel gas system) or back to the cylinder suction.

A means should be available to provide easy detection of excessive packing clearances. Alternatives are:

- Packing line flow switch
- Packing line orifice and pressure switch (only if compressor pressures are controlled to be constant)

FUNCTION: MINIMIZES GAS LEAKAGE FROM CYLINDER

Figure 18.5 Cylinder packing

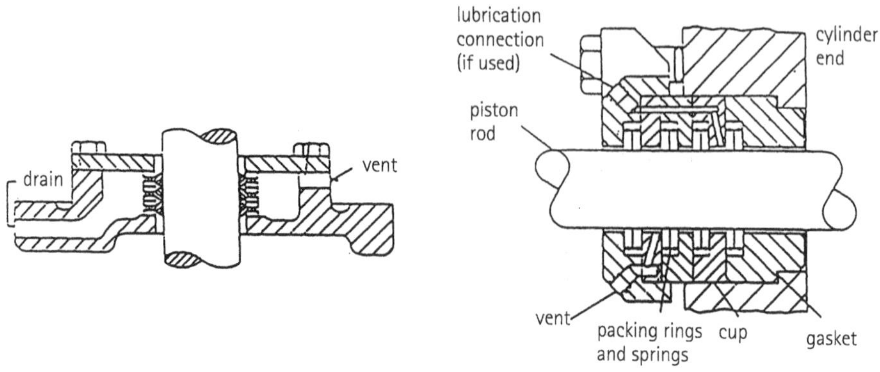

Figure 18.6 Left: Oil scraper ring arrangement for 9" & 11" ESH/V, HSE units; Right: Piston rod packing (Courtesy of Dresser Rand)

- Visual detection of gas flow (vent). Note: Flammable or toxic process gas must be purged with N_2 to attain a non-flammable mixture if the gas is to be vented to atmosphere.

Figure 18.6 shows additional packing assembly details.

The figure on the left side of the drawing is typical of a packing arrangement used between sections of a distance piece. Mounted horizontally, the assembly is equipped with a gravity drain and top vent.

The figure on the right side of Figure 18.6 shows a four (4) ring piston rod packing assembly. The upper part of the drawing shows the lubrication connections that are used when lubricated packing is required. Lube packing is normally used if the lubricant is compatible with the process stream.

If dry packing is used, piston rod speeds are usually slower and PTFE materials are usually employed. In the lower half of the drawing, the vent connections are shown and perform as previously discussed. The cup supports and positions an individual packing ring. Not shown is a purge connection which is inserted between the last and next to last packing ring (rings closest to the distance piece).

Cylinder and liner

Shown in Figure 18.7 are the two most common cylinder arrangements; double acting and single acting.

Most process reciprocating compressors are supplied with a replaceable cylinder liner. All cylinders are either jacked for cooling H_2O or finned for air cooling. Some older design cylinders use gaskets to isolate

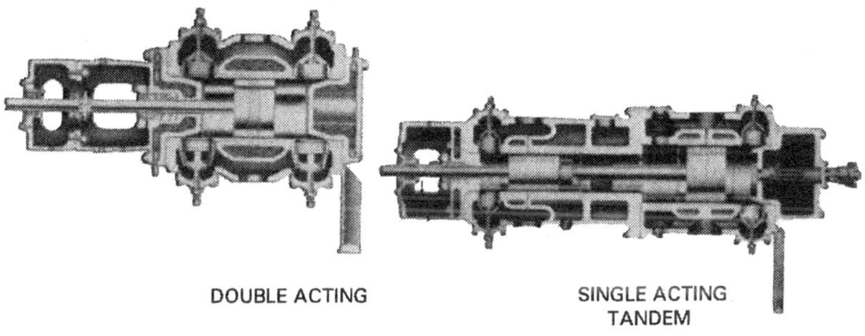

Figure 18.7 Cylinder and liner (Courtesy of Dresser Rand)

cooling water jackets from the cylinder. This design exposes the user to breakage from excessive cylinder H_2O entrainment if the gasket fails. Most reciprocating specifications today do not allow gaskets to be used in the cylinder. A double acting cylinder is designed to compress gas on both ends of the cylinder. (crank end and cylinder head end) while a single acting cylinder is designed for compression only on one end of the cylinder.

Reciprocating compressor cylinder valves

There are many different types of reciprocating compressor valves. Regardless of their design, all valves perform the same function … they allow gas to enter the cylinder, prevent recirculation flow back to the suction piping and allow gas to pass into the discharge system when the process discharge pressure at the compressor flange is exceeded. Valves are the highest maintenance item in reciprocating compressors. Their life is dependent on gas composition and condition, gas temperature and piston speed. Typical valve lives are:

- Process gas service in excess of one (1) year
- H_2 gas service – 8 – 12 months

In hydrogen service, particular attention should be paid to cylinder discharge temperature in order to obtain maximum valve life, cylinder discharge temperature for service with > 60% H_2 should be limited to 250°F. Recently, light weight, non metallic valves (PEEK) have been used successfully to increase the valve life in H_2 service above one (1) year.

Figure 18.8 shows a typical channel valve assembly.

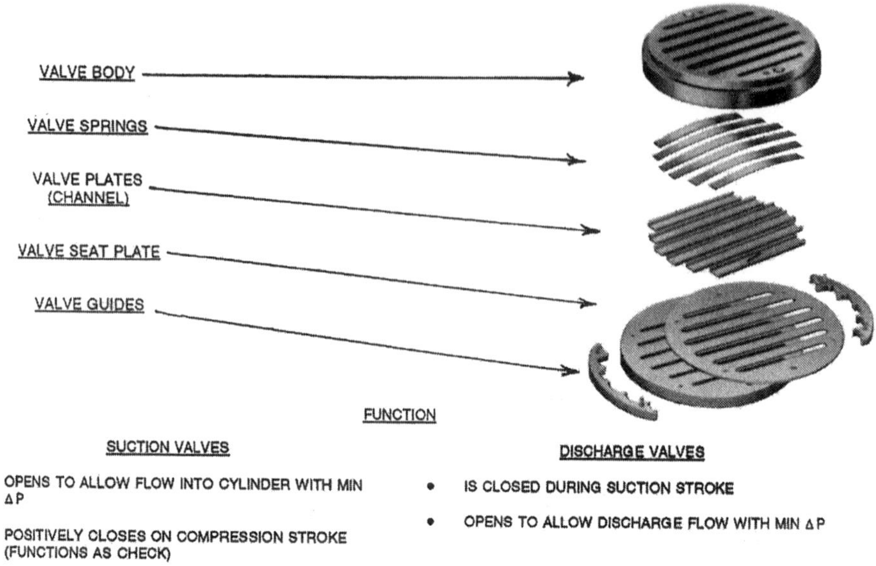

Figure 18.8 Reciprocating compressor valves – (Channel type, Suction and Discharge) (Courtesy of Dresser Rand)

The life of the channel valves shown is controlled by the spring force of the valve springs. The channel arrangement reduces the forces on the valve seal and usually results in increased valve life.

Figure 18.9 depicts a ring or plate valve assembly. This type of valve is most widely used.

Regardless of the type of valve, condition monitoring of valves is important to the profitability of any operation. The following parameter should be monitored.

Type valve

Suction
- Valve body temperature
- Compressor volume flow rate

Discharge
- Interstage process gas temperature
- Compressor volume flow rate

Changes in these parameters in excess of 10% from original (baseline) values should be cause for component inspection and replacement.

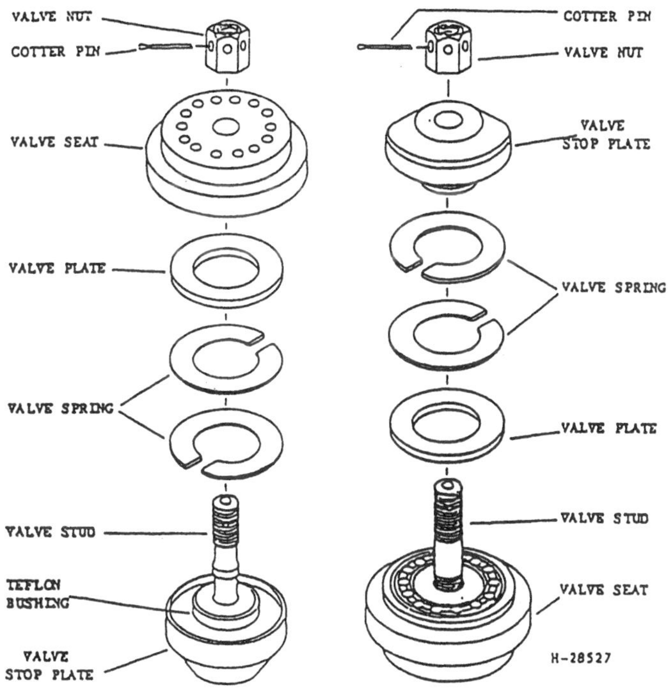

Figure 18.9 HDS suction and discharge valves (Courtesy of Dresser Rand)

Piston assembly

Figure 18.10 presents a typical piston assembly consisting of the piston rod nut, piston rod, piston and piston nut.

Piston rod materials are hardened steel and can include metal spray in packing areas to extend rod life. Piston materials can be steel, cast nodular iron or aluminum. The most common being cast iron due to its durability. Aluminum pistons are used in large cylinder applications (usually 1st stage) to minimize piston rod assembly weight.

Figure 18.11 shows piston rider bands (2) items 111 and piston rings (3) items 113.

Rider band and ring material is dependent on cylinder lubrication. If the cylinder is lubricated, carbon materials or compounds are used. If non-lubricated service is required, PTFE materials or other Teflon derivatives are used. Note also the piston hollowed area for piston weight control. Rider band and ring life is a function of piston speed, cylinder gas temperature and cleanliness of the process gas. In many process applications, a strainer is required upstream of the compressor to prevent excessive ring wear.

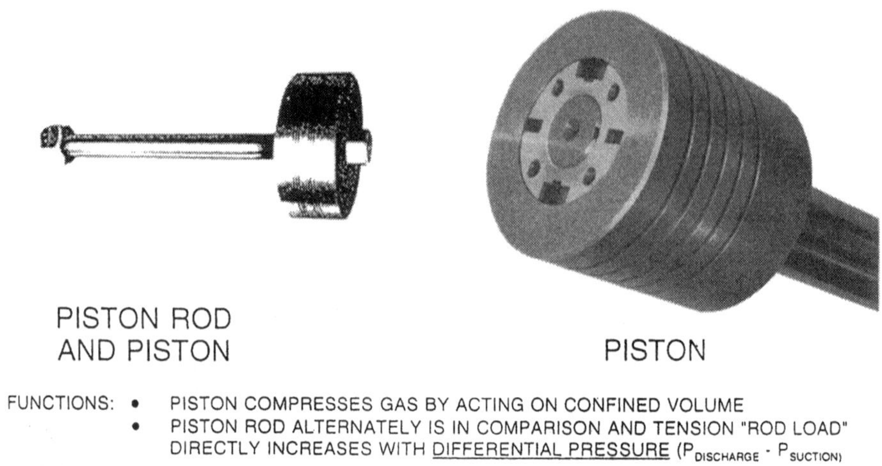

Figure 18.10 Rod and piston

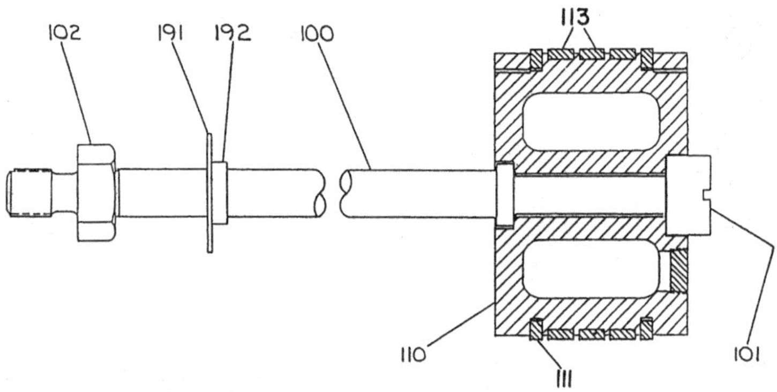

Figure 18.11 HDS off-gas-piston rod and piston (Courtesy of Dresser Rand)

Condition monitoring of rider band and piston ring wear can be accomplished by measuring and trending the vertical distance between each piston rod and a fixed point (known as rod drop). This can be accomplished either by mechanical or electrical (Bentley Nevada proximity probe) means. Of importance in piston assembly design is rod loading and rod reversal.

Rod loading is the stress (tension or compression) in the piston road and crosshead assembly caused by the ΔP across the piston. Rod load limits the maximum compression ratio that a cylinder can tolerate. This is the reason that many first stage cylinders are supplied with a suction pressure switch. Rod reversal is necessary so that the piston rod reaction forces on the crosshead pin will change allowing oil to enter the pin bushing. If the position of the pin in the bushing did not change (reverse) with each stroke, the bushing could not be sufficiently lubricated and would prematurely fail.

Pulsation dampeners

Since the action of the piston is non continuous, pressure pulsations will be generated. Depending upon the piping arrangement, these pulsations can be magnified to destructive levels. The use of pulsation dampeners, as shown in Figure 18.12 can reduce pulsations to 2% or lower.

There are methods available to evaluate and simulate the effect of pulsation dampeners prior to field operation. However, the correlation between predicted and actual results can be large and field modifications (installation of orifices or pipe modifications) may be necessary.

FUNCTION: TO REDUCE GAS PULSATION (EXCITATION FORCE THAT CAUSES PIPE MOVEMENT)

NOTE: EXCITATION FORCES VARY WITH PISTON SPEED AND CYLINDER LOADING

Figure 18.12 Pulsation dampeners (Courtesy of Dresser Rand)

Cylinder and packing lubricators

Whenever mineral oil is compatible with the process, lubricators will be used. Lubricators can be either positive displacement or dynamic type. Attendees are asked to review lubrication details in the appropriate instruction book. Lubricators will increase piston ring and packing life by reducing friction.

Figure 18.13 presents a typical lube oil system and its function.

ALL instruments in the lube oil system should be continuously monitored (baseline and current conditions). Remember, component (bearing) failure will occur if any major component in the system fails to function.

Figure 18.14 shows a lube oil system containing a shaft driven main lube oil pump with an internal relief valve.

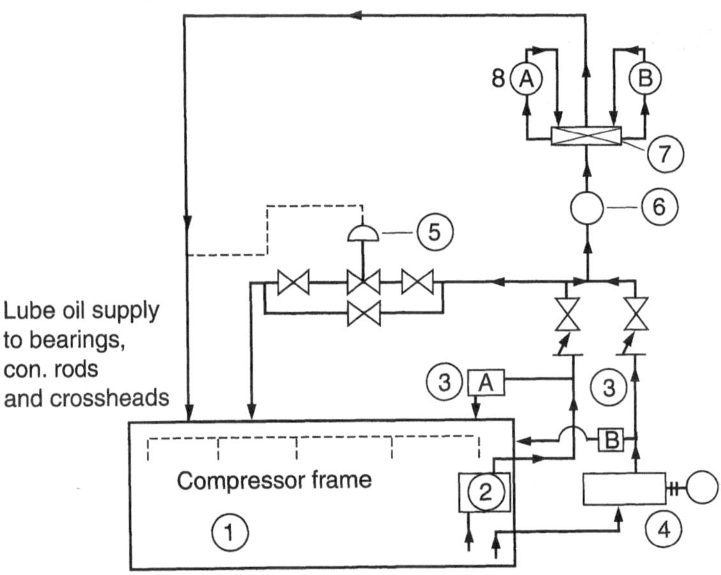

1. Crankcase (oil reservoir)
2. Main gear pump - shaft driven
3. Relief valves (A & B)
4. Aux pump - motor driven
5. Back pressure control valve (controls lube oil pressure)
6. Oil cooler
7. Transfer valve
8. Oil filters (A & B)

Note: Component condition instrumentation & auto starts not shown

Figure 18.13 Lube oil system

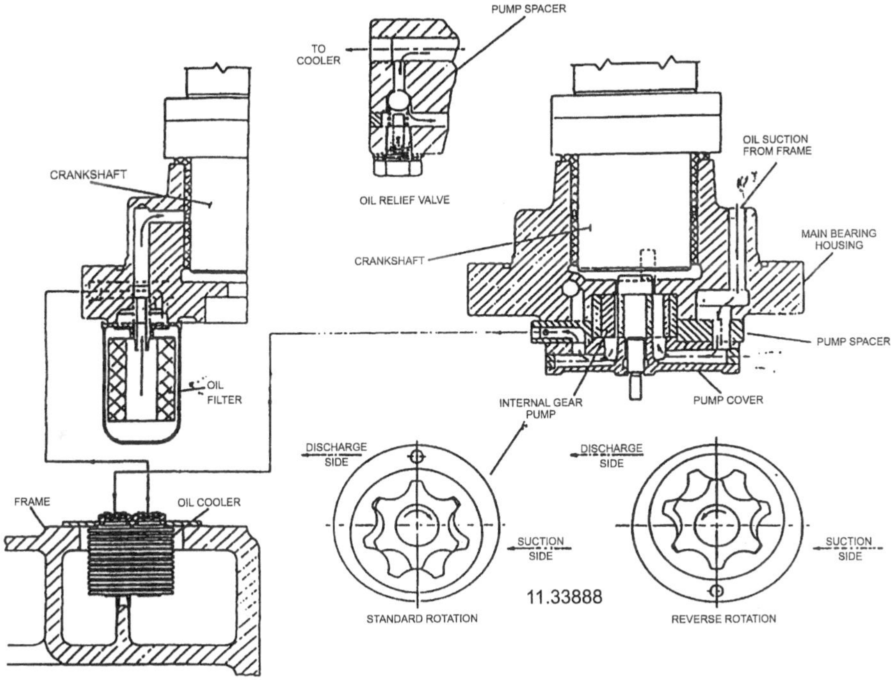

Figure 18.14 HDS off-gas-lube oil system (Courtesy of Dresser Rand)

This arrangement is a common one. Failure of the relief valve to seat can cause a low lube oil pressure trip.

Cooling system

The final topic to be covered is the cooling system. The cylinders, packing and process gas must be cooled to extend run time and minimize maintenance. Figure 18.15 presents a typical water cooled circuit.

In addition to cooling, the temperature of the cooling water must be regulated so that moisture (condensate) will not form in the cylinder in wet gas applications. It is recommended that the tempered water system temperature in the cylinder be maintained a minimum of 10–15°F above the cylinder inlet gas temperature.

Careful monitoring of the cooling circuit is essential in determining cooler, jacket, cylinder maintenance (cleaning) requirements.

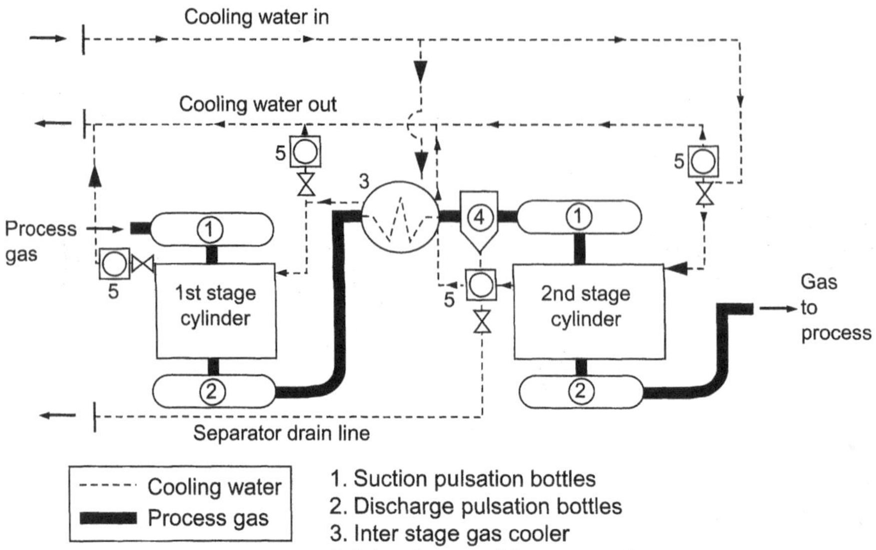

1. Suction pulsation bottles
2. Discharge pulsation bottles
3. Inter stage gas cooler
4. Inter stage moisture separator
5. Water flow sight glass

Note: Jacket/cooling system is designed to provide water to cylinder jackets 10-15 F above inlet gas temperature

Figure 18.15 Cylinder, packing and intercooler cooling water system

Flexible coupling design, installation and operation

- Introduction
- The coupling function
- Types
- The coupling system
- Coupling installation and removal
- Enclosed coupling guards
- Field retrofits from lubricated to dry couplings

Introduction

In this chapter, the subject of couplings or constant speed ratio transmission devices will be discussed. Every equipment train contains one or more couplings. The function of couplings will be defined. The types of couplings most frequently used will be presented and their limitations discussed.

In the writers' experience, the root cause of most coupling failures has not been the coupling itself, but the 'coupling system' (shaft mounting, coupling guard design, etc). The coupling system will be discussed with emphasis on coupling installation, removal and coupling guard design.

The coupling function

The function of a flexible coupling is to transmit torque from the driver to the driven machine while making allowances for minor shaft misalignment and shaft end position changes between the two

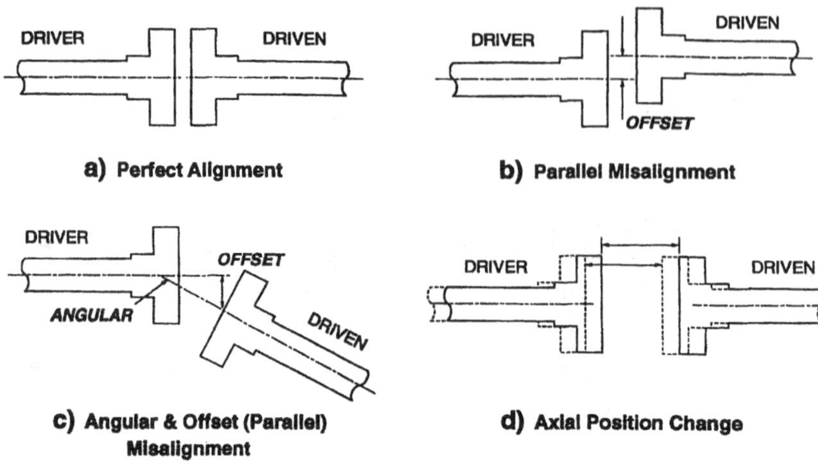

Figure 19.1 Shaft misalignment and axial position

machines. The design of the coupling should provide for transmission of the required torque at the required speed with a minimum of extraneous forces and perturbations exerted on either the driver or driven shaft. Shaft misalignment exists when the centerlines of two shafts joined by a coupling do not coincide. Figure 19.1 shows the various types of misalignment and shaft end position changes that can occur.

Each coupling type has a maximum tolerance of misalignment and axial position change that is noted on the coupling drawing. Regardless of coupling type, misalignment tolerance is stated in degrees and is usually 1/4°. Axial position change tolerance varies with coupling type. Gear type couplings have a large axial position change tolerance compared to flexible element types.

Types

The following is a list of various types of flexible couplings:

I Gear couplings
- A. Continuous lubrication
- B. Grease packed

II Flexible membrane or flexible disc couplings
- A. Single membrane type
- B. Multiple membrane or multiple disc type

III Couplings with elastomer insert flexible drive members.

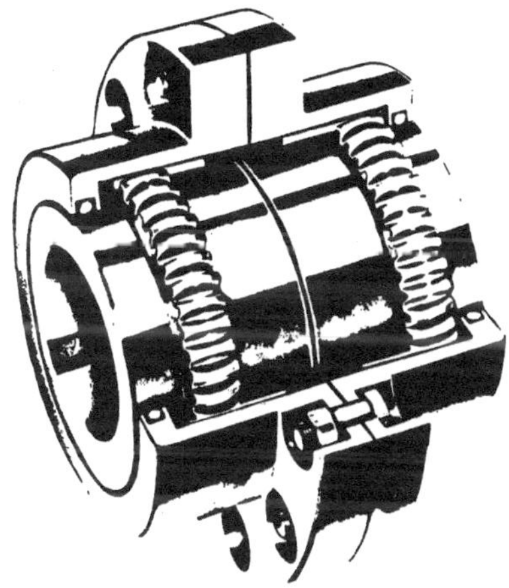

Figure 19.2 Gear tooth coupling (grease packed) (Courtesy of Zurn Industries)

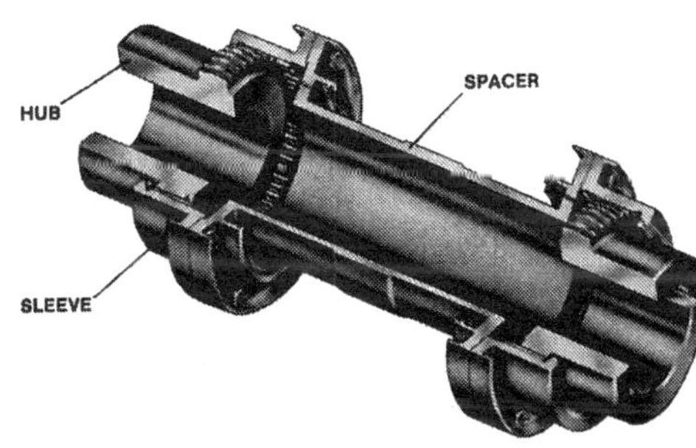

Figure 19.3 Continuously lubricated gear type coupling with spacer (Courtesy of Zurn Industries)

Gear couplings

Gear type couplings are shown in Figures 19.2 and 19.3.

Gear couplings usually include two separate gear mesh units. Each gear mesh unit consists of an external gear which fits closely into an internal gear. The internal gear can either be part of the coupling hub assemblies or mounted on each end of the coupling spacer assembly. If the internal gears are hub mounted, then the external gears are spacer mounted and vice-versa.

Grease pack couplings (see Figure 19.2) are normally designed with hub mounted external gears and the internal gears are part of a sleeve type spacer which serves as a retainer for the grease lubrication. The flange joint of the sleeve is either precision ground to avoid lubrication leaks or has a gasket between the two flange faces. The sleeve ends are fitted with 'O' ring seals to keep dust out and lubrication in.

In recent years, flexible element couplings have been used almost exclusively. However, many gear type couplings are still in use. They are the most compact coupling for a given amount of torque transmission of all the coupling designs. For this reason, they also have the least overhung weight. In addition, the gear coupling can adapt more readily to requirements for axial growth of the driver and driven shafts. Axial position change tolerances are on the order of 1/2" or greater.

There is a common disadvantage in all gear type flexible couplings. Any gear mesh has a break-away friction factor in the axial direction. This is caused by the high contact force between the two sets of gear teeth. The result is that the forces imposed on the driver and driven shafts are not totally predictable and are sometimes higher than desired due to the quality of the tooth machine surfaces and the inevitable build up of sludge or foreign material in the tooth mesh during extended service. These forces are detrimental to the ability of the coupling to make the required corrections for misalignment but, more importantly, can have a disastrous effect on the ability of the coupling to correct for thermal or thrust force changes between the driver and driven machines.

Both coupling manufacturers and user have long been aware of this problem and have used many methods to minimize the effect. Some of these methods are:

- Reduction of the forces between the gear teeth by increasing the pitch diameter of the gear mesh. This is often self defeating in that it results in increased size of the coupling and the coupling weight.
- Reduction of the break-away friction factor by the use of higher quality gear tooth finish and better tooth geometry and fit.
- Reduction of sludge and foreign material build up in the gear mesh by finer filtration of the coupling lubricant.
- Reduction of sludge and foreign material build up in the gear mesh by incorporating self flushing passages and ports in the coupling to allow any contaminants to pass through in the lubricant without being trapped in the gear mesh area.

These steps have been only partially successful and the problem still exists in many applications.

Coupling manufacturers are asked to quote the design break-away friction factor of their coupling as built and shipped from the factory. Machinery train designers then use this figure to calculate the maximum axial force that the coupling would be expected to exert on the connected shafts. From this information, the designers can decide if the thrust bearings adjacent to the coupling are adequate to handle the axial loads within the machine plus the possible load from the coupling resistance to any external forces.

There has been much discussion and some disagreement regarding the friction factor to be used when calculating possible thrust forces which can be transmitted by the coupling. When the coupling is in reasonably good condition, factors from .15 to .30 have been considered reasonable. Since the factor reflects the total force relationship, the coupling design can have a significant effect on the factor used. The factor is a function of the number of teeth in contact and the contact areas of each tooth plus the quality of the tooth contact surface. If we assume that the factor to be used is .30, then the axial force which must be exerted in order to allow the coupling to correct for axial spacing changes can be calculated as:

$$F_a = \frac{0.30 \times T}{Dp/2}$$

Where: F_a = Required axial force in pounds
T = Design torque in in/pounds
Dp = Pitch diameter of gear mesh in inches

We can assume then, that if we use a coupling with a six inch pitch diameter gear mesh transmitting 25,000 in/pounds of torque and a break-away friction factor of .30, the axial force required to move the gear mesh to a new axial position would be 2,500 pounds. Adjacent thrust bearings must be capable of handling this force in addition to the machine's normal calculated thrust forces. Machinery train designers and users must be aware of this and make provisions for it in the built-in safety factors of thrust bearings and machinery mounting design.

The machinery user must know that the same phenomena has an effect on machinery vibration when machinery is operated with excessive misalignment. The gear mesh position must change with each revolution of the shaft to correct for the misalignment. This results in counter axial forces on a cyclic basis since the mesh is moving in opposite directions at each side of the coupling. Vibration detection and monitoring instrumentation will show that the resulting vibration will occur at twice the running frequency of the shafts. Although the primary force generated is axial, the resultant can show up as a radial vibration due to the lever arm forces required on the coupling spacer to make the gear meshes act as ball and socket connections. **Axial or radial vibration in rotating machinery which occurs at twice the frequency of the shaft rotational speed will normally be an indication of misalignment between the two machines.**

Figure 19.3 shows a continuously lubricated, spacer gear type coupling. Spacers are usually required for component removal (seals, etc). They also provide greater tolerance to shaft misalignment. A common spacer size used for unspared (critical) equipment is 18 inches.

Flexible membrane or flexible disc couplings

Couplings in these categories do not have moving parts and derive their flexibility from controlled flexure of specially designed diaphragms or discs. They do not require lubrication and are commonly known as 'dry couplings'. The diaphragms or discs transmit torque from one shaft to the other just as do the gear meshes in a gear coupling.

The following features are common to all flexible disc or flexible membrane type couplings:

1. None require lubrication.

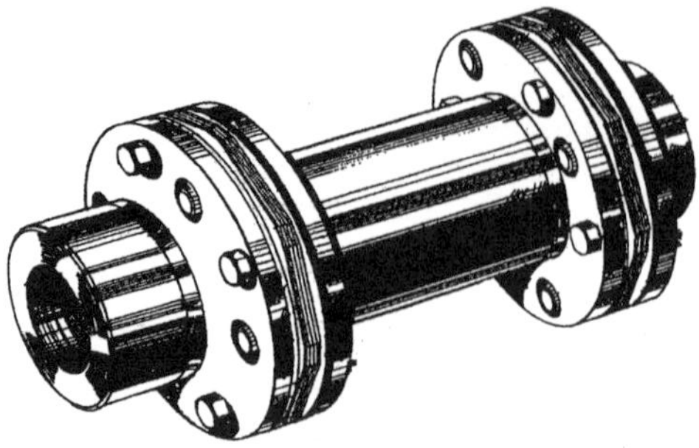

Figure 19.4 Flexible disc spacer coupling (Courtesy of Rexnord)

2. All provide a predictable thrust force curve for a given axial displacement range.
3. Properly applied, operated and maintained, none are subject to wear and have an infinite life span.
4. All provide smooth, predictable response to cyclic correction for minor misalignment.

It should be noted that **none** of the above comments can be applied across the board to gear type flexible couplings. For this reason, more and more special purpose machinery trains are being supplied with flexible metallic element couplings in their design. Many users do not allow the use of gear type coupling for critical (unspared) applications.

The following is a discussion of the various types of 'dry' couplings with comments pertaining to their application ranges and limitations.

Figure 19.4 shows a typical flexible disc coupling.

This is the most common type and is generally used for general purpose applications (pumps, fans, etc). The major consideration with this type of coupling is assuring the shaft end separation (B.S.E.) is within the allowable limits of the couplings. This value is typically only 0.060" for shaft sizes in the 1" – 2" range. At shaft sizes over 4" the maximum end float can be 0.150". Exceeding the allowable end float will significantly increase the axial load on the thrust bearings of the equipment and can fail the coupling discs. A single diaphragm, spacer type coupling is shown in Figures 19.5 and 19.6. Figure 19.5 is a cutaway view and Figure 19.6 presents a two dimensional assembly drawing.

This type of coupling is commonly used for critical (unspared) applications where axial end float values are less than .125". This limit is based on an approximate axial float of ± 0.062". If end float is greater than 0.125", a convoluted (wavy) diaphragm or multiple type diaphragm must be used. During disassembly, care must be taken when

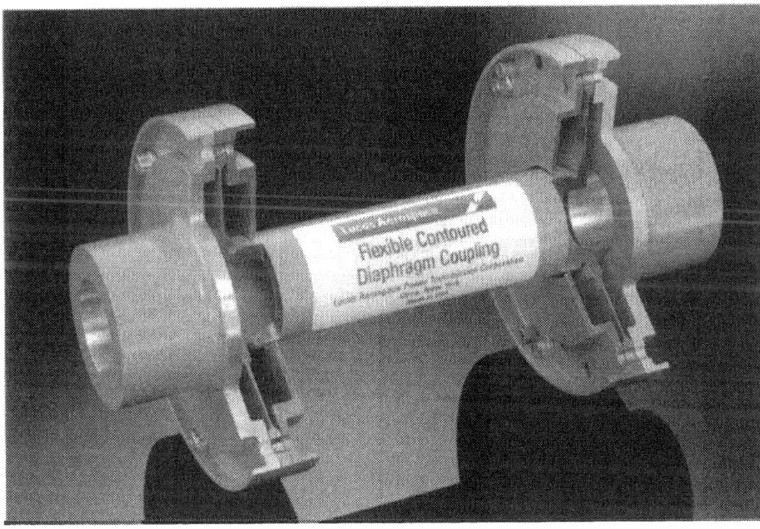

Figure 19.5 Single diaphragm spacer coupling (Courtesy of Lucas Aerospace)

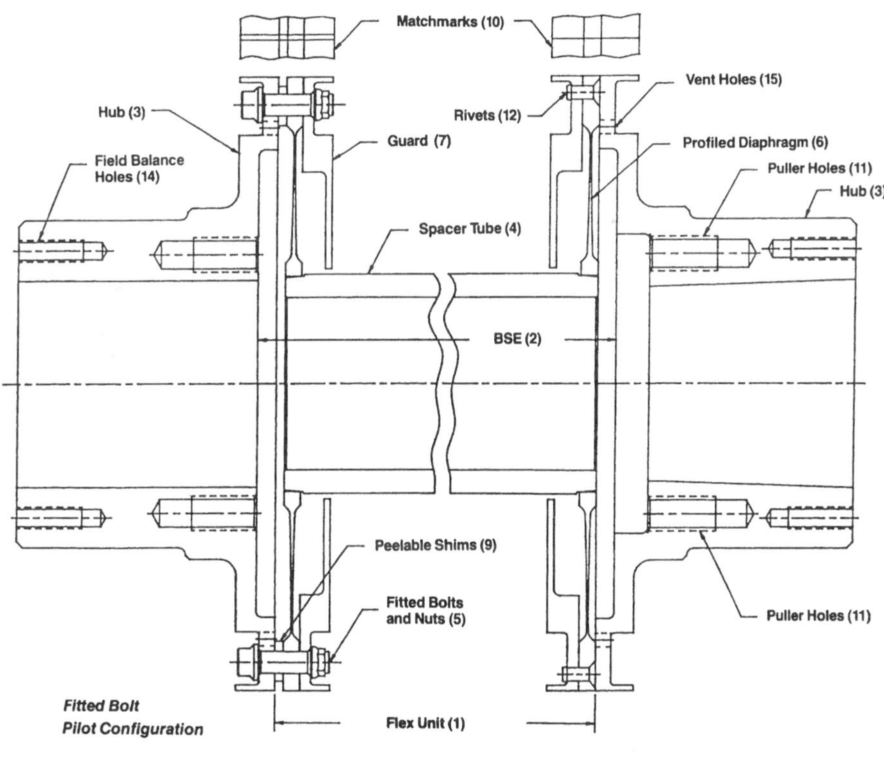

Figure 19.6 Single diaphragm spacer coupling (Courtesy of Lucas Aerospace)

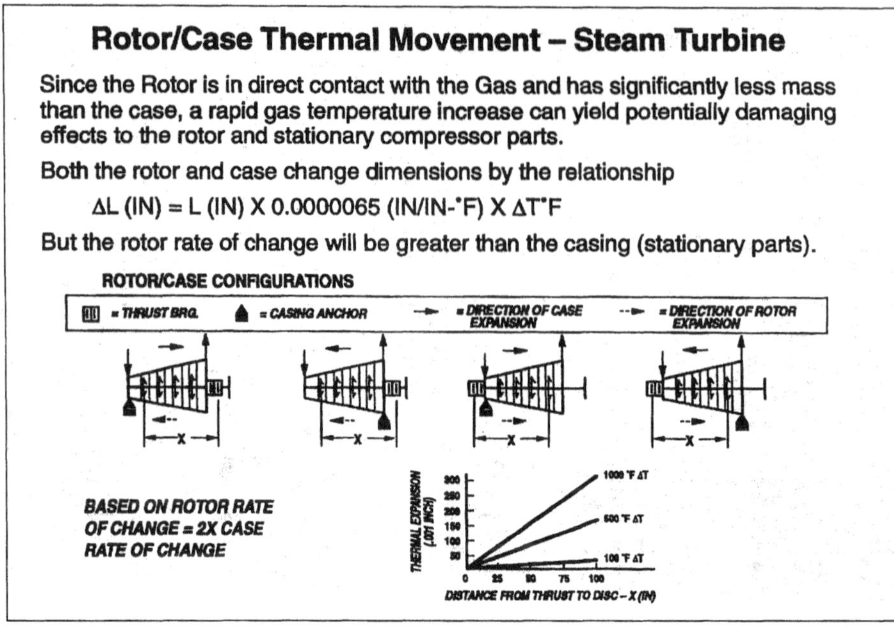

Figure 19.7 Rotor/case thermal movement – steam turbine

removing the spacer to not scratch or dent the diaphragm element. A dent or even a scratch that penetrates the protective coating can cause a diaphragm failure.

Regardless of the type of diaphragm couplings, it is common practice to ‘pre-stretch’ these couplings to take full advantage of the maximum available end float. Readers are cautioned to always require equipment vendors provide axial shaft movement calculations in order to confirm that the coupling maximum end float is not exceeded. Figure 19.7 graphically displays the various combinations of end shaft movement and the calculation method.

Figure 19.8 is a picture of a multiple, convoluted (wavy) diaphragm spacer coupling.

This type coupling is used whenever large values of axial end float exist. Axial end float values as high as ± 0.875" are attainable with this type of coupling.

As previously mentioned, gear type couplings provide the lowest value of overhung weight (coupling moment) on the bearing. However, a dry type coupling will usually have a higher coupling moment because the flexible assembly is farther from the bearing centerline than the gear teeth in a gear coupling. An excessive coupling moment will reduce the second natural frequency (N_{c2}) of a turbo-compressor and could move it close to or within the operating speed range. A solution in these cases

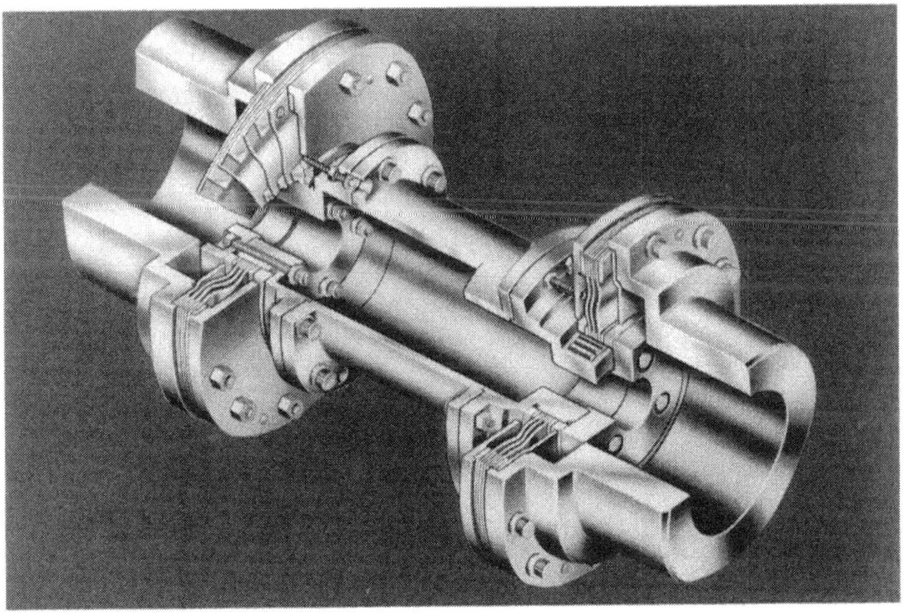

Figure 19.8 Multiple, convoluted diaphragm-spacer coupling (Courtesy of Zurn Industries)

can be to use a reduced moment diaphragm coupling as shown in Figure 19.9.

In this design, the diaphragm is moved to the back of the hub and the flange diameter is reduced thus significantly reducing the coupling moment. The reduced moment coupling approaches the gear coupling in term of coupling moment value.

Couplings with elastomer insert flexible drive members

This type coupling is normally used only for low horsepower, general purpose applications. Their limitations are based primarily on the wear factor and the difficulty in maintaining shape and concentricity of the elastomer insert. These items have a tendency to limit the maximum design speed at which such couplings can be operated. A typical 'Jaw and Spider' type is shown in Figure 19.10.

One exception is a special design used for synchronous motor driven compressor trains. A characteristic of synchronous motors is a variable oscillating torque that decreases linearly in frequency from 2X line frequency (50 HZ or 60 HZ) at 0 RPM to 0 frequency at rated RPM. Figure 19.11 shows a plot of motor RPM vs transient torsional excitation frequency.

The excitation frequency inherent in all synchronous motors will excite all torsional natural frequencies present between 2x line frequency and 0 RPM.

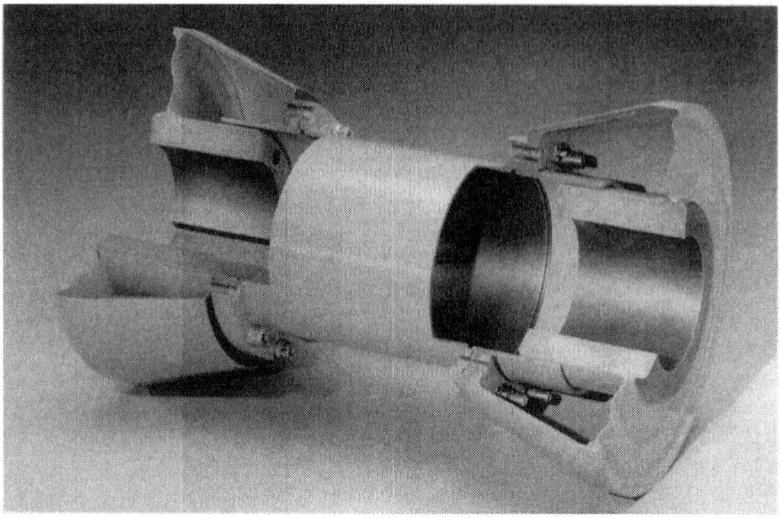

Figure 19.9 Reduced moment convoluted (wavy) diaphragm spacer coupling (Courtesy of Lucas Aerospace)

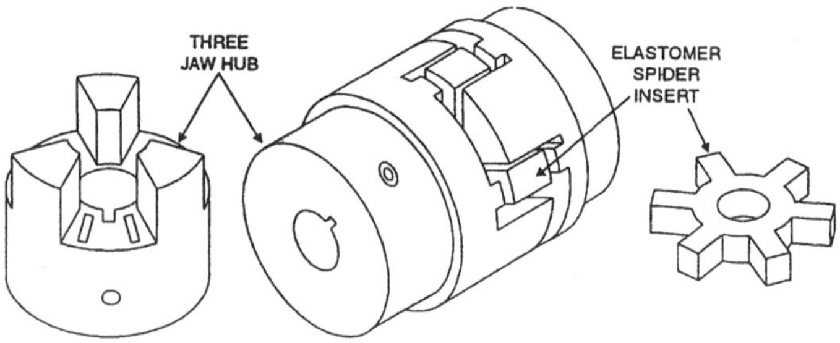

Figure 19.10 Jaw and spider coupling

When the motor torsional excitation frequency briefly coincides with a torsional natural frequency, torque values can amplify to as much as 5 or 6 times full load torque. The 'Holset' or elastomeric coupling shown in Figure 19.12 significantly reduces the torque amplification by dampening out the response in the elastomeric elements. The hardness of these elements is controlled to limit the maximum amplification factor to an acceptable value (usually 2–3 X rated torque).

The coupling system

It has been the writers' experience that if couplings are properly selected, the root cause of failure, if it occurs, is in the coupling system.

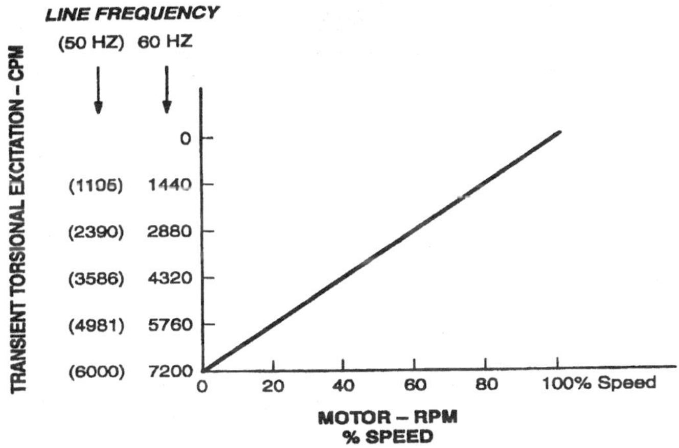

Figure 19.11 Transient torsional excitation – Frequency vs. motor speed

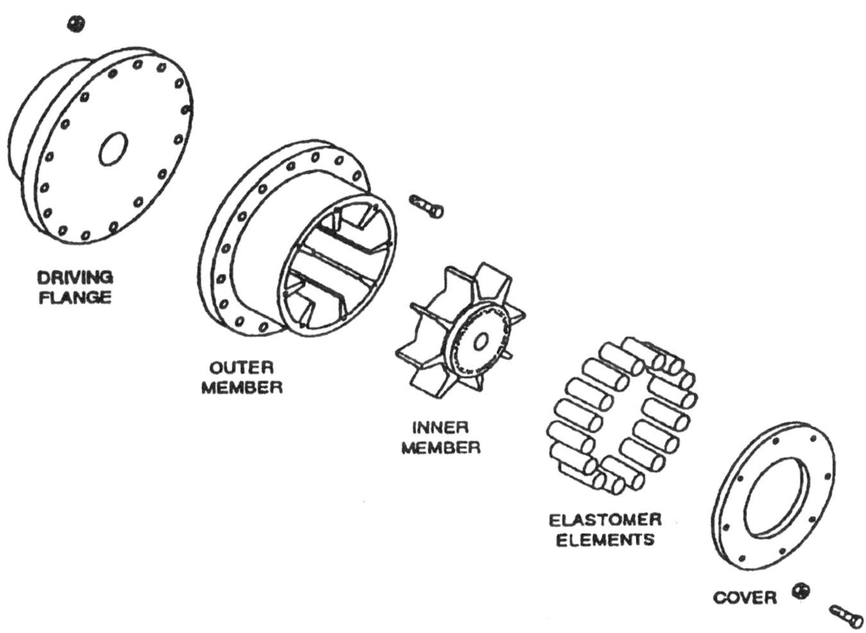

Figure 19.12 Holset coupling (exploded view) non-spacer type

The 'coupling system' must continuously transmit torque safely between the driver and driven equipment and must allow for changes in shaft misalignment and axial movement. The components that make up the shaft system are:

- The driver shaft
- Driver shaft/coupling fit
- Coupling
- Driven shaft/coupling fit
- The driven shaft
- The coupling spacer system
- Lubrication (if required)
- Cooling system (if required)

A schematic of a coupling system is shown in Figure 19.13.

The reliability of the coupling is a function of the coupling system design and assembly. If any of the items noted above are not properly designed or assembled a coupling failure can occur.

Coupling assembly/disassembly errors and enclosed coupling guard design are two important areas that are critical to coupling system reliability.

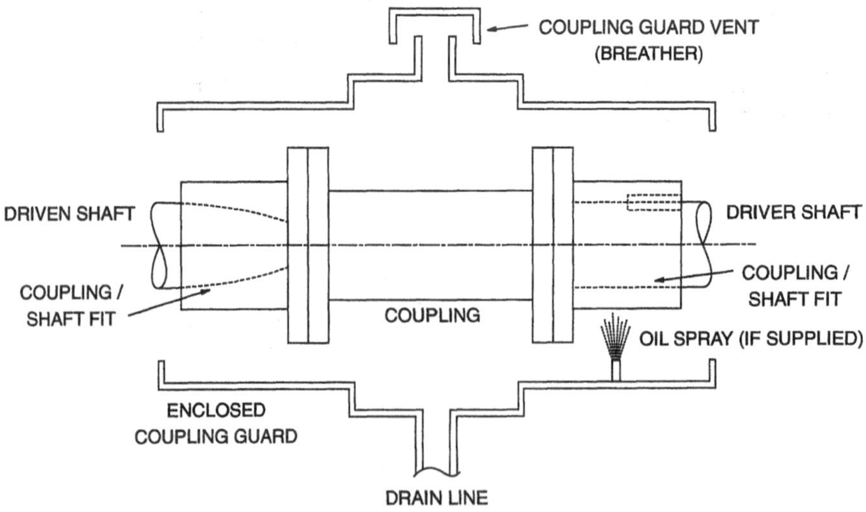

Figure 19.13 The coupling system (Courtesy of M.E. Crane Consultant)

Coupling installation and removal

The most common methods of coupling attachment are:

- Key fit
- Spline fit
- Hydraulic fit

Key fits are used whenever possible. They are the most common method of shaft fit. It is important to assure keys and keyways are properly manufactured to avoid problems with removal or breakage. Key fits will be used on equipment that does not require coupling removal to remove shaft components (seals, bearings, etc). Keyed fits are usually used on motors, gearboxes and most pumps and small steam turbines. Since heat is usually required to remove keyed couplings, they will not be used where removal in the field is necessary.

In these applications either spline or hydraulic fits are used.

Spline fits consist of a male (on the shaft) and female (in coupling hub) finely machined mating gear teeth with line to line fit (no backlash). When assembled on the shaft, the fit is rigid and provides no flexibility. Spline fits are commonly used in the gas turbine industry. They do not usually require heat for removal.

Hydraulic fits are used where heat to remove the coupling hub is either not available or not permitted. Usually, turbo-compressors will utilize hydraulic fits for this reason, since hydrocarbon gas, usually present, requires a flame free environment. Figure 19.14 shows a typical coupling hydraulic shrink fit arrangement. Note that the entire torque load is transmitted by the shrink fit and that no keys are used!

The equipment vendor calculates the required shrink fit based on the shaft and coupling dimensions. Typical values of hydraulic shrink fit are 0.002 inch/inch of shaft diameter.

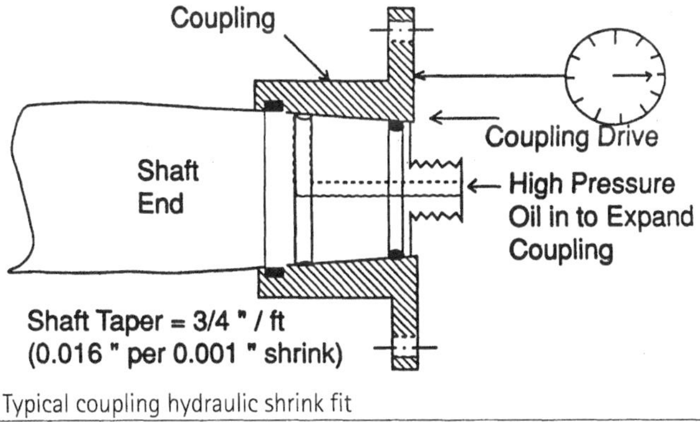

Figure 19.14 Typical coupling hydraulic shrink fit

For ease of hydraulic fit assembly and disassembly, all shafts and coupling hubs are tapered. Different shaft/coupling hub matching tapers are used. The most common are:

- ½° taper
- ½" per foot taper
- ¾" per foot taper (shafts above 4" diameter)

Once the shrink fit is calculated, the value appears on the coupling drawing and is usually expressed as 'drive' or 'push' based on the shaft taper. This is the axial distance the coupling must be moved up the shaft. The coupling drive per 0.001" of shrink fit for the most common shaft tapers is noted below:

Shaft taper	Drive per 0.001 shrink
½°	0.057"
½" per foot	0.024"
¾" per foot	0.016"

As an example, a hydraulic fit coupling with a 4" bore requires a 0.008" shrink fit (i.e. the bore diameter is 0.008 less than the shaft). To expand the coupling bore 0.008", what is the drive if the shaft taper is:

Taper	Drive
½°	0.456"
½" per foot	0.192"
¾" per foot	0.128"

Since the load torque is completely transmitted by the shrink fit, one can see the importance of assuring that the correct shrink fit (or drive) is obtained. The shrink fit amount is directly proportional to torque load capability. If the shrink fit is 50% of the specified value, so is the torque capability! However, industry specifications require that the shrink fit at minimum tolerances be a minimum of 125% greater than the driver maximum torque. Observing the calculated drives in the example above it can be seen that the smaller the shaft, the more critical the correct drive becomes for a given shaft taper. The coupling drive is measured by positioning a dial indicator on the coupling hub and measuring the axial distance traveled during coupling assembly.

Figure 19.15 shows a typical hydraulic coupling mounting arrangement used by Dresser-Rand.

All turbo-compressor manufacturers use similar arrangements. There are some slight differences which are:

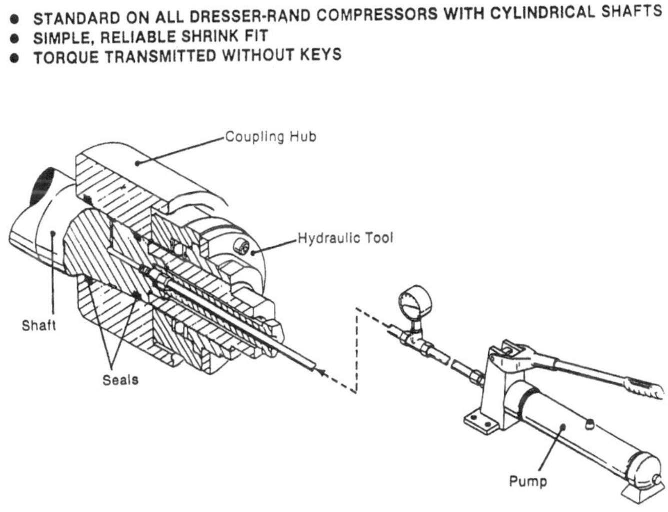

Figure 19.15 Hydraulic fit coupling (Courtesy of Dresser-Rand Corp)

- Hydraulic oil enters the coupling hub and not the shaft.
- An additional pump is used to move the hydraulic tool axially (Figure 19.15 shows a nut which is manually turned to push the coupling axially).

The basic coupling mounting procedure is as follows: (readers must refer to the specific vendors' instruction book for the exact procedure)

1. Clean shaft end and coupling bore with light oil.
2. Remove all 'O' rings from shaft end and coupling.
3. Lightly blue the coupling hub.
4. Push coupling on shaft **without 'O' rings** and tap hub with wood to assure tight fit.
5. While hub is on shaft, index coupling hub axial position relative to a machined surface on shaft (usually shaft end).
6. Remove hub and confirm contact area of blue is a minimum of 85%. If not, correct as required.
7. When coupling contact of 85% is confirmed, clean shaft and coupling hub and install shaft and coupling 'O' rings.
8. Hand push coupling on shaft to **indexed position in step 5. NOTE: It may be necessary to use pump since 'O' rings can provide significant resistance to movement.**
9. **With hub at indexed (zero drive) position**, use hand pump to push coupling axially to value noted on coupling drawing.

Coupling drive must be within tolerances noted. **NOTE: Pump pressures will be high. Be extremely careful when connecting pump and tubing. Be sure to secure pump so that hand jacking cannot break tubing.** Pressures typically required range from 15,000–30,000 PSI depending on shaft dimensions, coupling dimensions and shrink fit.

10. When coupling is on shaft correct amount, do not remove dial indicator but reduce pump pressure to zero and **back off hydraulic tool slightly. Observe that dial indicator does not move before removing tool.**
11. Promptly assemble shaft end coupling nut.
12. Measure between shaft end dimension (B.S.E.) to assure it is as stated on coupling drawing before assembling coupling spacer. If this dimension is not correct, consult instruction book and O.E.M. if necessary before taking corrective action. **Under no circumstances should coupling spacers be added unless allowed by the coupling manufacturer or should equipment axial shaft position be changed without O.E.M. consent.**
13. When coupling is properly assembled check alignment using 'reverse dial indicator procedure'. **NOTE: For coupling removal, consult vendor's instruction book. Under no circumstances should coupling be pulled or heated. Usually, hydraulic pressure required for removal will be higher (5–10%) than that required for assembly. If the value required exceeds 35,000 PSI, do not proceed until consulting O.E.M. for additional options concerning removal.**

Incorrectly mounting a hydraulic coupling can cause catastrophic coupling and/or shaft end failure.

Enclosed coupling guards

Most turbo-compressor couplings are completely enclosed by a spark proof (usually aluminum) coupling guard. This is because the couplings are continuously lubricated gear type or to prevent oil siphoned from the bearing brackets by the windage action of the dry couplings. In either case, proper design of the coupling guard is essential to maintaining coupling reliability. Many coupling failures have resulted from high coupling enclosure temperatures, enclosures full with oil and debris educted into the coupling guard from the atmosphere.

As a minimum the following must be checked by the O.E.M. and coupling vendor during equipment design or field coupling retrofit from gear to dry type:

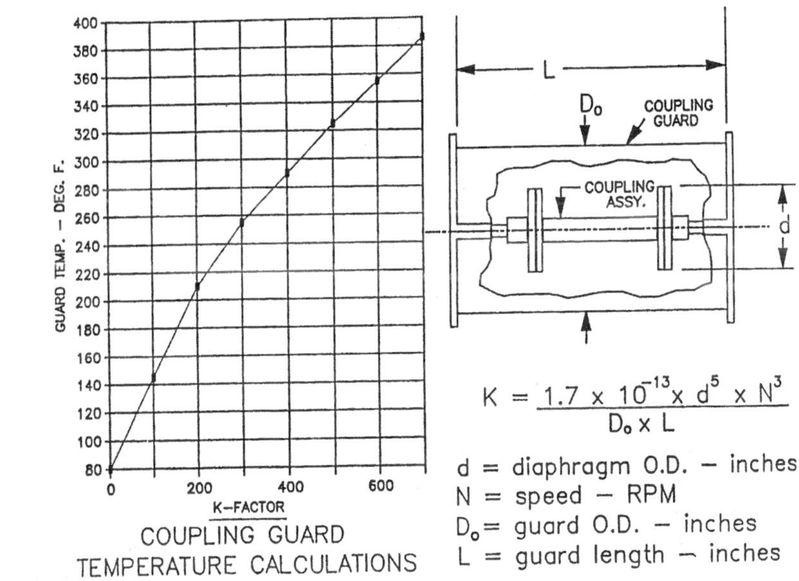

Figure 19.16 (Courtesy of M.E. Crane Consultant)

- Proper coupling O.D. to guard and/or bearing bracket I.D. clearance.
- Proper coupling guard baffle design to allow proper drainage. **NOTE: All enclosed coupling guards must be supplied with vent and drain.**
- Proper vent breather sizing and design.

Figure 19.16 presents coupling guard dimensional design criteria for dry type couplings operating in enclosed coupling guards. Note that in some designs, D_o may be the I.D. of the bearing bracket.

Recommend coupling guard skin operating temperatures should be below 200°F to avoid coupling and coupling guard leakage problems. Under no circumstances should coupling guard skin temperatures approach the flash point of lubricating oil (400°F for new mineral oil).

Field retrofits from lubricated to dry couplings

Considering the advantages of dry couplings, many users are retrofitting their older style lubricated couplings to dry couplings. Whenever considering a retrofit, the following action should be taken to maintain or increase coupling system reliability.

1. Consult equipment or coupling vendor for proper selection of new coupling.

2. Consult equipment O.E.M.(s) (each affected vendor) to confirm:
   - Critical speeds will not be affected
   - Coupling guard design is acceptable
3. Advise coupling vendor or any environmental considerations that may affect dry flexible element life (environmental gases, temperatures, excessive dust, etc).

20 Steam turbine function and types

- Introduction
- Steam turbine types
- Steam turbine applications

Introduction

In this section we will examine several types of Expansion Turbines commonly installed in Refineries, Petrochemical Plants and other installations.

All types of Expansion Turbines regardless of their design, perform similar duties. That is, they extract usable energy from a vapor and provide sufficient power to operate their load (driven equipment) at rated conditions. An Expansion Turbine performs the opposite duty of a turbo-compressor. A turbo-compressor requires power to increase the energy of a vapor while an expansion turbine obtains power to drive the turbo-compressor from the potential energy of the vapor. In other words, in a turbo-compressor the blades work on the gas and in an expansion turbine, the gas performs work on the blades.

There are many different types of expansion vapors. Steam, because of its ease of generation and comparative cost effectiveness, is the most widely used expansion vapor. However, many cryogenic (ethylene, hydrogen, etc.) and fired vapors (gas generated) are also used. Fired vapors are used in a gas turbine. It is extremely important to understand that the operation of a steam turbine and an expansion turbine in a gas turbine are identical. We will examine the expansion of steam on a Mollier Diagram and also a hydrocarbon gas (ethylene) on its Mollier Diagram.

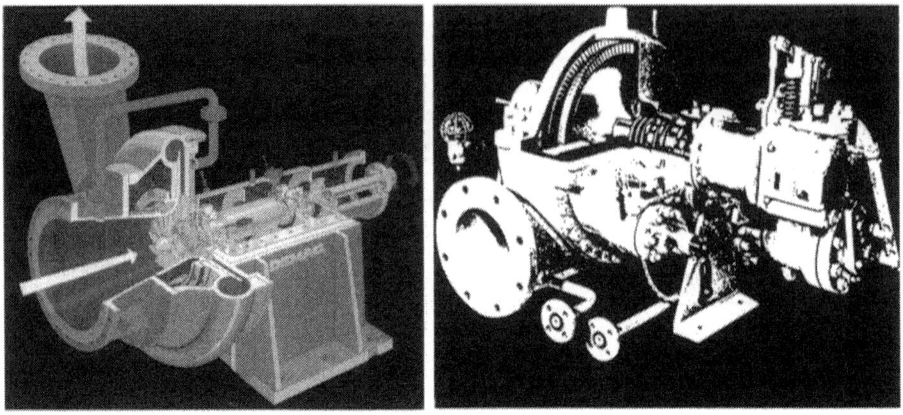

Figure 20.1 Turbo-compressors/expansion turbines Left: A turbo-compressor requires power to increase the energy (head) of a vapor (Courtesy of Mannesmann Demag); Right: An expansion turbine obtains power to drive the turbo-compressor from the potential energy of a vapor (Courtesy of Westinghouse Canada, Inc.)

The definition of single-stage, multi-stage, single valve, multi valve, back pressure and condensing turbines will be discussed along with the reason for using these various types of steam turbines.

In addition, we will discuss extraction and admission steam turbines and explain why double and triple flow exhaust ends are used in large steam turbines.

We will conclude this section by discussing the various applications of the different types of turbines discussed here. This will provide the reader with a working knowledge of when to use various types of steam turbines and why those types are used for specific applications.

Figure 20.1 defines the functions of a turbo-compressor and expansion turbine.

The turbo-compressor requires power and is termed a 'driven' machine. The expansion provides power and is called the driver or 'prime mover'.

A function diagram of one row of impulse and reaction blading is shown in Figure 20.2.

The two basic types of blading sets used to extract energy from a vapor and produce power are shown. In general, impulse blading has been widely used in the steam turbine industry and reaction blading has been widely used in the gas turbine industry. In recent years, the steam turbine industry has been designing a 'Hybrid Turbine' utilizing rugged impulse blading in the initial stages and high efficiency reaction blading in the final stages.

The advantages and disadvantages of each blading type are noted

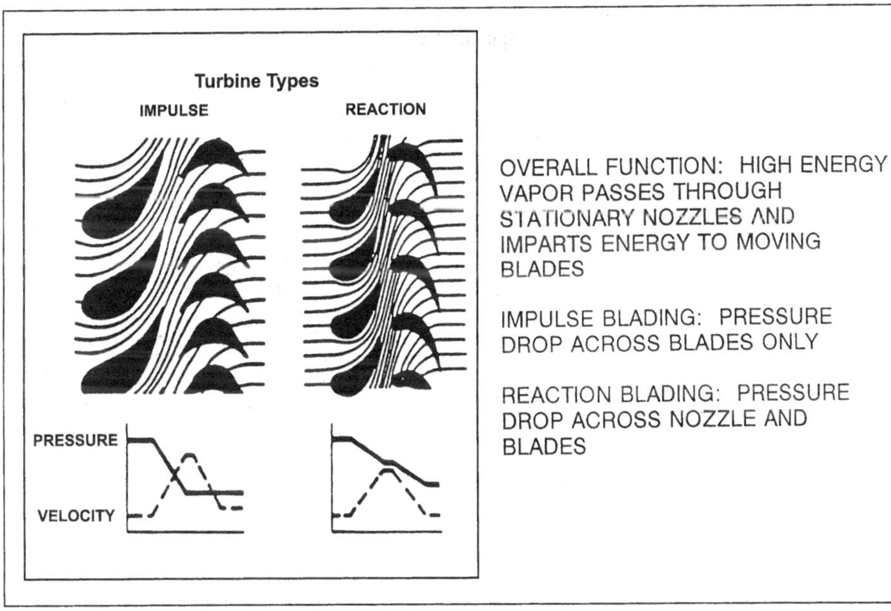

Figure 20.2 Expansion turbines simple function diagram

below.

Type	Advantages	Disadvantages
Impulse	■ Rugged	■ Lower efficiency
	■ No thrust forces	■ Heavy weight per stage
Reaction	■ Higher efficiency	■ More stages required
	■ Light weight per stage	■ Significant thrust force per stage

Figure 20.3 shows an example of a hybrid turbine design that incorporates impulse blading in the first stage and reaction blading in all other stages.

The principle of operation of any type of expansion turbine is shown in Figure 20.4. The thermodynamic expansion process is exactly the opposite of the ideal thermodynamic compression process used in turbo-compressors. An Isentropic (Adiabatic) reversible expansion is commonly used to determine steam turbine performance.

There are many different types of expansion vapors used. Steam is the most common. Figure 20.5 presents some other commonly used expansion vapors.

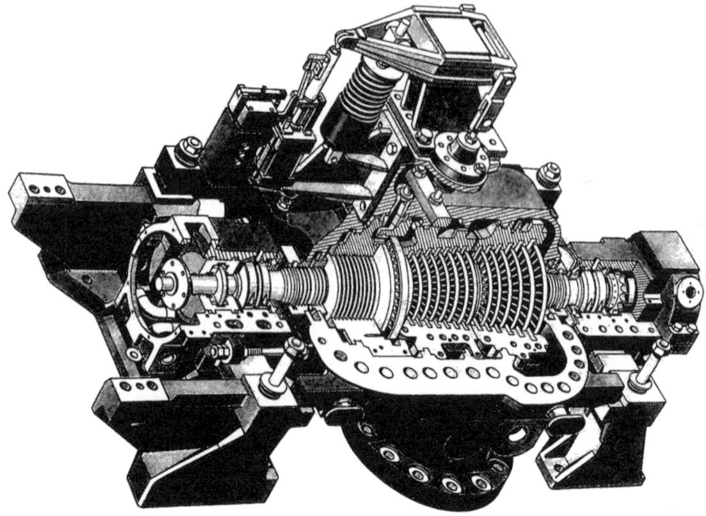

Figure 20.3 Backpressure steam turbine 1st stage impulse. All other stages reaction. (Courtesy of Siemens Corporation)

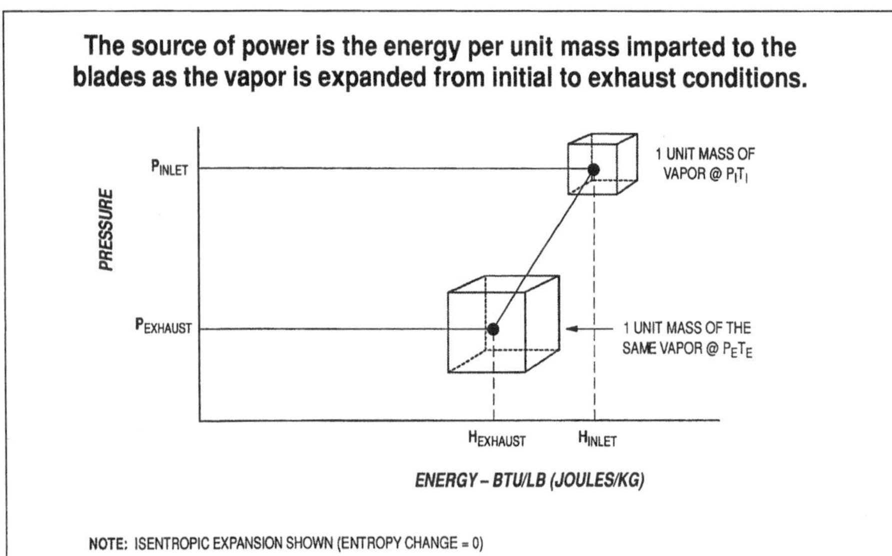

Figure 20.4 Expansion vapors

Types of expansion vapors

The types of expansion vapors available are many. They can be grouped as follows:

Steam	Cryogenic	Fired vapors (gas generated)
	■ Ethylene	■ Diesel fuel & air
	■ Hydrogen	■ Natural gas & air
	■ Nitrogen	■ Refinery gas & air

Naturally, the vapors with the highest amount of energy (BTU's or joules) per unit mass will be used.

Figure 20.5 Types of expansion vapors

Steam turbine types

Figure 20.6 contains an assembly drawing of a Single and Multi-stage steam turbine.

Note that the single stage turbine shown actually has two blade rows. This arrangement is known as a 'Curtis' stage and is used in single stage and some older design multi-stage turbines to reduce the blade loading. All single stage steam turbines contain one Curtis stage.

The limiting factors for a single stage expansion turbine are shown in Figure 20.7.

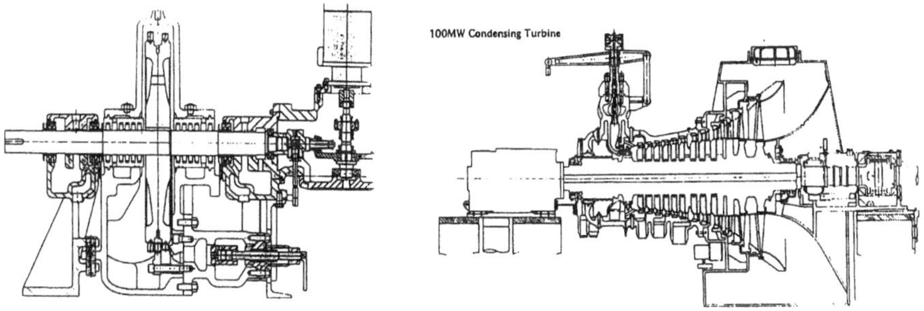

Figure 20.6 Single stage and Multi-stage steam turbines. Left: Single stage (Curtis blading) (Courtesy of Westinghouse, Canada, Inc.) and multi-stage (Condensing) (Courtesy of Mitshibushi Industries)

Single vs. multi-stage

The energy extracted from the vapor is limited by aerothermal and mechanical factors:

Aerothermal	Mechanical
■ Exhaust moisture content	■ Blade bending stresses
■ Nozzle and blade velocities	■ Blade attachment (root) stresses
■ Blade incident angles	■ Blade disc stresses

Figure 20.7 Single vs. multi-stage

The approximate power limitations for single and multi-stage turbines are noted in Figure 20.8

Single vs. multi-stage (continued)

As a result:

- Single stage - limited to approximately 2000 H.P. (1500 KW)
- Multi-stage - above 2000 H.P. (1860 KW)

Figure 20.8 Single vs. multi-stage (Continued)

The objective of any expansion turbine is to extract the maximum possible energy from each pound of vapor to produce power. The inlet throttle valves control the amount (mass flow) of steam admitted to the expansion turbine to meet the power requirements of the driven equipment. However, to maximize energy extraction, the losses across the inlet throttle valves must be minimum. Pictured in Figure 20.9 is a Single Valve and a Multi-Valve steam turbine.

Figure 20.9 Left: Single valve steam turbine and Right: Multi-valve steam turbine (Courtesy of Westinghouse, Canada, Inc.)

The factors that determine the choice of a single valve or a multi-valve steam turbine are presented in Figures 20.10 and Figure 20.11.

Single valve and multi-valve steam turbines

The choice depends on:

- Steam flow requirements
- Operating requirements
- Cost of steam (efficiency)

Figure 20.10 Single valve and multi-valve steam turbines

Schematics depicting a backpressure and condensing steam turbine are shown in Figure 20.12. In a backpressure turbine, the exhaust pressure is greater than atmospheric pressure. A condensing turbine's exhaust pressure is equal to or less than atmospheric pressure. Most condensing turbines operate at a high vacuum (3–4" Hg Vac) for maximum efficiency and energy extraction.

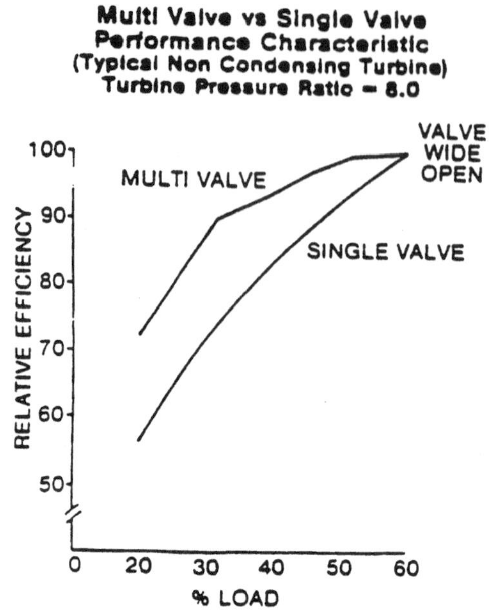

Figure 20.11 Why multi-valve turbines are used (Reprinted with permission of Gas Processors Supplies Association. GPSA Engineering Data Book 10th edition)

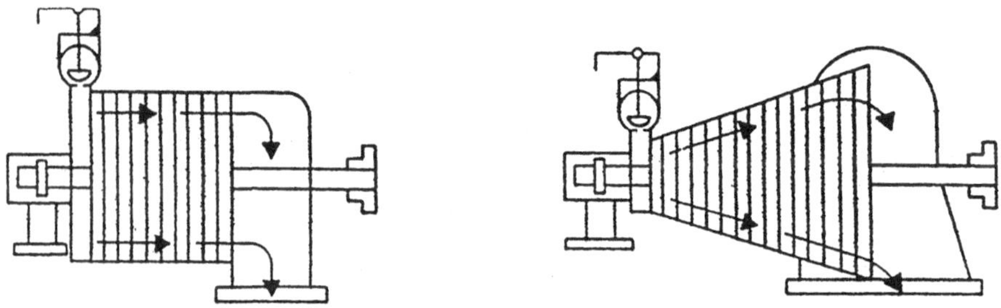

Figure 20.12 Backpressure and condensing steam turbines

Figures 20.13 and 20.14 present facts concerning the definition and selection of backpressure and condensing steam turbines.

Backpressure/condensing steam turbines

- The greater inlet to exhaust ΔP, the greater amount of potential energy per pound of vapor
- Backpressure turbine exhaust pressures are above atmospheric pressure
- Condensing turbine exhaust pressures are below atmospheric pressure

Figure 20.13 Backpressure/condensing steam turbines

Backpressure/condensing steam turbine

The choice depends on

- The cost of steam
- The plant steam balance
- The capital cost of auxiliaries (condenser, condensate system, etc.)

Figure 20.14

Extraction and Admission turbines are used to optimize the cycle efficiency for a given plant steam balance. Figure 20.15 shows a Single Extraction and Single Admission steam turbine.

By using an Extraction or Admission turbine steam pressure can be efficiently reduced to a desired pressure level or excess steam can be utilized to produce power. Either type of turbine is actually a series combination of either backpressure or condensing type turbines. Each turbine section is supplied with a throttle valve.

In Figure 20.16 a Single Flow and Double Flow steam turbine are shown. In large steam turbines the volume of exhaust steam is too large for one stage of blading to accommodate due to excessive blade length. As a result, the flow is divided into two stages. In some very large turbines, a triple flow back end can be used for the same purpose.

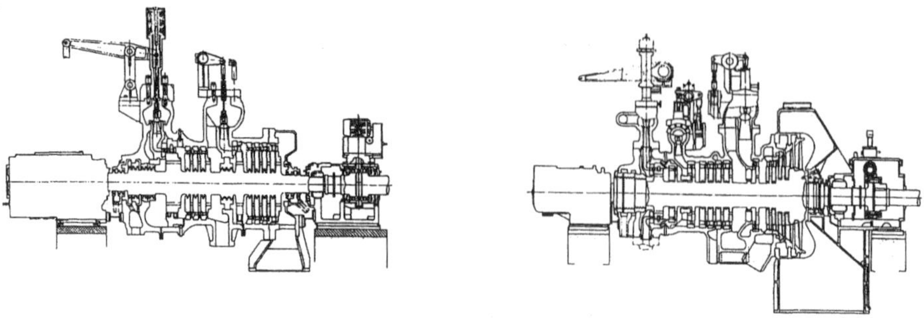

- EXTRACTION OR ADMISSION LOW PRESSURE TURBINES ARE ACTUALLY MULTIPLE TURBINES MOUNTED ON THE SAME SHAFT.

Figure 20.15 Extraction and/or admission steam turbines. Left: Single extraction; Right: Single admission (Courtesy of M.H.I)

- WHEN EXCESSIVE STEAM VOLUME FLOW IN THE EXHAUST SECTION OF A STEAM TURBINE RESULTS IN EXCESSIVE BLADE HEIGHTS OR VELOCITIES, THE STEAM PATH IS SPLIT INTO DOUBLE FLOW:

Figure 20.16 Single and double flow condensing steam turbines. Left: Single flow (Courtesy of Siemens); Right: Double flow (Courtesy of General Electric Co.)

Steam turbine applications

Figures 20.17, 20.18, 20.19 and 20.20 present lists of typical applications arranged according to number of stages, number of valves, exhaust end design and extraction/admission design.

Steam turbine applications

Single stage

- Process pump drive
- Boiler fan drive
- Generator drive
- Lube/seal oil pump drive

Multi-stage

- Generator drive
- Boiler feed pump drive
- Cooling water pump drive
- Turbo-compressor drive

Figure 20.17 Steam turbine applications

Steam turbine applications (continued)

Single valve

- Process pump drive
- Boiler fan drive
- Lube/seal oil pumps
- Generator drive (small co-gen HRSG)
- Cooling water pump drive
- Compressor drive < 3000 H.P. (1860 KW)

Multi valve

- Boiler feed pump drive > 3000 H.P. (746 KW)
- Compressor drive > 3000 H.P. (1860 KW)
- Generator drive

Figure 20.18 Steam turbine applications (continued)

Steam turbine applications (continued)

Backpressure	Condensing
■ Process pump drive	■ Compressor drive*
■ Boiler fan drive	■ Generator drive*
■ Lube/seal oil pump drive	■ Boiler feed pump drive*
■ Cooling water pump drive*	
■ Compressor drive*	
■ Boiler feed pump drive*	

* choice depends on power level (approximately > 5000 BHP (3700 KW) and plant steam balance

Figure 20.19 Steam turbine applications (continued)

Steam turbine applications (continued)

Extraction/admission

- Generator drive*
- Compressor drive*
- Boiler feed pump drive*

* approximately > 15,000 H.P. (11,000 KW)

Figure 20.20 Steam turbine applications (continued)

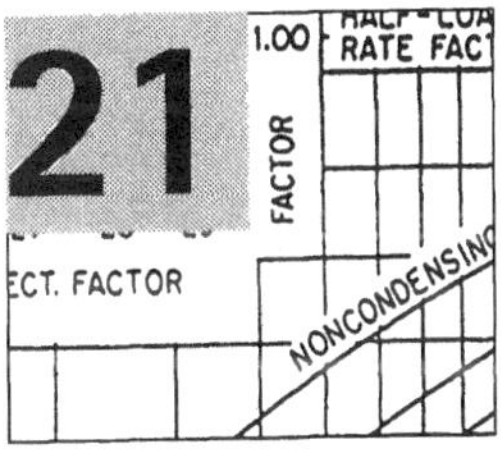

Steam turbine performance

- Steam turbine performance characteristics
- Steam conditions
- Theoretical steam rate
- Actual steam rate
- Turbine efficiency
- Why steam turbines are not performance tested
- Performance curves

Steam turbine performance characteristics

In this section we will discuss how energy is extracted from the vapor (steam) in a steam turbine. Power output is determined by the following factors in a steam turbine:

- The energy available from the vapor
- The external efficiency of the turbine
- The steam flow rate

Note: The external efficiency is equal to the internal efficiency (steam path efficiency) of the turbine factored by the mechanical losses and steam leakage losses from the turbine.

The energy available from the vapor is determined by the steam conditions of the particular application. That is, the pressures and temperatures at the turbine inlet and exhaust flanges. To determine the energy available from the vapor, steam tables or a Mollier Diagram is used. We have included a Mollier Diagram for steam in this section. We will review in practical terms the use of a Mollier Diagram to determine the following:

- Theoretical steam rate
- Actual steam rate
- Turbine efficiency

An example will be presented for both non-condensing and condensing turbines. It should be noted that the exercises in this section will deal with overall steam turbine performance only. In the next section, we will discuss individual blade performance and efficiencies and determine as an exercise, the number of stages required for a given turbine application.

In this section we will also observe the effect of design steam conditions on steam turbine performance and reliability.

Finally, typical turbine efficiencies and performance curves will be presented and discussed.

Steam conditions

Steam conditions determine the energy available per pound of steam. Figure 21.1 explains where they are measured and how they determine the energy produced.

Steam conditions

- The steam conditions are the pressure and temperature conditions at the turbine inlet and exhaust flanges.
- They define the energy per unit weight of vapor that is converted from potential energy to kinetic energy (work).

Figure 21.1 Steam conditions

Frequently, proper attention is not paid to maintaining the proper steam conditions at the flanges of a steam turbine. Failure to maintain proper steam conditions will affect power produced and can cause mechanical damage to turbine internals resulting from blade erosion and/or corrosion. Figure 21.2 presents these facts.

Steam condition limits

Inlet steam conditions should be as close as possible (+/- 5%) to specified conditions because:

- Power output will decrease
- Exhaust end steam moisture content will increase, causing blade, nozzle and diaphragm erosion.

Figure 21.2 Steam condition limits

Mollier Diagram or steam tables allow determination of the energy available in a pound of steam for a specific pressure and temperature. Figure 21.3 describes the Mollier Diagram and the parameters involved.

The Mollier Diagram

Describes the energy per unit mass of fluid when pressure and temperature are known.

- Enthalpy (energy/unit mass) is plotted on Y axis
- Entropy (energy/unit mass degree) is plotted on X axis
- Locating P_1, T_1 gives a value of enthalpy (H) horizontal and entropy (S) vertical
- Isentropic expansion occurs at constant entropy ($\Delta S = 0$) and represents an ideal (reversible) expansion

Figure 21.3 The Mollier Diagram

Refer to Figure 21.4 which is an enlarged Mollier Diagram.

As an exercise, plot the following values on the Mollier Diagram in this section and determine the corresponding available energy in BTUs per pound.

1. $P_1 = 600$ PSIG, $T_1 = 800°F$ $\qquad h_1 = \frac{BTU}{LB_M}$

2. $P_2 = 150$ PSIG, $T_2 = 580°F$ $\qquad h_2 = \frac{BTU}{LB_M}$

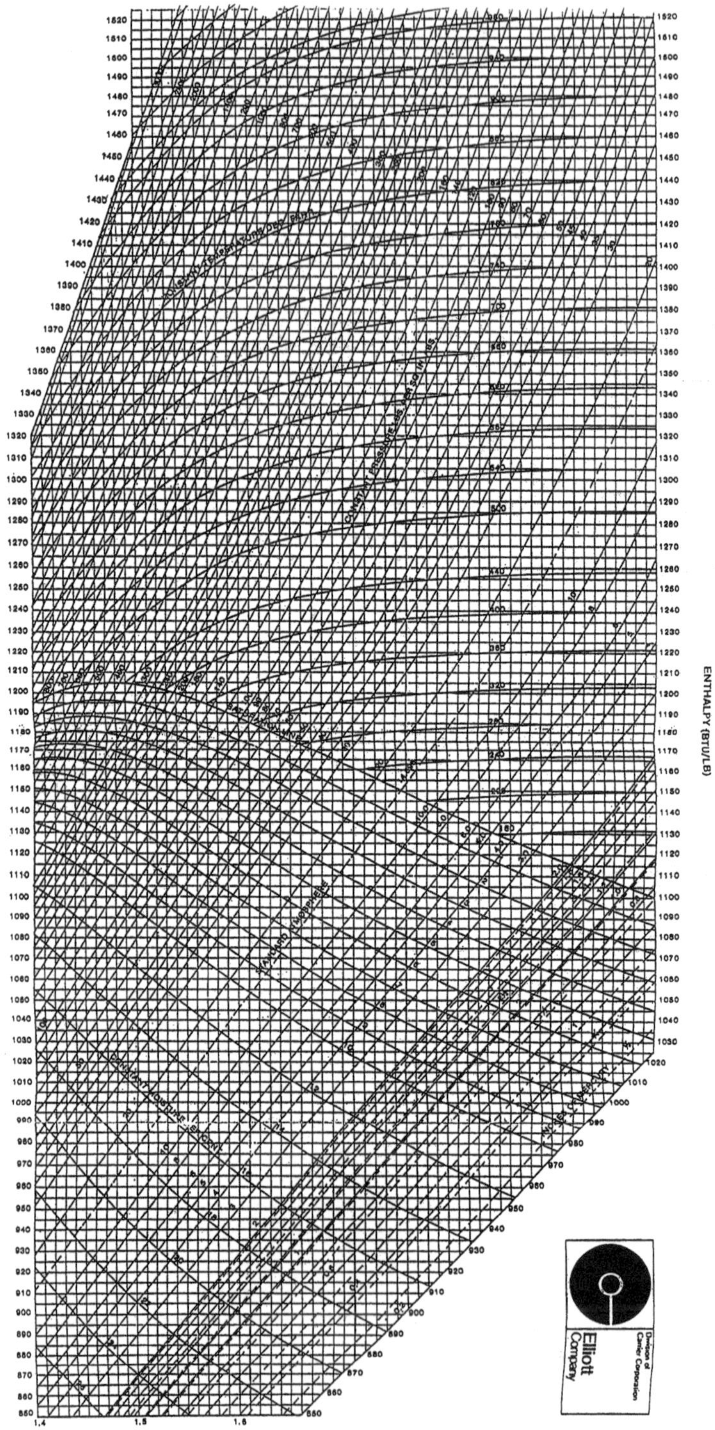

Figure 21.4 Mollier steam diagram (Courtesy of Elliott Company)

3. $P_1 = 1500$ PSIG, $T_1 = 900°F$ $\qquad h_1 = \frac{BTU}{LB_M}$

4. $P_2 = 2$ PSIG, % moisture = 9% $\qquad h_2 = \frac{BTU}{LB_M}$

Having plotted various inlet and exhaust conditions on the Mollier Diagram to become familiar with its use, please refer to Figure 21.5 which presents the definitions and uses of steam rate.

Determining steam rate

Uses:

- Determine the amount of steam required per hour
- Determine the amount of potential KW (horsepower)

Required:

- Steam conditions
- Theoretical steam rate table or Mollier Diagram
- Thermal efficiency of turbine

Formula:

- Theoretical steam rate

$$\text{T.S.R. (LB/HP-HR)} = \frac{\text{2545 BTU'S/HP-HR}}{\Delta H_{\text{ISENTROPIC}}}$$

- Actual steam rate

$$\text{A.S.R. (LB/HP-HR)} = \frac{\text{T.S.R.}}{\text{Efficiency}} = \frac{\text{2545 BTU'S/HP-HR}}{\Delta H_{\text{ACTUAL}}}$$

- Turbine efficiency

$$\text{Efficiency} = \frac{\text{T.S.R.}}{\text{A.S.R.}} = \frac{\Delta H_{\text{ACTUAL}}}{\Delta H_{\text{ISENTROPIC}}}$$

Figure 21.5 Determining steam rate

Theoretical steam rate

The theoretical steam rate is the amount of steam, in LBS per hour required to produce one (1) horsepower if the isentropic efficiency of the turbine is 100%. As shown in Figure 21.5, it is determined by dividing the theoretical enthalpy $\Delta h_{\text{isentropic}}$ into the amount of BTU'S/HR in horsepower.

Actual steam rate

The actual steam rate is the amount of steam, in LBS per hour, required to produce one (1) horsepower based on the actual turbine efficiency. As shown in Figure 21.5, it is determined by dividing the theoretical steam rate (T.S.R.) by the turbine efficiency. Alternately, if the turbine efficiency is not known and the turbine inlet and exhaust conditions are given (P_2, T_2 or % moisture), the actual steam rate can be obtained in the same manner as theoretical steam rate but substituting ΔH_{actual} for $\Delta H_{isentropic}$.

Turbine efficiency

As shown in Figure 21.5, turbine efficiency can be determined either by the ratio of T.S.R. to A.S.R. or Δh_{actual} to $\Delta H_{isentropic}$.

It is relatively easy to determine the efficiency of any operating turbine in the field if the exhaust conditions are superheated. All that is required are calibrated pressure and temperature gauges on the inlet and discharge and a Mollier Diagram or Steam Tables. The procedure is as follows:

1. For inlet conditions, determine h_1
2. For inlet condition with $\Delta S = 0$, determine h_{2ideal}
3. For outlet conditions, determine $h_{2actual}$
4. Determine $\Delta h_{ideal} = h_1 - h_{2ideal}$
5. Determine $\Delta h_{actual} = h_1 - h_{2actual}$
6. Determine efficiency

$$\text{Efficiency} = \frac{\Delta H_{actual}}{\Delta H_{ideal}}$$

However, for turbines with saturated exhaust conditions, the above procedure cannot be used because the actual exhaust condition cannot be easily determined. This is because the % moisture must be known. Instruments (calorimeters) are available, but results are not always accurate. Therefore the suggested procedure for turbines with saturated exhaust conditions is as follows:

1. Determine the power required by the driven equipment. This is equal to the power produced by the turbine.
2. Measure the following turbine parameters using calibrated gauges:
   - Pin
   - $P_{exhaust}$
   - Tin
   - Steam flow in (LBS/HR)

3. Determine the theoretical steam rate by plotting P_{in}, T_{in}, $P_{exhaust}$ @ $\Delta S = 0$. and dividing $\Delta h_{isentropic}$ into the constant.
4. Determine the actual steam rate of the turbine as follows:

$$\text{Actual Steam Rate (A.S.R.)} = \frac{\text{Steam Flow (LB/HR)}}{\text{BHP required by driven equipment}}$$

5. Determine efficiency

$$\text{Efficiency} = \frac{\text{T.S.R.}}{\text{A.S.R.}}$$

Figures 21.6, 21.7 and 21.8 present the advice and values concerning steam turbine efficiencies. The efficiencies presented can be used for estimating purposes.

Typical steam turbine efficiencies

- Quoted turbine efficiencies are external efficiencies, they include mechanical (bearing, etc.) and leakage losses
- Turbine efficiency at off load conditions will usually be lower than rated efficiency
- Typical efficiencies are presented for impulse turbine:
  - Condensing multi-stage
  - Non condensing multi-stage
  - Non condensing single state

Figure 21.6 Typical steam turbine efficiencies

Why steam turbines are not performance tested

When purchasing large steam turbines that do not use proven components, keep in mind that it will not be cost effective to performance test the turbine prior to field installation. If the turbine does not meet predicted output horsepower values, the field modifications will be lengthy and costly in terms of lost product revenue resulting from reduced output horsepower. In some cases, the output power predicted may never be attained. Figure 21.9 presents the reasons why steam turbines are not performance tested.

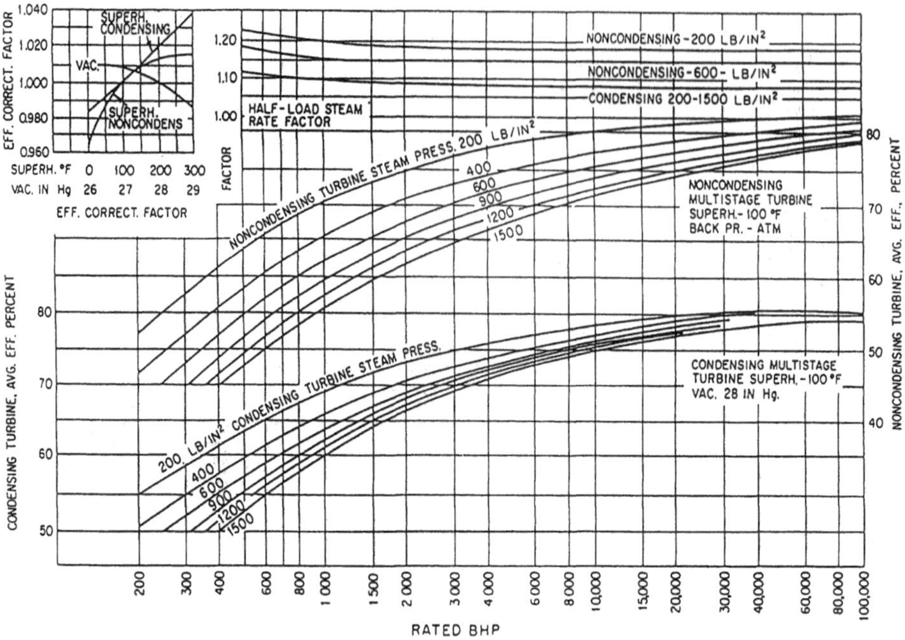

Figure 21.7 Efficiency of multistate turbines (Courtesy of IMO Industries)

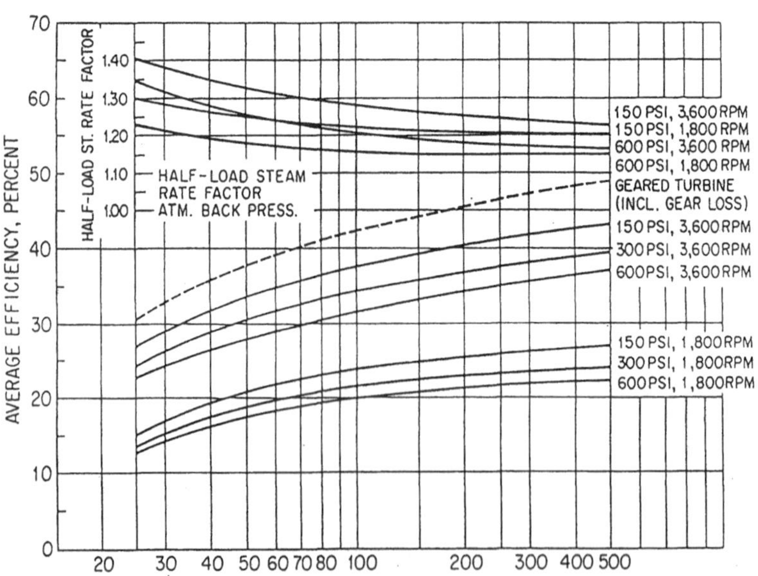

Average efficiency of single-stage turbines (non-condensing dry and saturated steam).

Figure 21.8 Efficiency of single-stage turbines (Courtesy of IMO Industries)

Steam turbine performance testing

- Steam turbines are not usually shop performance tested
- They are only tested mechanically (vibration, bearing temperatures etc.) at no load

Because

- Performance (steam rate, efficiency) varies directly with steam velocity
- Steam velocity varies directly with steam flow
- Steam flow varies directly with power requirement because:
  Power = ΔH (energy per unit mass) x mass flow rate
- Testing at full load is not cost effective

Figure 21.9 Steam turbine performance testing

Performance curves

The performance curve format for steam turbines is to plot steam flow on the y axis and produced shaft horsepower on the x axis. Figure 21.10 presents important facts concerning steam turbine performance curves.

Steam turbine performance curves

- Non-extraction turbines present performance on a Willians-line
- A Willians-line describes the amount of steam flow (throttle flow) required for a given load at a given speed
- An extraction map describes throttle flow for a given load and extraction flow

Note: all curves are for a specific set of steam conditions

Figure 21.10 Steam turbine performance curves

In Figure 21.11, a typical performance curve is presented for a single stage turbine with manual hand valves.

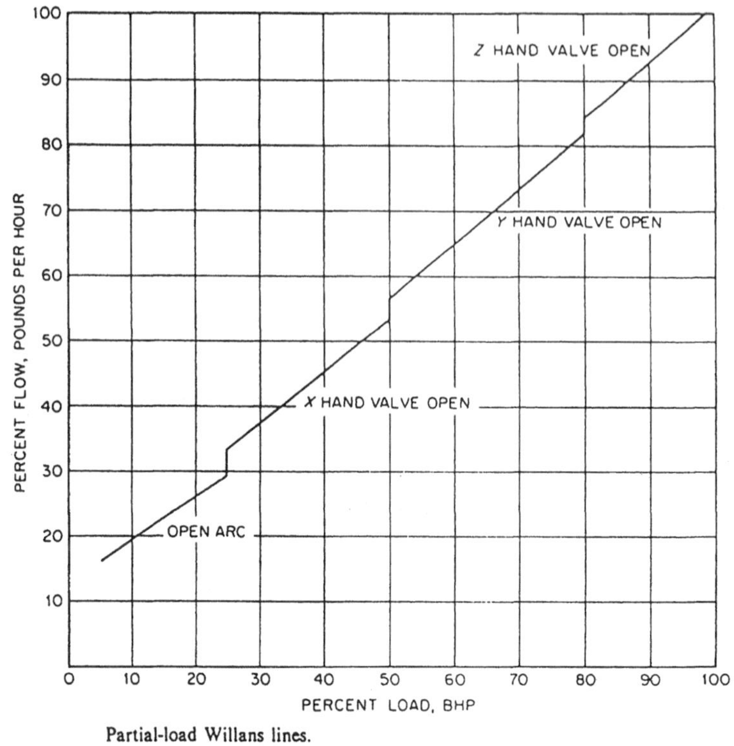

Figure 21.11 Typical performance curve for a single stage turbine with manual hand valves (Courtesy of IMO Industries)

Note that this turbine contains three (3) manual hand valves (x, y, z). Closing hand valves for low horsepower loads increases the efficiency of the turbine. However please note that closed hand valves limit the steam flow through a turbine and therefore the horsepower produced. Hand valves are not modulating. That is, they are either full open or full closed. Throttling a hand valve will destroy the valve seat and may damage the valve stem thus rendering it immovable. Normally hand valves are manually actuated. However, modern electronic governor systems provide outputs to open or close hand valves based on power requirements.

Figure 21.12 shows a performance curve for a typical extraction steam turbine.

This performance curve plots inlet flow and extraction flow vs produced turbine horsepower. When selecting an extraction turbine, care must be taken to be sure the turbine produces the horsepower required during the start-up of the process. The cost of an extraction steam turbine can be significantly reduced if the exhaust section (L.P.

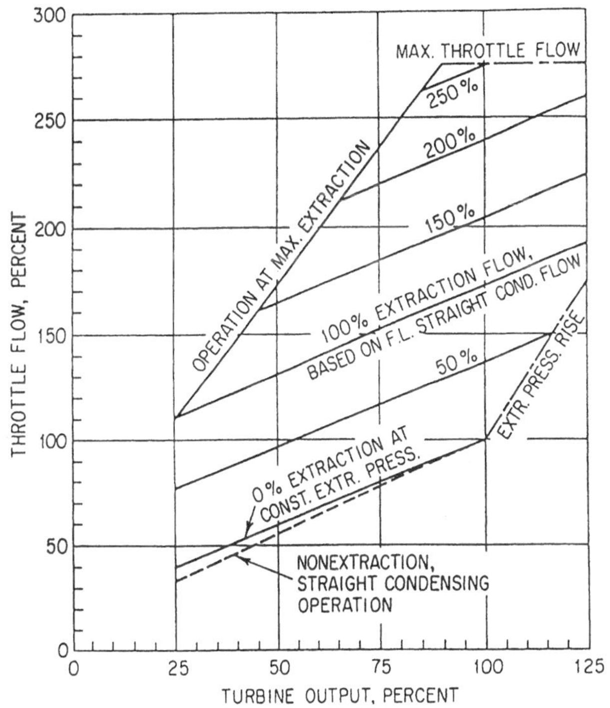

Figure 21.12 Performance curve for a typical extraction steam turbine (Courtesy of IMO Industries)

steam section) size is reduced. Figure 21.12 shows an extraction turbine capable of producing 100% power with 0% extraction flow. Usually, extraction turbines are sized to only provide the process start-up horsepower with 0% extraction. These values may be as low as 50–60% of full load horsepower.

Steam turbine mechanical design overview

- Introduction
- Steam turbine casings
- Throttle valves
- Nozzle ring
- Rotor
- Nozzle/diaphragm
- Shaft end seals
- Bearings

Introduction

In this section we will discuss the function of the major components of a steam turbine. Each major component will be presented, its function discussed, and the most common problems associated with each component reviewed.

After discussing each major component, we will focus on blade design and we'll present information concerning the advantages and disadvantages of different blade types and stage types. We will then proceed to discuss blade efficiency considerations and will present an example to determine the number of blade rows used for both an impulse and reaction turbine application.

We will conclude this section by discussing the mechanical aspects of blade design, namely blade root types, blade natural frequencies and blade loading considerations.

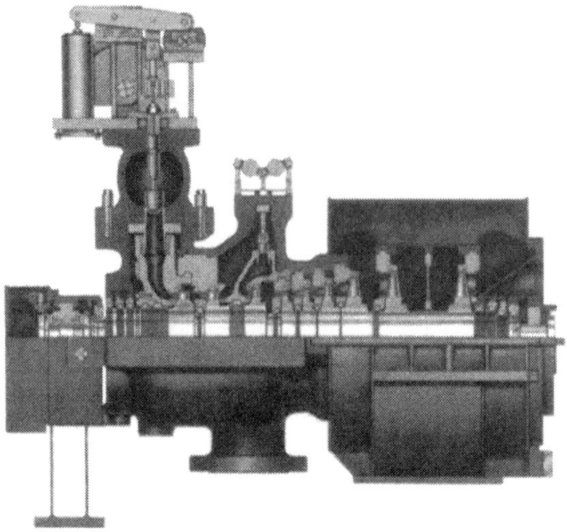

Figure 22.1 Multi-stage extraction condensing turbine (Courtesy of General Electric Company)

Figure 22.1 presents an assembly of a multi-valve, multi-stage, extraction condensing turbine with the major mechanical components noted.

The components can be divided into two basic groups, gas path and mechanical as noted below:

Gas path components

- Steam chest
- Throttle valve(s)
- Nozzle ring
- Blade rows
- Interstage diaphragms (Nozzles and assemblies)

Mechanical components

- Rotor
- Shaft seals
- Bearings
- Casing

Details concerning the function and problems encountered with each major component are as follows:

Steam turbine casings

The function of any expansion turbine casing is defined in Figure 22.2.

All expansion turbine casings, with the exception of single stage cryogenic expanders are horizontally split. As with turbo-compressor casings, expansion turbine casings can be distorted by external piping

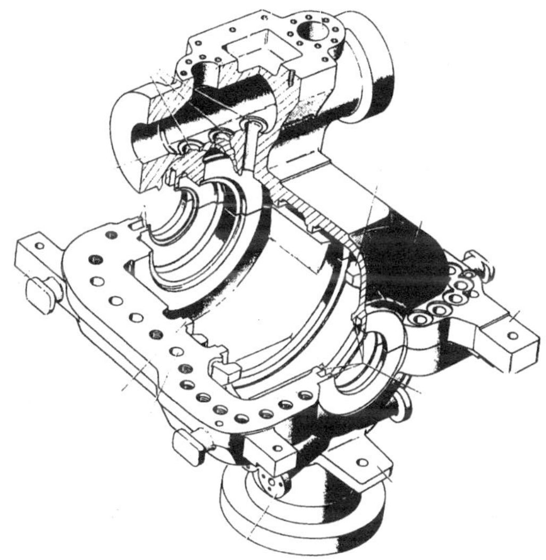

THE FUNCTION OF THE TURBINE CASING IS TO CONTAIN THE EXPANSION VAPOR AND PROVIDE RIGID SUPPORT FOR THE BEARINGS AND SEALS. MAJOR PARTS ARE: ADMISSION SECTION, INTERMEDIATE SECTION AND EXHAUST SECTION.

Figure 22.2 Steam turbine casing (Courtesy of Siemens)

and foundation forces which can result in excessive bearing loads and rotor misalignment. All of the precautions discussed for turbo-compressor casings apply to expansion turbine casings.

The only significant operating condition difference between turbo-compressor and expansion turbine casings is temperature. The maximum turbo-compressor operating temperature is usually limited to 350°F. Steam turbine casings can operate at inlet temperatures above 900°F. Therefore, the casings must incorporate special design features to allow for case thermal growth during start-up and shut down. Special features include; horizontal joint relief (see turbo-compressor case module) and casing joint bolt tensioning devices (hydraulic or thermal). Steam turbine joint leaks, particularly on the H.P. end, should be corrected as soon as possible to prevent permanent joint face damage from H.P. steam.

Most steam turbine inlet casings are cast while intermediate sections and either cast or forged. Exhaust sections of back pressure turbines can be cast, forged or fabricated. Exhaust sections of condensing turbines are usually fabricated.

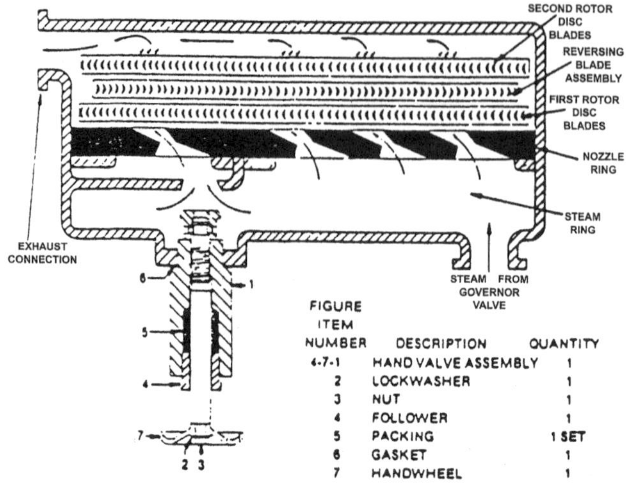

FIGURE ITEM NUMBER	DESCRIPTION	QUANTITY
4-7-1	HAND VALVE ASSEMBLY	1
2	LOCKWASHER	1
3	NUT	1
4	FOLLOWER	1
5	PACKING	1 SET
6	GASKET	1
7	HANDWHEEL	1

Hand Nozzle Valve Assembly

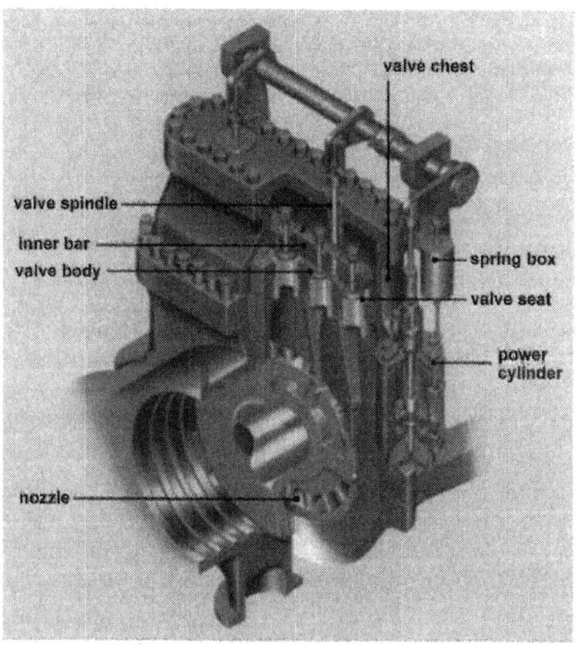

Function: To adjust steam throughput to desired amount with the minimum of efficiency loss

Figure 22.3 Throttle valves. Top: Single throttle valve (Courtesy of Elliott Co.), Bottom: Multiple throttle valve (Courtesy of M.H.I.)

Throttle valves

The function of single and multiple throttle valves is presented in Figure 22.3. All inlet throttle valves are specially designed for minimum pressure drop. Single throttle valves are designed for the normal operating condition. As a result, pressure drop significantly increases at low steam flows (horsepower loads). In order to increase turbine efficiency at low loads, one or more hand valves are used (see Figure 22.3).

Multiple throttle valves are selected such that the pressure drop across any valve will be minimum.

Problems encountered with throttle valves include valve seat breakage. Broken valve seat parts can block the nozzle ring reducing first stage pressure to the turbine. All multi-stage turbines are provided with a pressure gauge that monitors first stage pressure. Noting the baseline first stage pressure, for a given load, and trending this parameter will provide valuable information concerning pressure drop in the throttle valves and steam passages to the first stage of the turbine.

Nozzle ring

The function of the nozzle ring is shown in Figure 22.4.

In single valve turbines, all the steam is admitted to one dedicated set of nozzles. If hand valves are provided, an additional set of nozzles is furnished for each hand valve. In multi-valve turbines, each valve has a specific passage to a set of nozzles.

The arc of admission is defined as the degrees of a circle that the first stage nozzles (or control nozzle) occupies. For single stage turbines, a typical value would be 30°. Multi-valve turbines can have admissions of 180° or full admission (360°) depending on the steam flow requirements.

The nozzle ring, often called the control nozzle, is the only nozzle in a turbine that is not designed for full admission. All other nozzle areas are designed on the assumption that the steam will enter the entire arc.

There have been cases where the axial distance between the control stage and the second stage was not sufficient to allow full arc admission of steam to the second stage. This resulted in higher nozzle velocities than anticipated and lower turbine efficiency.

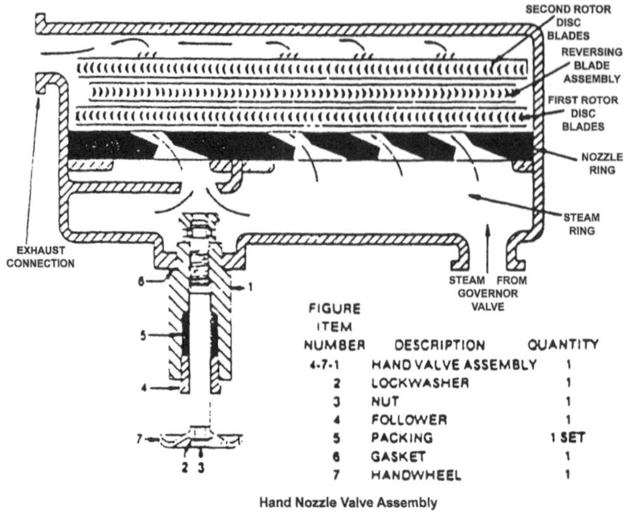

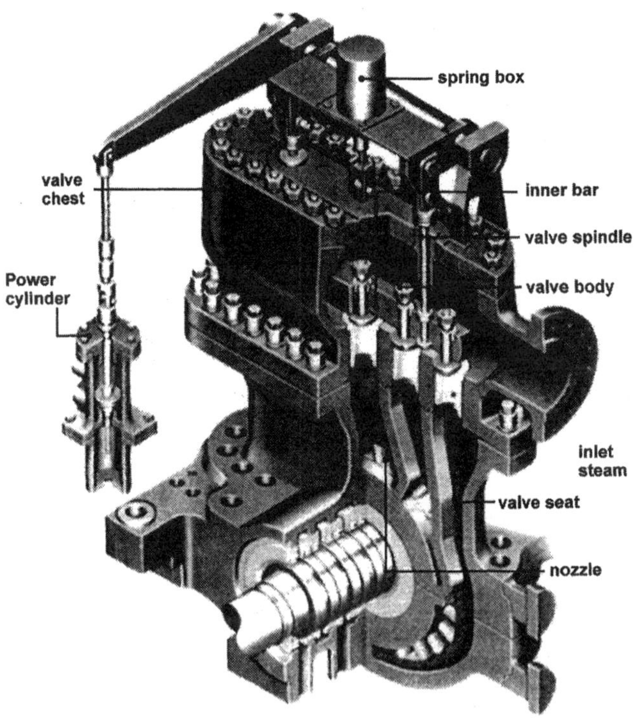

Function: Distributes the steam from the throttle valve to the first stage (control stage) with a minimum of losses

Figure 22.4 Nozzle ring. Top: Single valve turbine with hand valves (Courtesy of Elliott Co.). Bottom: Multi-valve turbine (Courtesy of M.H.I.)

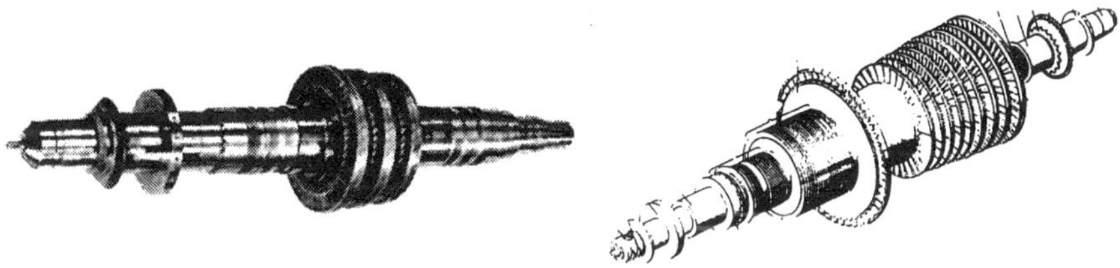

Function: Secures all blades and converts thermal energy from the expansion vapor to mechanical energy to drive the load

Figure 22.5 Rotor. Left: Impulse turbine rotor (Courtesy of Elliott Co.). Right: Reaction turbine rotor (Courtesy of Siemens)

Rotor

An example of an impulse bladed and reaction bladed rotor and their function are presented in Figure 22.5.

Unlike turbo-compressor rotors, most turbine rotors are integral design. That is, the discs which hold the blades are integral with the shaft. Integral designs are preferred because the disc shaft fit will not be affected during transient conditions (start-up and shutdown). Built up rotors are manufactured for single stage and multi-stage rotors that are used for general purpose (spared) applications. Typical applications would be pump and fan drives. One note of caution, in general purpose automatic start applications, vendors must be advised that the turbine will rapidly start so that the interference (shrink) fit between the shaft and disc will be properly designed. This action will avoid any looseness in the interference fit during transient conditions.

Blade design

The three types of expansion turbine blading have already been discussed and are shown in Figure 22.6.

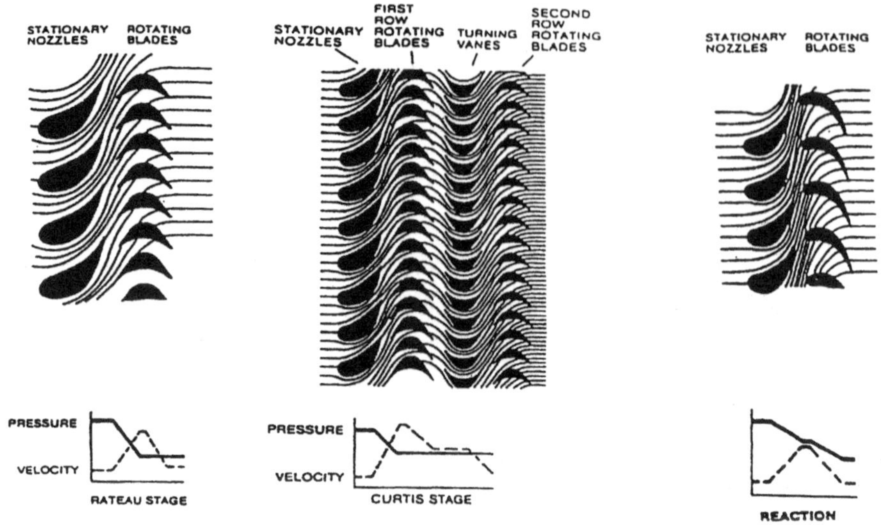

Figure 22.6 Steam turbine blade types (Reprinted with permission of GPSA)

The advantages and disadvantages of impulse and reaction blading are shown in Figure 22.7 and Figure 22.8 respectively.

Impulse blade

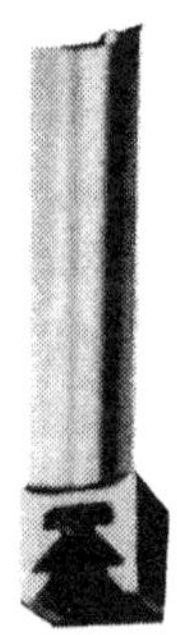

Advantages

- can recover greater energy per stage
- can withstand higher stresses
- do not generate significant thrust forces

Disadvantages

- lower blade efficiency
- larger stage spacing required

Figure 22.7 Blading reaction vs impulse advantages/disadvantages

The different staging alternatives are described in Figure 22.9.

Shown is a multi-stage turbine with a Curtis control stage (first stage). As previously mentioned, this arrangement is used on older design multi-stage turbines and lowers the overall efficiency of the turbine. Also shown is a multi-stage Rateau bladed rotor. Most turbine rotors today use one row of blading for each turbine stage.

Reaction blade

Advantages

- higher blade efficiency
- smaller stage spacing

Disadvantages

- recovers less energy per stage
- higher stresses than impulse for some vapor load
- generates a thrust force
- blade tip seal losses

Figure 22.8 Blading (continued)

The expansion turbine blading design objective is to design a given nozzle row and blade row set for optimum gas velocity to obtain maximum stage efficiency. Figure 22.10 presents the important facts regarding blade design.

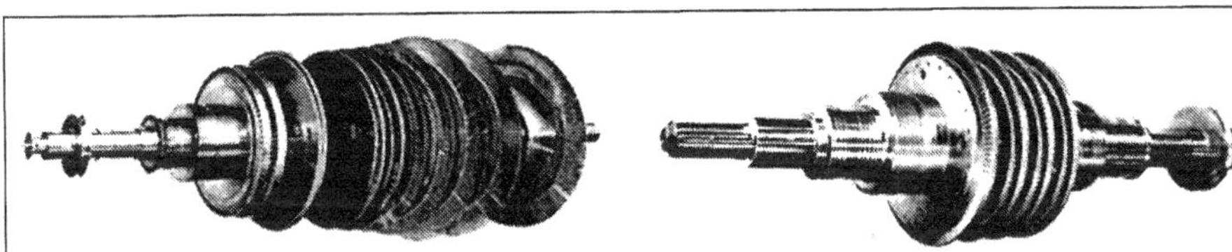

Curtis

- two rows of blades
- used on first (control stage) (older designs)
- reduced blade loading

Rateau

- one row of blades
- used on stages other than control stage
- used on control stage in place of Curtis stage (newer designs)

Figure 22.9 Stage types (Courtesy of Elliott Co.)

Blade and vapor velocities efficiency considerations

- Maximum energy is extracted from a vapor when the blade tip velocity is 50–75% of the vapor velocity
- Note this value varies with blade design

Figure 22.10 Blade and vapor velocities efficiency considerations

Figures 22.11 and 22.12 present the information required to optimize expansion turbine stage efficiency.

Blade efficiency

- Turbine blade efficiency is a function of:
  - **Blade tip speed 'U'**
    U = RPM x nozzle pitch diameter
  - **Theoretical vapor velocity 'C'**
    C = constant x energy/unit mass
- Actual blade tests have shown there is an optimum value of u/c for each blade type
- 'U' is limited by mechanical design stresses (blade shape, blade root, material, etc.)

Figure 22.11 Blade efficiency

Every turbine vendor has a blade row efficiency curve plotted as a function of blade tip speed to theoretical steam velocity ratio for each blade row design. As can be seen in Figure 22.12, there exists an

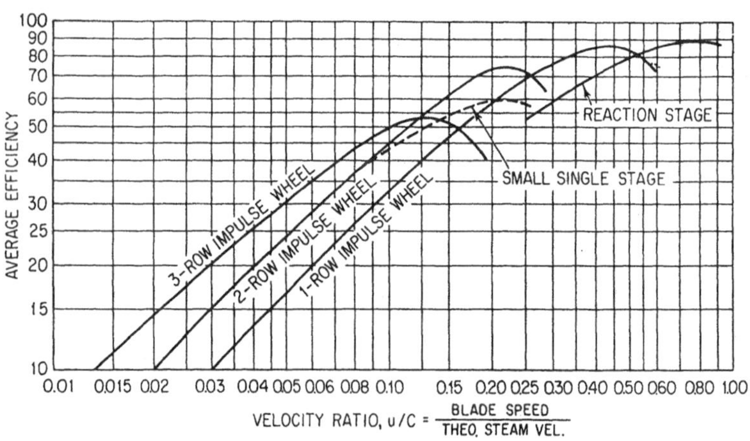

Average efficiency of turbine stages.

U = BLADE TIP VELOCITY - FT/SEC

$$= \frac{\text{BLADE DIAMETER X RPM}}{229}$$

C = THEORETICAL STEAM VELOCITY - FT/SEC

$$C = 223.8 \times \sqrt{\text{BTU PER STAGE}}$$

Figure 22.12 Blade efficiency vs velocity ratio (u/c) (Courtesy of IMO Industries)

optimum U/C ratio for each blade row design. Blade tip speed (blade diameter × rotor speed) is varied within the mechanical stress limits of the blading and nozzle area is adjusted to attain the maximum blade row efficiency for the rated steam flow conditions. Study of Figure 22.12 shows that once the blade is designed, turbine efficiency will usually decrease for a change in stream flow (turbine load).

Refer to Figure 22.13.

The number of stages required

- Based on the optimum value of u/c and mechanical limits, there is a maximum value of energy that each blade row can extract from a vapor.

Figure 22.13 The number of stages required

The number of expansion turbine stages is determined by proportioning the total theoretical enthalpy available to attain the optimum turbine efficiency. For a given blade speed, based on mechanical limitations, the theoretical steam velocity 'C' can be obtained from Figure 22.12 for a known stage design. Once 'C' is known, the theoretical BTU's per stage can be obtained. Dividing the total theoretical enthalpy ($\Delta H_{ISENTROPIC}$) by the theoretical BTU/per stage will yield the number of stages:

$$\text{Number of stages} = \frac{\Delta H_{\text{ISENTROPIC}}}{\Delta H_{\text{ISENTROPIC PER STAGE}}}$$

The actual BTU's per stage is obtained by multiplying the theoretical enthalpy by the actual predicted stage efficiency.

Blade root design must allow for acceptable blade root stresses at 110% of turbine trip speed. Turbine trip speed is 110% of turbine maximum continuous speed which is 105% of rated turbine speed. Figure 22.14 describes the factors that contribute to blade stress.

Each blade row has many modes of natural frequency. Figures 22.15 and 22.16 describe the important facts concerning natural frequency.

- blade stress is produced by:
  - steady state stresses
    - centrifugal stress
    - blade bending stress
  - alternating stresses
    - blade vibratory stresses
- many different types of blade attachments (roots) are used

Figure 22.14 Blade root types

Blade Frequency

- Each blade has many modes of natural frequency
- If a natural frequency is excited by any external frequency, a significant amount of alternating stress will be transferred to the blade which can be destructive.
- Every blade row is analyzed and tested to determine all of its natural frequencies.
- Every effort is made to eliminate all excitation frequencies in the operating range.

Figure 22.15 Blade frequency

Each blade row must be analyzed for blade interference. IE, to assure that blade natural frequencies will not be excited so as to cause blade failure. This is achieved in two steps. First, each blade row is analyzed using a Campbell Diagram which will show any possible interferences. A margin of ± 10% excitation outside the turbine speed range is required to assure interference does not exist. If this requirement cannot be met and blade interference exists, a blade loading diagram

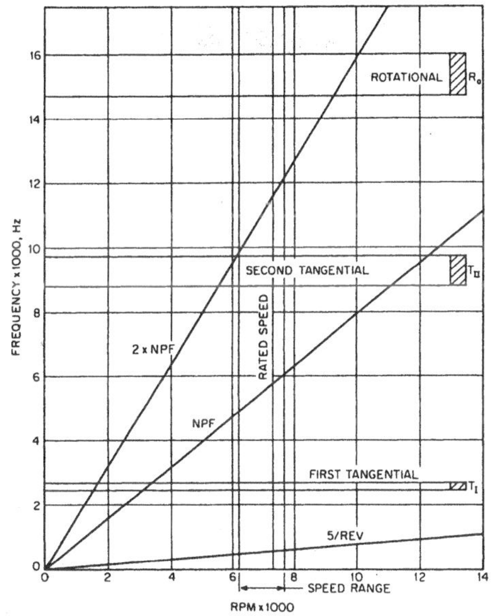

Figure 22.16 The blade interference diagram (Courtesy of IMO Industries)

must be developed to determine if the blades are excessively loaded. Figure 22.17 describes the facts concerning blade loading and Figure 22.18 presents a modified Goodman Diagram for a blade row with interference.

Blade Loading

To determine if blades are excessively loaded, a blade loading diagram is plotted (modified Goodman diagram) for the operating temperature, material and steady state stresses.

- The resulting allowable alternating stress must not exceed a known % of maximum allowable value

Figure 22.17 Blade loading

As long as the intersection of the blade steady state and alternating stresses are within the vendors' limit, continuous operation of the blade will not result in failure.

The correlation between predicted Campbell and Goodman values and actual values has been generally good. However, there have been

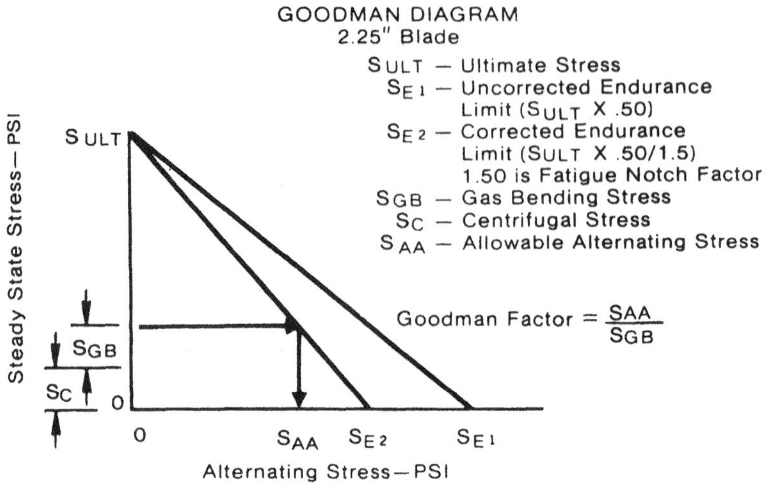

Figure 22.18 Modified Goodman diagram (Courtesy of Elliott Co.)

instances where blades predicted to be in a safe operating range have failed. It should be noted that the traditional Campbell and Goodman approach does not consider the effect of natural frequency of the blade discs. Some turbine vendors now include these items in their analysis and have reportedly achieved good correlation.

Nozzle/Diaphragm

The function of nozzles is defined in Figure 22.19.

Shown are examples of reaction and impulse nozzles. Reaction nozzles are inserted directly into an inner casing shell known as a blade carrier. Impulse nozzles are inserted into a carrier plate known as a diaphragm. Both blade carriers and diaphragms are horizontally split. When operating condensing turbines, care must be taken to assure the vendors' maximum allowable % moisture is not exceeded in the exhaust blading. Operating for continuous periods can severely erode blading, blade carriers and diaphragms and cause turbine failure.

Shaft end seals

Facts concerning shaft end seals and sealing systems for critical service non-condensing and condensing turbines are shown in Figures 22.20, 22.21, 22.22 and 22.23.

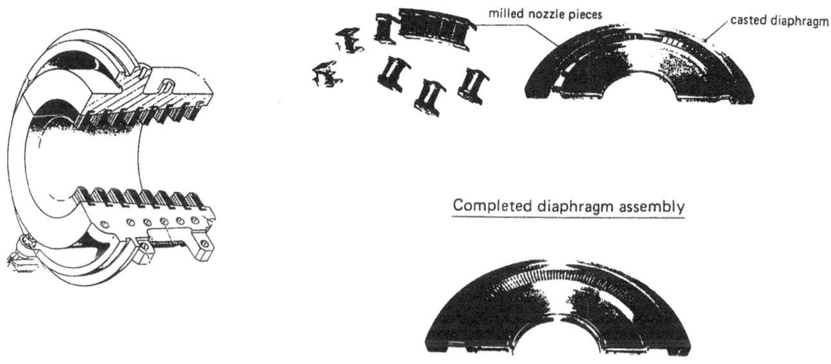

FUNCTION: THE NOZZLES EXPAND THE VAPOR AND DIRECT IT TO THE SUCCEEDING STAGE AT A GAS ANGLE (INCIDENCE) THAT RESULTS IN A MINIMUM EFFICIENCY LOSS

Figure 22.19 Nozzles/blade carriers/Diaphragms Left: Reaction nozzle and blade carrier (Courtesy of Siemens) Right: Impulse nozzle and diaphram (Courtesy of M.H.I.)

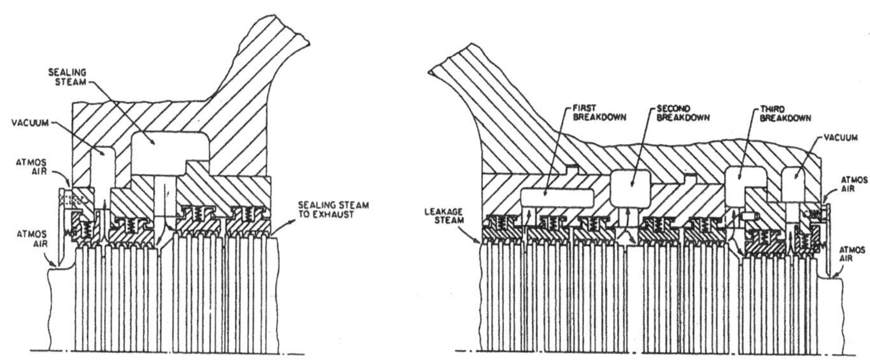

Function: The labyrinth type shaft end seals safely direct the expansion vapor to a defined location and draw in a buffer gas (air) to prevent expansion vapor contact with the bearings

Figure 22.20 Expansion turbine shaft end seals (Special purpose (unspared) turbines) Left: Typical exhaust end seal. Right: Typical inlet end seal (Courtesy of IMO Industries)

Steam turbine shaft sealing systems

- Function: prevent steam from escaping to atmosphere along the shaft and entering the bearing housing
- Special purpose turbines usually employ a low vacuum (5–10' H_2O vacuum) to buffer atmospheric end labyrinth with air
- General purpose turbines usually do not employ a vacuum system and do not totally prevent moisture from entering bearing housing

Figure 22.21 Steam turbine shaft sealing systems

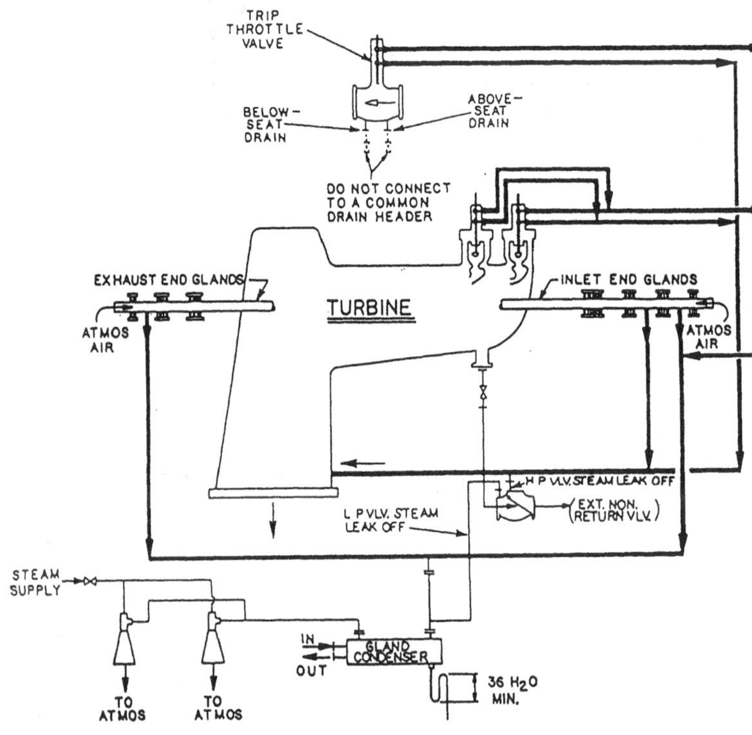

Figure 22.22 Gland seals and drains: noncondensing automatic-extraction turbine (Courtesy of IMO Industries)

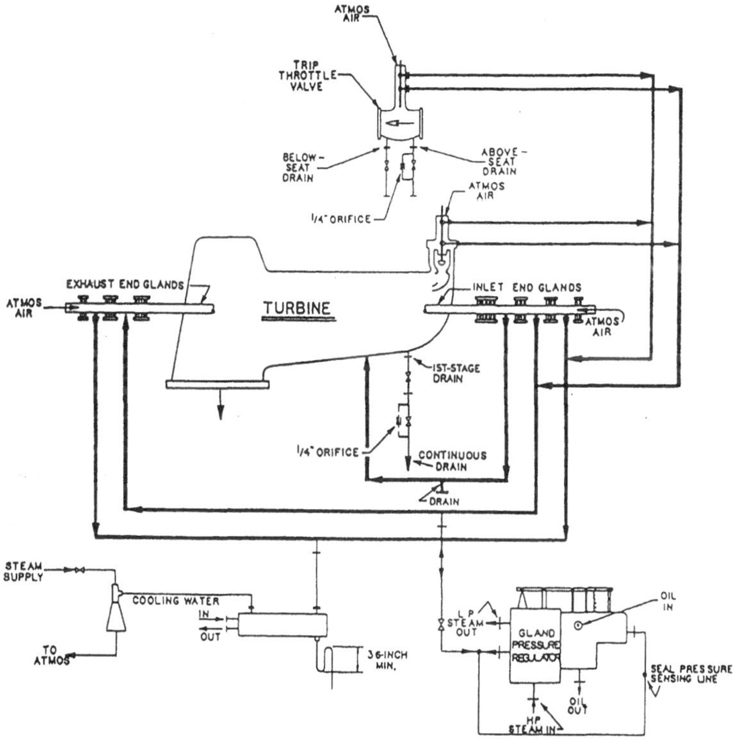

Figure 22.23 Grand seals and drains: condensing turbine (Courtesy of IMO Industries)

The key to successful shaft end seal operation is to continuously maintain a slight (5–10" H_2O) vacuum in the last chamber of the seal. By maintaining a vacuum at this location, atmospheric air will enter the seal thus assuring that steam (moisture) will not enter the bearing bracket and contaminate the oil system.

Condition monitoring of the system vacuum is essential to maintaining moisture free lubrication oil. Many a turbine bearing has failed because of poor seal system preventive and predictive maintenance practices. 'THINK SYSTEM' and check all components of the seal system frequently.

Bearings

The last components to be covered in this chapter are the turbine radial and thrust bearings. The function and types of bearings used are the same as for a turbo-compressor (please refer to the radial and thrust chapters).

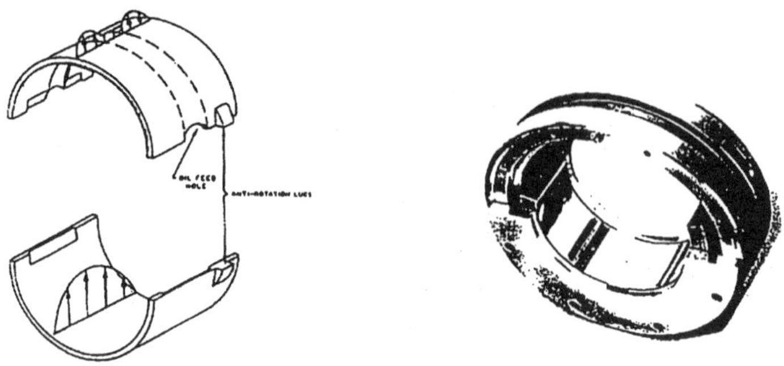

FUNCTION: JOURNAL BEARINGS REGARDLESS OF TYPE CONTINUOUSLY SUPPORT THE ROTOR WITH AN OIL FILM OF LESS THAN ONE THOUSANDTH OF AND INCH.

Figure 22.24 Journal bearings. Left: Sleeve bearing. Right: Tilt-pad bearing

Figures 22.24 and 22.25 show the types of radial (journal) and thrust bearings used.

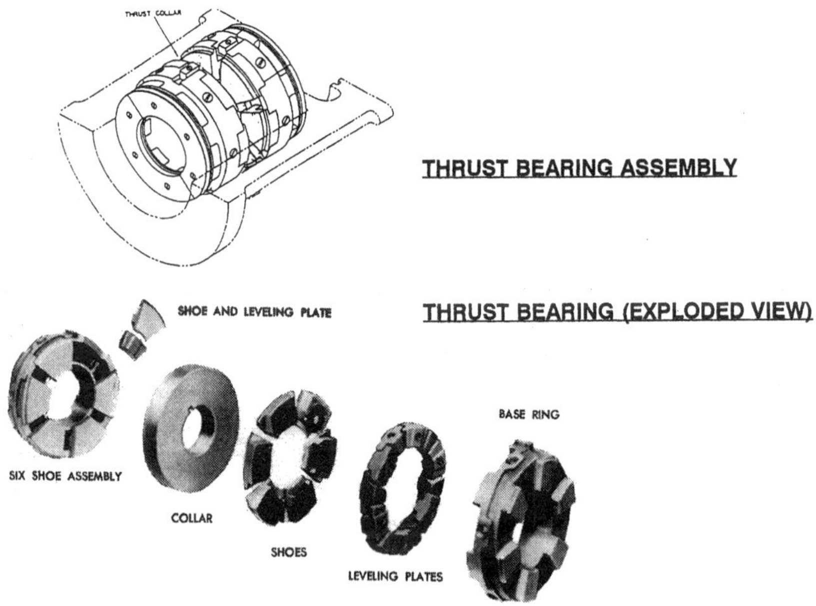

FUNCTION: THE THRUST BEARING POSITIONS THE ROTOR AXIALLY IN THE CASING AND TAKES THE LOAD GENERATED BY THE BLADES AND BALANCE DRUM*

* NOTE: ONLY INSTALLED ON REACTION TURBINES

Figure 22.25 Thrust bearing assembly (Courtesy of Kingsbury, Inc.)

23 Steam turbine inlet steam regulation

- Introduction
- Requirements
- Single valve design and admission path
- Multi-valve design and admission path

Introduction

In this section we will discuss the method of steam regulation as it applies only to the inlet throttle valves and their lift mechanisms. We will discuss the total control system in the next chapter.

We will begin this section by discussing the requirements of any steam turbine inlet regulation system. We will then discuss single valve design used in low horsepower general purpose steam turbines. These applications are usually pump drives, fan drives and low horsepower compressor drives. We will review single valve design and discuss its limitations. We will also cover hand valves and review their proper usage. Single valve modulation will also be discussed and methods to increase mechanical linkage reliability will be covered.

Multi-valve design will also be covered in this section. We will discuss valve design, how valves are set up and timed. The admission path will be discussed in detail and design considerations will be covered.

Finally, the modulation of multiple valve designs will be reviewed dealing with various lift mechanisms and servomotors.

Requirements

The requirements for an efficient expansion turbine admission path are shown in Figure 23.1.

Steam turbine inlet valve and admission path requirement

- Lowest possible energy loss (pressure drop)
- Non-oscillating
- Non-binding (minimum actuator friction)
- Directs steam to rotor with a minimum of external transmitted force

Figure 23.1 Steam turbine inlet valve and admission path requirement

Regardless of the type or size of expansion turbine, these requirements are the same.

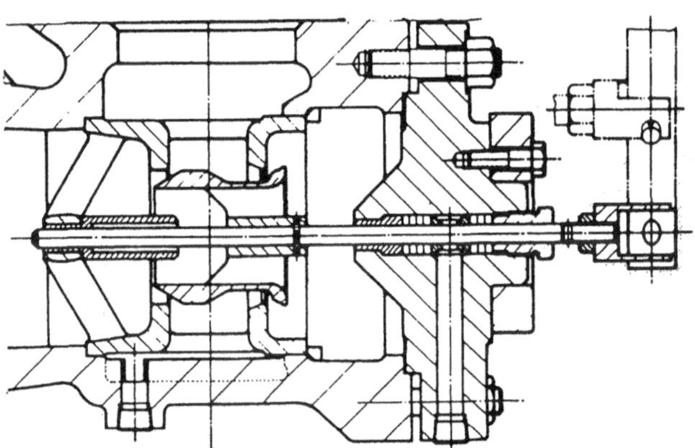

FEATURES

- DOUBLE SEATED
- PRESSURE BALANCED
- REQUIRES MINIMUM EXTERNAL FORCE FOR ACTUATION
- USUALLY ACTUATED BY MECHANICAL LINKAGE (NORMAL ACTUATOR TORQUE 10-25 LBS - FT)

Figure 23.2 Single valve

Single valve design and admission path

The features noted are essential for reliable operation. Since single throttle valves are not hydraulically actuated, the force available for valve actuation is limited. Therefore, the valve must be pressure balanced and designed for minimum actuation force.

From a performance point of view, a single throttle valve is inefficient at reduced flow rates (turbine load) as shown in Figure 23.3.

As a result, hand valves are normally used to accommodate off design conditions thus keeping the single throttle valve close to full open under varying steam flows. Please refer to Figure 23.4 which shows a hand valve installed in a single valve turbine.

In this turbine, as an example, the throttle valve might be designed for 75% of the rated steam flow (turbine load). Note that for a set of steam conditions and turbine efficiency,

Mass Flow (Lb/Hr) $\propto$ Horsepower

$$\text{Since, Horsepower} = \frac{\Delta H \text{ X Mass Flow}}{\text{Constant X Efficiency}}$$

Where ΔH is the ideal (Isentropic) energy available in the steam.

Therefore below 75% of load, the hand valve would be closed. Above 75% of load, the hand valve would be open. If two hand valves were used, the throttle valve may be designed for 50% of load. Below 50% of load, both hand valves are closed. Between 50% and 75% load one hand

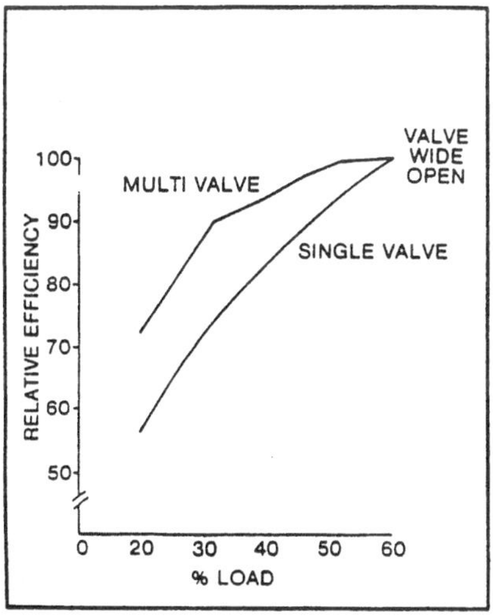

Figure 23.3 The effect of multi-valve vs single valve turbines on relative efficiency (Reprinted by permission of GPSA)

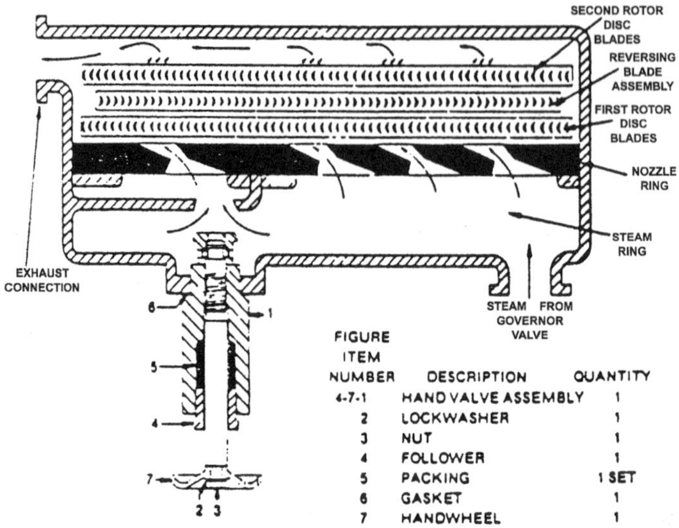

FIGURE ITEM NUMBER	DESCRIPTION	QUANTITY
4-7-1	HAND VALVE ASSEMBLY	1
2	LOCKWASHER	1
3	NUT	1
4	FOLLOWER	1
5	PACKING	1 SET
6	GASKET	1
7	HANDWHEEL	1

Hand Nozzle Valve Assembly

FEATURES:

- NOZZLE BLOCK IN LOWER HALF OF CASE
- ONE OR MORE HAND VALVES FOR EFFICIENCY AND INCREASED LOAD
- HAND VALVES NOT MODULATING!
  (FULL OPEN OR FULL CLOSED)

Figure 23.4 Single valve turbine admission path (Courtesy of Elliott Co.)

valve is open and above 75% load, both hand valves are open. Most single valve turbines utilize two (2) or more hand valves. Hand valves are normally manually actuated. As noted in Figure 23.4, hand valves are not modulating. They are either full open or closed.

Throttling a hand valve will cause damage to the valve and valve seat.

In some single valve applications, such as generator drives which normally operate at full load, pneumatic or hydraulic operated hand valves open or close at off design conditions as required by the electronic control system.

While hand valves increase single valve turbine efficiency at off design conditions, operators must constantly be aware of their position (open or closed) and of the power requirements of a turbine. If one or more hand valves is closed and the inlet steam pressure to a turbine decreases, less energy per pound of steam will be available and more steam flow will be required. However, with hand valves closed, the turbine will not be able to admit the required steam flow and output power and turbine speed will decrease. Readers are cautioned to be sure turbine hand valves are kept in the open position in critical applications where steam

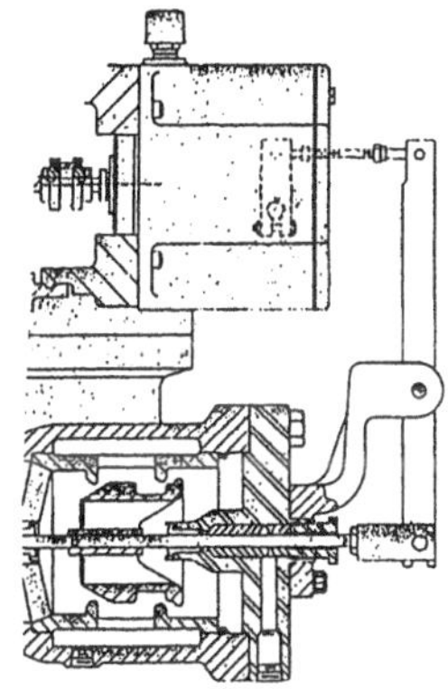

Figure 23.5 Single valve actuation reliability considerations

conditions can vary. Example applications are: lube and seal oil main pumps, boiler feed pumps and critical process pumps.

Single valve turbines less than 500 horsepower are usually actuated directly by the governor output and are not supplied with an actuator amplifier (Servomotor). Since the output power (torque) of the governor is limited, preventive maintenance is essential in maintaining single throttle valve system reliability. Figure 23.5 shows a single valve and its actuation system.

Reducing friction in all linkage bushings is the main objective. Bushing friction will cause valve instability (hunting) and reduce control system response. The use of a high temperature lubricant (molycote or equal) is recommended. On line inspection of linkages involve checking linkage movement during turbine operation. When linkages and bushings are functioning properly, a slight movement can be felt. If linkages are not moving, bushing lubrication, bushing or valve stem packing should be inspected at the first opportunity.

Multi-valve design and admission path

When steam flow and power requirements are large and will change frequently, multi-valve throttling arrangements are used. Figure 23.6 shows a cutaway view of a multi-valve nozzle block arrangement with 180° arc of admission.

The components of a multi-valve rack assembly are shown in Figure 23.7.

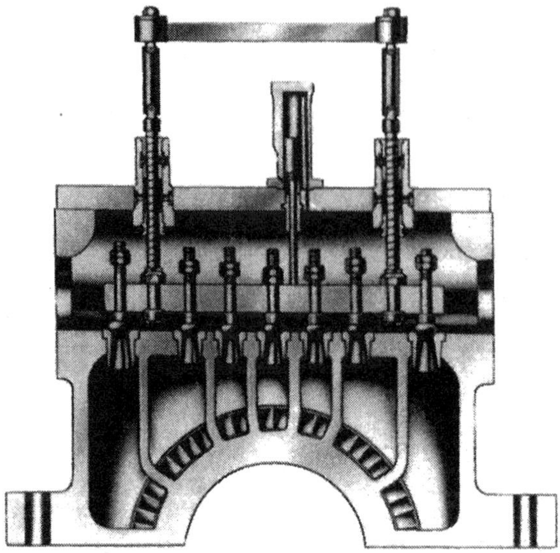

PURPOSE: USED ON SPECIAL PURPOSE TURBINES TO MINIMIZE LOSSES (INCREASED EFFICIENCY) AT "OFF DESIGN" CONDITIONS

FUNCTION: VALVES ARE POSITIONED ("TIMED") TO OPEN AS REQUIRED THUS MINIMIZING THROTTLING

Figure 23.6 Multi valve (Courtesy of Westinghouse Canada)

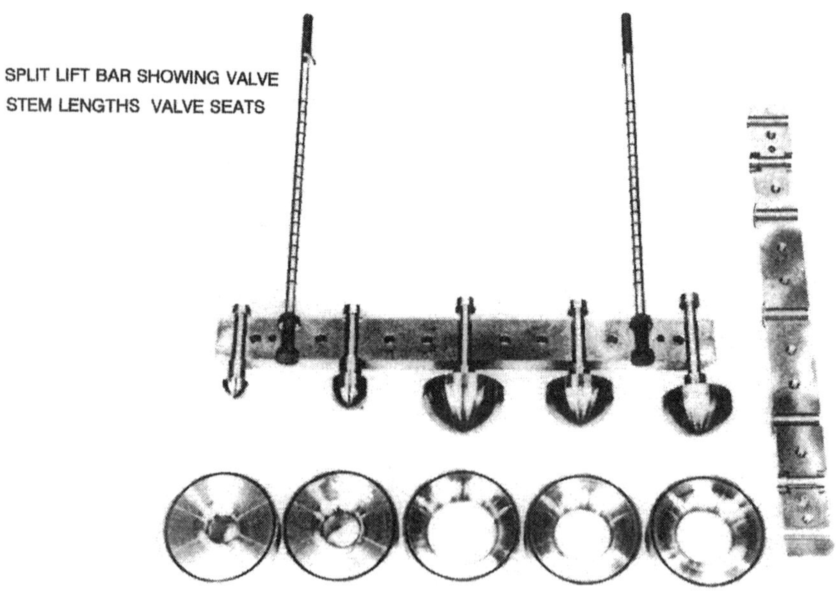

FUNCTION: THE VALVE RACK ACCURATELY POSITIONS VALVE STEMS ALLOWING FREE MOVEMENT VERTICALLY. STEM TRAVEL IS ADJUSTED BY DOUBLE NUTS.

Figure 23.7 The valve rack (Courtesy of Dresser (Terry))

Each valve is selected for a specific maximum flow range. The valve stems, as shown in Figure 23.7, are adjusted such that the valve with the shortest stem will lift first. From the left of the picture the sequence of valve opening will be: 1, 2, 3, 4, 5.

Referring back to Figure 23.6, observe that each valve supplies steam to a dedicated passage and set of nozzles. Valve location is designed to assure that steam flow at all operating conditions will be balanced. "Balanced" flow means that the valve opening sequence in Figure 23.6 is arranged so that the stem flow will be equally distributed as much as possible to avoid large forces on the rotor that can cause rotor vibratory instability (gas whirl). In Figure 23.6, the opening sequence from left to right will be: 4, 5, 3, 6, 2, 7, 1. This arrangement will result in "balanced" steam flow distribution under all operating conditions.

Figure 23.8 is a valve timing plot for a five (5) valve multi-valve steam turbine.

Valves will be adjusted to open at a specific value of valve bar lift (x-axis) to accommodate the required steam flow (y-axis) while maintaining minimum valve pressure drop.

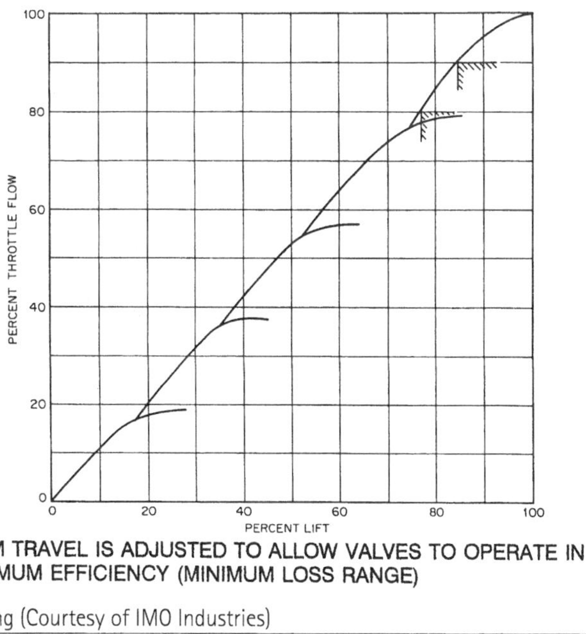

Figure 23.8 Valve timing (Courtesy of IMO Industries)

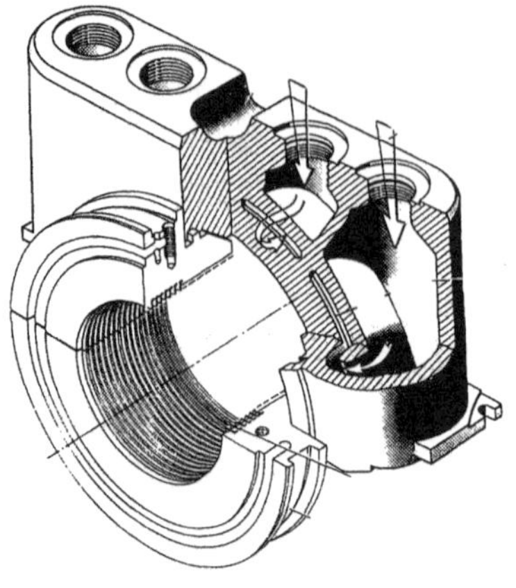

FUNCTION: TO DIRECT STEAM TO FIRST STAGE NOZZLE WITH:

- MINIMUM ENERGY (PRESSURE) LOSS
- MINIMUM OF EXTERNAL FORCES ON THE ROTOR

Figure 23.9 Multi-valve admission path (Courtesy of Siemens)

As an exercise, note the % of valve bar lift where each valve begins to open.

Valve	% of valve bar lift
1	________________
2	________________
3	________________
4	________________
5	________________

If valve timing is not correct, steam turbine efficiency and power output will be affected. Figure 23.9 shows a 3-dimensional view of a multi-valve admission path.

Figure 23.10 presents facts concerning admission to the turbine first stage (control stage).

Admission

- Admission is usually expressed in degrees of admission:

  180° = Half arc admission (usually in top half of case)

  360° = Full arc admission

Note: the first row of blades takes admission into consideration. Subsequent rows usually *assume* full admission.

Figure 23.10 Admission

Multi-valve arrangements, unlike single valve arrangements, require large lift forces. Figure 23.11 states important facts concerning multi-valve lift mechanisms. A typical bar lift system is shown in Figure 23.12.

Multi-valve lift mechanism

- Large actuation forces are required because:
  - Valve lift mechanism is large and heavy
  - Valves cannot be pressure balanced
- Valve lift is accomplished using:
  - Bar lift
  - Cam lift
    - Integral
    - Separate

Figure 23.11 Multi-valve lift mechanism

A power cylinder, hydraulically or pneumatically operated is required to overcome steam forces and lift bar assembly weight to move the lift bar as required. Note that valve timing modifications require disassembly of the valve chest.

In Figure 23.13 a cam lift multi-valve assembly is shown.

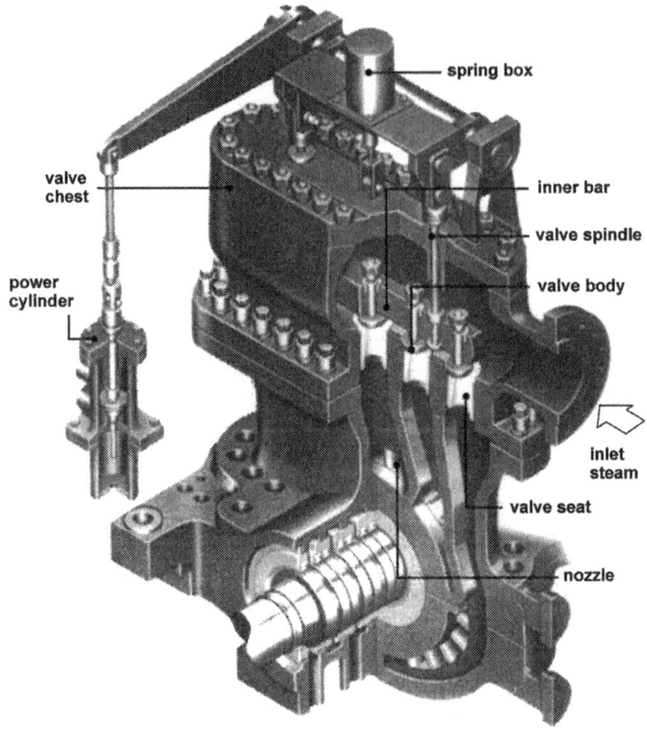

Figure 23.12 Bar lift (Courtesy of M.H.I.)

Instead of using a lift bar and valve steam adjustment to control valve opening time, this arrangement utilizes individual cams to control valve opening. Cam lifts are externally adjustable and therefore do not require valve chest disassembly for valve timing.

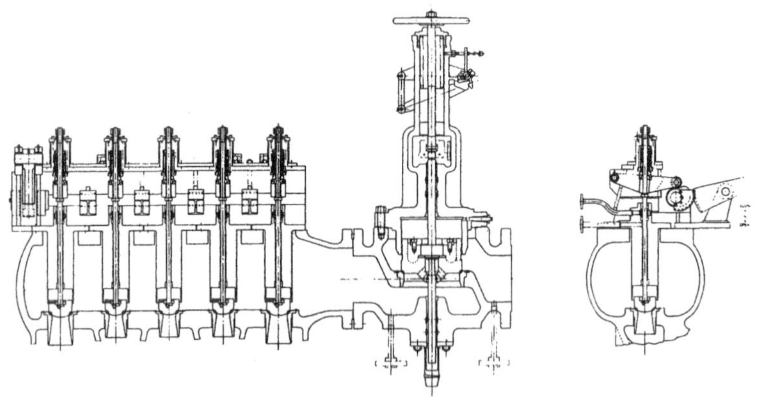

Figure 23.13 Cam lift: five valve individual cam lift mechanism (Courtesy of IMO Industries)

- THE SERVOMOTOR IS A FORCE AMPLIFIER THAT INCREASES THE GOVERNOR OUTPUT SIGNAL TO PROVIDE SUFFICIENT FORCE TO REGULATE THE THROTTLE VALVES.

- THE SERVOMOTOR CONSISTS OF:

    - RELAY VALVE

    - POWER PISTON

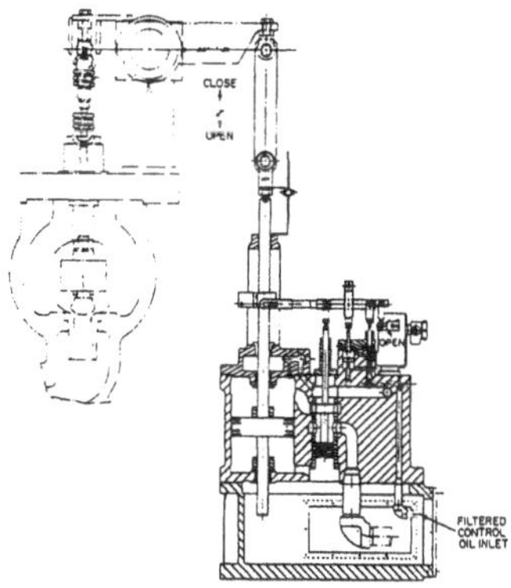

Figure 23.14 The Servomotor (Courtesy of IMO Industries)

Regardless of the type of multi-valve arrangement, a force amplifier is required to accurately open valves upon command from the control (governor) system. Figure 23.14 shows an assembly drawing and defines the function of a servomotor.

Note that external linkages and bushings, which need preventive maintenance, like single valve turbines are also required for multi-valve turbines.

24 Steam turbine control/protection systems

- Introduction
- Total train control and protection objectives
- Control
- Protection

Introduction

We will now discuss steam turbine driven total train control and protection.

The total train control and protection objectives are as follows:

- Meet driven equipment control requirements
- Meet above objectives when operating in series or parallel
- Continuously protect the entire train from damage

We will cover these aspects by examining a simple process system and explaining the total train control and protection objectives.

The control of a steam turbine driven train will be presented by first overviewing a simple control system functionally and comparing it to that of an automotive cruise control system. This approach will simplify the complex nature of the subject matter and allow the student to relate to common everyday terms.

The function of the major components of a governoring system will be defined and the various types will be presented, namely, mechanical governoring systems, mechanical hydraulic governing systems and electro-hydraulic governoring systems. Each system will be examined in detail and the function of the major components of each system will be discussed.

Extraction control systems will be presented for both mechanical hydraulic and electrical hydraulic systems. Also the application of the types of governor systems will be presented along with the advantages and disadvantages of each system.

Steam turbine protection systems will be covered as well in this section. We will begin by presenting a protection system overview and discussing the major objectives of any steam turbine protection system. We will then discuss the various component functions of each system and the types of different protection systems. Finally the applications of systems will be discussed and emphasis will be placed on protection systems philosophy both domestically and overseas.

Total train control and protection objectives

Figure 24.1 presents the total train control and protection objectives.

Total train control/protection objectives

- Meet driven equipment control requirements
    - Compressor – pressure or flow
    - Pump – pressure, flow or level
    - Generator – load
- Meet above objectives when in series or parallel with other trains
- Continuously protect entire train from damage due to:
    - Overspeed
    - Loss of auxiliaries
    - Component mechanical failure
    - Driven equipment upsets (surge, minimum flow, high load, etc.)

Figure 24.1 Total train control/Protection objectives

Regardless of the type of driven equipment, the objective of the control and protection system is to assure that the required quantity of product or generated power is continuously supplied maintaining the highest possible total train efficiency and reliability.

Figure 24.2 presents a process diagram for a steam turbine driven compressor train.

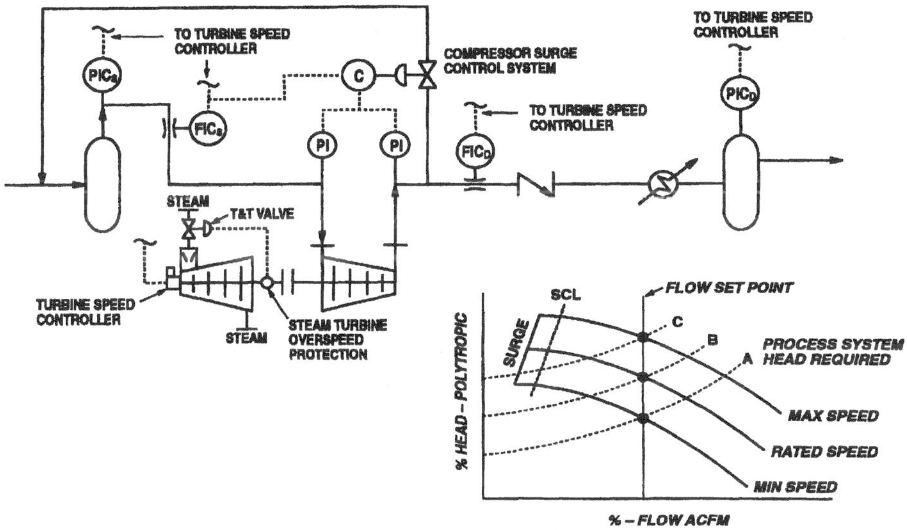

Figure 24.2 Total train control

Depending on the selected process variable and location, any PIC or FIC will continuously monitor the selected process variable sending its signal as an input signal to the turbine speed controller. For this example, assume the set point is a flow controller located in the discharge line of the turbo-compressor (FIC_D). The process system head (energy) requirements A, B, C are shown. These different energy requirements can represent either increased pressure ratio requirements (suction strainer blockage exchanger ΔP etc.) and/or gas density changes (M.W. P or T). As the process head (energy) requirements increase from A to B to C, the input flow variable will decrease if the turbo-compressor speed does not change. However, as soon as the monitored process variable, $FIC_D \neq$ flow set point, the turbine speed controller output will open the turbine inlet throttle valves to provide more turbine power to increase the head (energy) produced by the compressor to meet the additional process system head requirements and therefore maintain the desired throughput.

Adjusting the speed of the driven equipment is the most efficient control method since there are no control valves required in the system. Therefore only the exact value of head required by the process system is produced by the turbo-compressor.

Also noted in Figure 24.2 are the two major protection systems for the compressor and steam turbine, the surge protection and turbine overspeed protection systems. The surge system has been previously discussed, the turbine overspeed system will be discussed later in this chapter. In addition to the two major protection systems mentioned above, other typical protection systems for a rotating equipment train are:

- Shaft vibration
- Bearing bracket vibration
- Axial thrust displacement
- Bearing temperature
- Process gas temperature
- Lube oil pressure
- Seal oil ΔP
- Suction drum high liquid level (compressors)

Control

A turbine governor is a speed controller. Important facts concerning expansion turbine governors are shown in Figure 24.3.

Control

- The governor is the heart of the control system
- The governor in simple terms compares input signal(s) to a set point and sends an output signal to achieve the desired set point.
- An example of a simple governor system is "cruise control" in a car

Figure 24.3 Control

Regardless of type, all controllers have three identical parameters:

- Input
- Set point
- Output

Some familiar controllers are:

- Pressure
- Flow
- Level

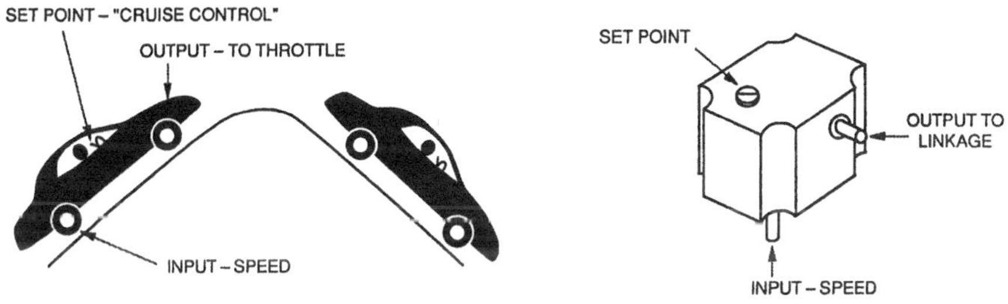

Figure 24.4 A control system analogy. Left: Cruise control. Right: Steam turbine governor. In both cases, load change is inversely related to speed change. The controller compares input to set point and changes output appropriately.

- Temperature
- Surge
- Speed

As an example, refer to Figure 24.4 which is a speed controller that may be familiar.

In Figure 24.4, we compare an auto "Cruise Control" to a steam turbine governor (typical single stage mechanical/hydraulic). Both are speed controllers and have an:

- Input
- Set point
- Output

The table below shows a comparison of these parameters.

Parameter	C.C. (Cruise control)	T.G. (Turbine governor)
Input	Actual speed from speedometer	Actual speed from speed pick-up
Set point	Selected by driver	Selected by operator
Output	To fuel control system	To steam throttle valve

Figure 24.5 is a schematic of a steam turbine governor system.

Note that the set point can either be a manual set point, similar to a driver setting a "speed" in a cruise control system or a process variable. Examples of process variable set points would be:

- Pressure
- Flow
- Level (pump applications)

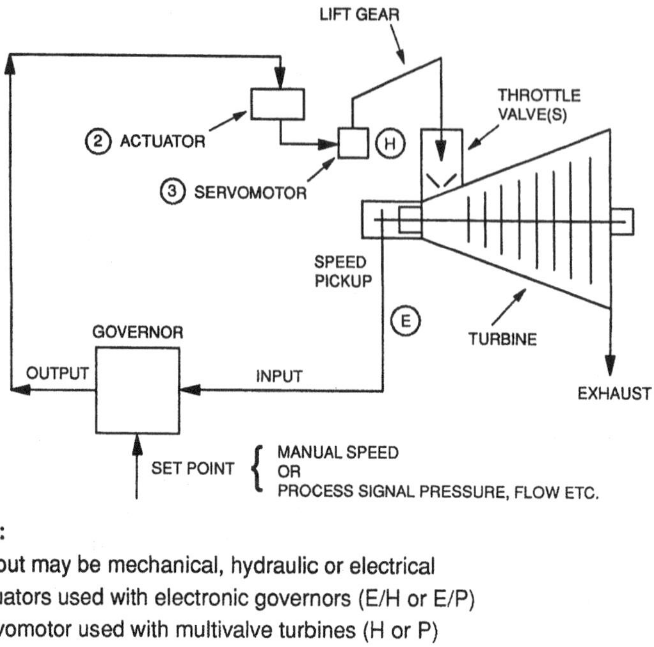

Figure 24.5 Steam turbine control (Courtesy of M.E. Crane, Consultant)

There are many controller designs. Historically, the first controllers were entirely mechanical. An example of a mechanical speed controller is shown in Figure 24.6.

Commonly called "Fly Ball Governors", the **input** shaft from the driver would rotate the weights through a gear set. As the weights rotated, centrifugal force would move the weights outward, compressing the spring and thus moving the **output** linkage. The tension on the spring from the speed changer **(set point)** would control the speed as the equilibrium point of the input and set point values.

Many mechanical governors are still in use today on older, small single valve steam turbines. The mechanical governor output force is limited and lead to the development of the mechanical hydraulic governor pictured in Figure 24.7.

The mechanical-hydraulic governor uses the same mechanical mechanism to determine the output signal. However, the output shaft moves a pilot valve which allows hydraulic fluid (usually oil) to provide the output signal to the throttle valve(s). The common Woodward "T.G." and "P.G." governors are examples of mechanical/hydraulic governors. These governors have internal positive displacement oil pumps driven by the governor input shaft.

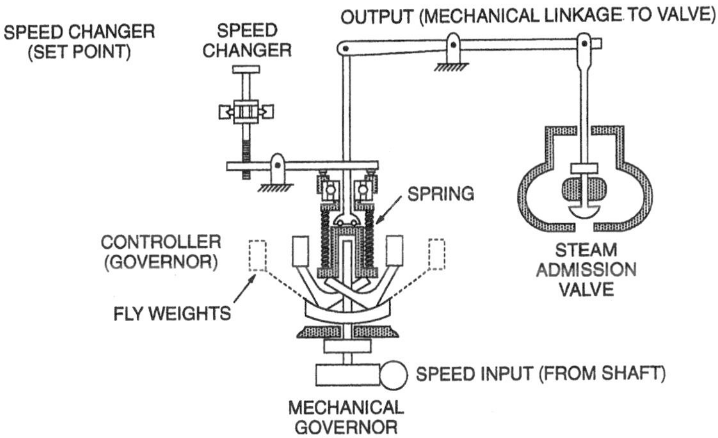

Component Function:

- **Input** – *Continuously Advises Value of Turbine Speed*
- **Set Point** – *Defines the Desired Value of Speed or Process Variable*
- **Controller** – *Produces an Output Force (Signal) Proportional to Weight Mass, Radius of Rotation and Rotation SPD²*
- **OutPut System** – *Accepts Controller Output Signal and Provides Sufficient Force to Modulate Steam Admission Valve*

Figure 24.6 A mechanical governor system

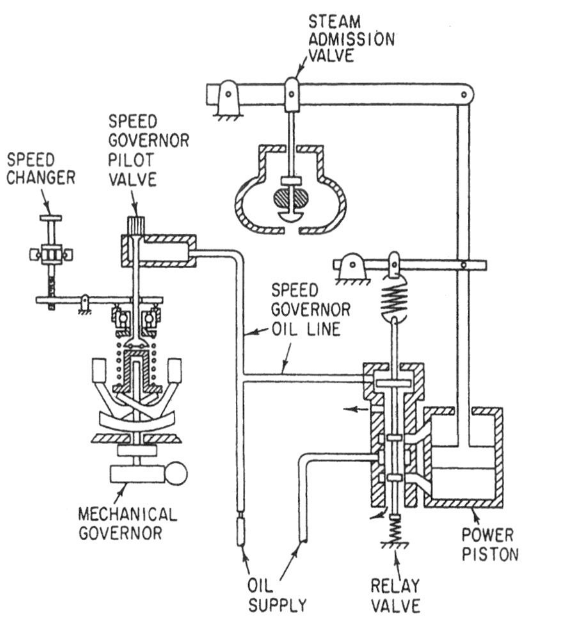

FUNCTION: OUTPUT OF CONTROLLER MODULATES RELAY VALVE WHICH PROVIDES HYDRAULIC SIGNAL TO MOVE OUTPUT LINKAGE. NOTE: HAS IT'S OWN INTERNAL PUMP

Figure 24.7 A mechanical hydraulic governor system

All mechanical-hydraulic governors require hydraulic fluid and site preventive maintenance practices must include these governors. They are provided with a sight glass to indicate the operating level of the hydraulic fluid. Typical fluids used are turbine oil and automatic transmission fluid "ATF". Governor instruction books must be consulted for specific hydraulic specifications. In larger systems, the governor hydraulic fluid reservoir may not be large enough to provide a sufficient fluid quantity to fill all of the speed governor oil lines. **Readers are cautioned that additional hydraulic fluid may have to be added during initial start-up and whenever work has been done on the governor system during a turnaround.**

Figure 24.8 is a representation of a mechanical-hydraulic governor system for a multi-valve steam turbine.

The system shows a Woodward "P.G. – P.L." governor system. These systems, common in the 1960s and 1970s are still in use today and have provided extremely reliable service. However, both mechanical and mechanical-hydraulic governors receive their input signal via a gear arrangement. Therefore, they cannot be repaired or removed while the

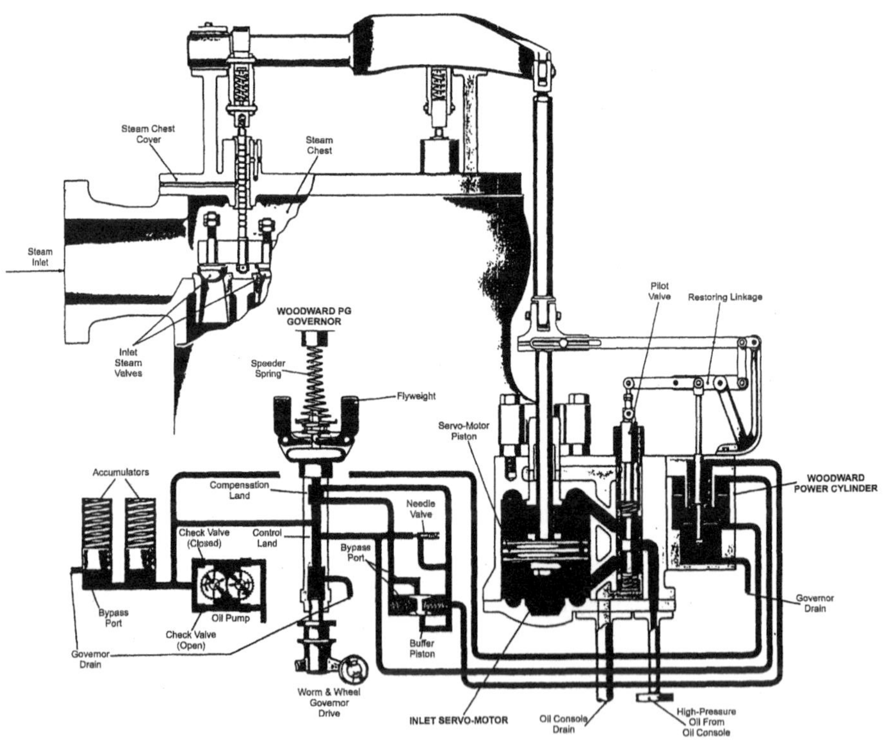

Figure 24.8 Typical mechanical-hydraulic governor for turbine drive (Courtesy of Elliott/Woodward)

turbine is operating. During the 1970s refinery, petrochemical and gas plant capacities increased significantly. As a result, the lost product revenue for one day downtime for governor repair became very large (typically $500,000 to over $1,000,000 U.S. dollars!). Therefore, there was an urgent need for a governor system that could be maintained without having to shut down the turbine. The electro/hydraulic governor met this need. Figure 24.9 presents the important facts concerning this system.

Electro/hydraulic governors

- Electro/hydraulic governors:
  - Do not require a mechanical input signal
  - Provide extremely accurate control
  - Provide self diagnostics, fault tolerance and auto-start capability
  - Require actuator to convert electric output signal to control signal (hydraulic or pneumatic)
- Types
  - Analog
  - Digital
- Either type can be:
  - Non-redundant
  - Redundant
  - Triple redundant

Figure 24.9 Electro-hydraulic governors

Since they did not require a mechanical (gear or shaft drive) input signal, these governors could be exchanged while the operators kept the turbine in the manual mode. As an analogy, exchanging automatic control valves is the same procedure. In this case, the operator maintains process conditions by manually throttling the bypass valve while the automatic control valve undergoes repair.

The first electronic governors were analog type which required significant maintenance to change out cards. Digital governors were introduced in the late 1970s and are the only type of speed control used today. As micro-processors became popular, digital governors also offered the great advantage of redundancy. Redundant and triple redundant governors became very popular because governors could now automatically transfer on line to allow control to be maintained while the other governor required maintenance. Operator assistance

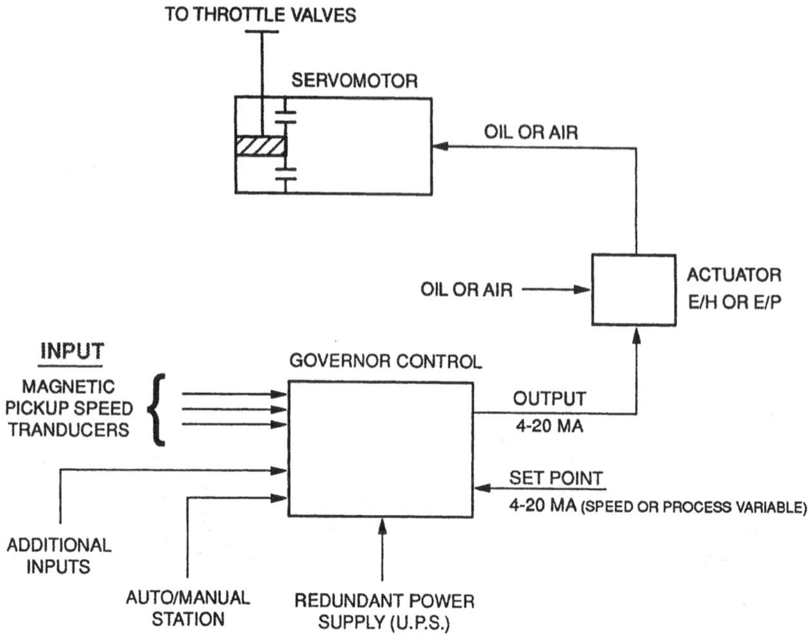

Figure 24.10 Electrohydraulic governor block diagram (Courtesy of M.E. Crane Consultant)

was no longer required. Figure 24.10 presents a block diagram for an electro-hydraulic governor systems.

In the 1990s, the trend is to control all process and machinery functions through the plant central distributed control system. A new chemical plant in South America is presently designing a D.C.S. system that will control all critical system functions:

- Turbine speed control
- Process control
- Surge protection
- E.S.D. systems
- On-line monitoring
- Emergency pump auto-start

In this design, all critical functions are actuated on the basis of a two out of three voting system.

As previously discussed, extraction turbines are used to optimize plant steam balance and overall steam cycle efficiency. Figure 24.11 defines the function of an extraction steam turbine control system.

Extraction control

Function: satisfy driven equipment control requirement and provide required extraction steam quantity at desired flow or pressure

An extraction control system consists of multiple governors with feed back

Figure 24.11 Extraction control

Both mechanical-hydraulic and electro-hydraulic extraction control systems are successfully operating in the field. Either design incorporates two (2) or more governors operating together to meet the control system objectives. Each governors' output controls a specific set of throttle valves. In addition, each governor in an extraction or admission system continuously receives an input signal from the other governors in the system. Each governor will respond to this input signal as required to meet all of the control objectives of the governor system.

Mechanical-hydraulic extraction or admission systems have proven to require a significant amount of adjustment and maintenance due to the high amount of friction in the systems. Please refer to Figure 24.12 which shows a mechanical-hydraulic single extraction governor system.

As a result, all new systems incorporate electro-hydraulic governor arrangements as shown in Figure 24.13.

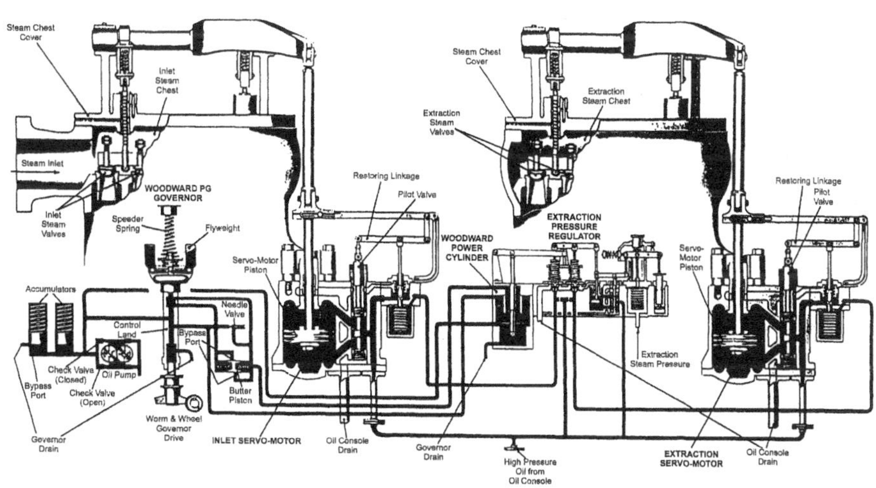

Figure 24.12 Mechanical/hydraulic extraction control (Courtesy of Elliott/Woodward)

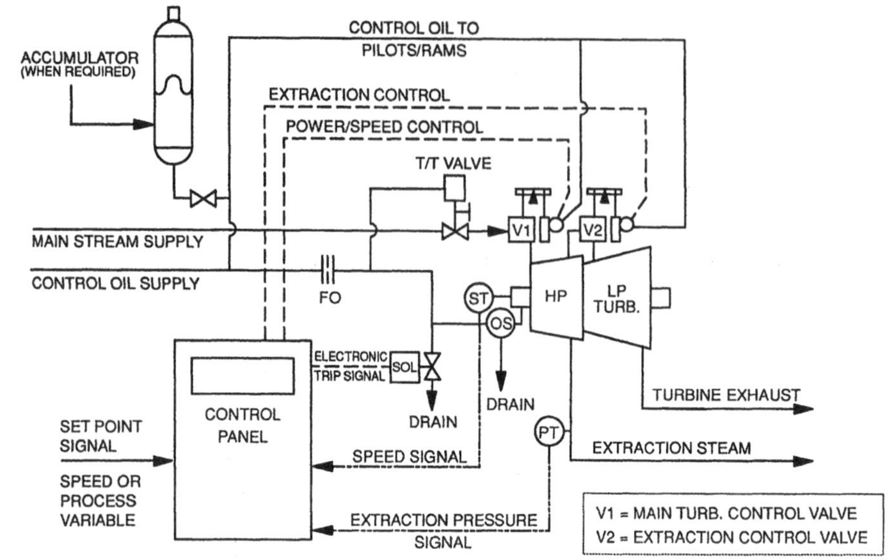

Figure 24.13 Electrohydraulic extraction control & protection system (Courtesy of M.E. Crane, Consultant)

Coupled with redundant features, these systems offer high reliability and efficient process control. Regardless of the type of governor utilized, mechanical-hydraulic and electro-hydraulic governors must be supplied with a reliable control oil system. Figure 24.14 presents the function and frequent problem areas of hydraulic control systems.

Control oil system

Function: Continuously provide cool, clean control oil to control and protection system at proper pressure, flow rate and temperature

Frequent problem areas:

- Main to auxiliary pump transfer
- Control oil valve instability
- Instantaneous flow requirement changes (need for accumulator)

Figure 24.14 Control oil system

Usually, the hydraulic control system is integral with the lubrication system. Typical pressure operating ranges for these systems are:

Low pressure	40–100 PSI
Medium pressure	120–600 PSI
High pressure	Above 600 PSI

Figure 24.15 is an application chart showing type of governor classification, speed regulation and type of governor used.

Steam turbine governor system application chart

Application-driven equipment	Speed regulation %	Type of governor system
Spared pump	NEMA A ± 10%	Mechanical (older applications) Mechanical (hydraulic) Elector-hydraulic (optional)* Non-redundant
Fan(s)	NEMA A ± 10%	Mechanical (older hydraulics) Mechanical hydraulic
Lube/seal oil pump(s)	NEMA A ± 10%	Mechanical/hydraulic
Turbo-compressor	NEMA D ± 0.5%	Electro-hydraulic (post 1980) Non-redundant Optional-redundant, triple redundant
Generator	NEMA D ± 0.5%	Isochronous (0% droop) Mechanical/hydraulic Present Electro-hydraulic

Figure 24.15 Steam turbine governor system application chart

In general, NEMA A governors are used in general purpose (spared) applications and NEMA D governors are used in special purpose (unspared) applications.

Protection

In the writers' experience the function of the steam turbine protection system is often confused with the control system. The two systems are entirely separate. The protection system operates only when any of the control system set point parameters are exceeded and steam turbine will be damaged if it continues to operate. Figure 24.16 defines the typical protection methods.

Protection

The protection system monitors steam turbine total train parameters and assures safety and reliability by the following action:

- Start-up (optional) provides a safe, reliable fully automatic start-up and will shut down the turbine on any abnormality
- Manual shutdown
- Trip valve exerciser allows trip valve stem movement to be confirmed during operation without shutdown
- Rotor overspeed monitors turbine rotor speed and will shut down turbine when maximum allowable speed (trip speed) is attained
- Excessive process variable signal monitors all train process variables and will shut down turbine when maximum value is exceeded

Figure 24.16 Protection

A schematic of a multi-valve, multi-stage turbine protection system is shown in Figure 24.17.

This system incorporates a mechanical overspeed device (trip pin) to shut down the turbine on overspeed (10% above maximum continuous speed). Centrifugal force resulting from high shaft speed will force the trip lever which will allow the spring loaded handle to move inward. When this occurs, the port in the handle stem will allow the control oil pressure to drain and drop to 0 PSI. The high energy spring in the trip and throttle valve, normally opposed by the control oil pressure will close suddenly (less than 1 second). In this system there are two other means of tripping the turbine (reducing control oil pressure to 0 PSI):

- Manually pushing spring loaded handle
- Solenoid valve opening

The solenoid valve will open on command when any trip parameter set point is exceeded. Solenoid valves are designed to be normally energized to close.

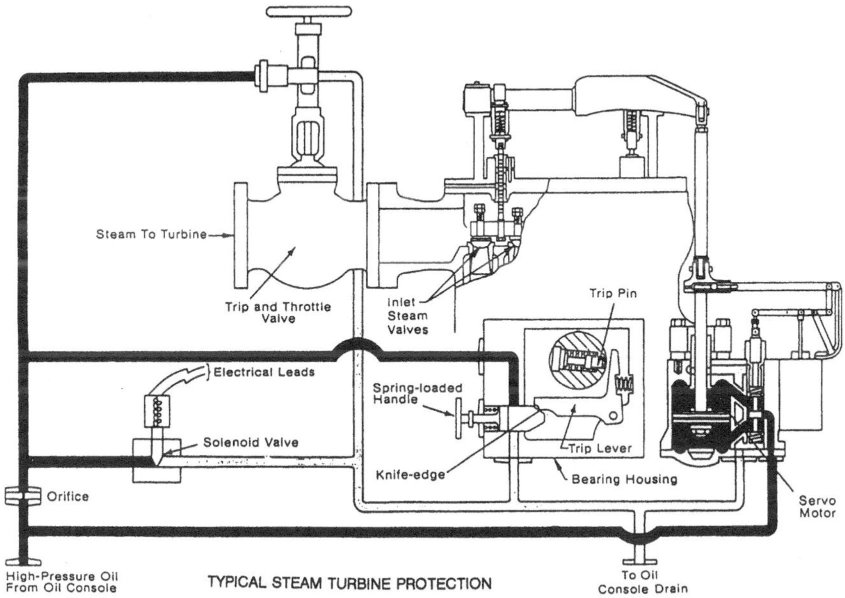

Figure 24.17 Typical steam turbine protection (Courtesy of Elliott Co.)

In recent years the industry has required parallel and series arrangements of solenoid valves to assure increased steam turbine train reliability.

Figure 24.18 shows two popular methods of overspeed protection used in the past.

Today, most speed trip systems incorporate magnetic speed input signals and two out of three voting for increased reliability. Figure 24.19 presents the devices that trip the turbine internally. That is, they directly reduce the control oil pressure causing a trip valve closure without the need of a solenoid valve (external trip method).

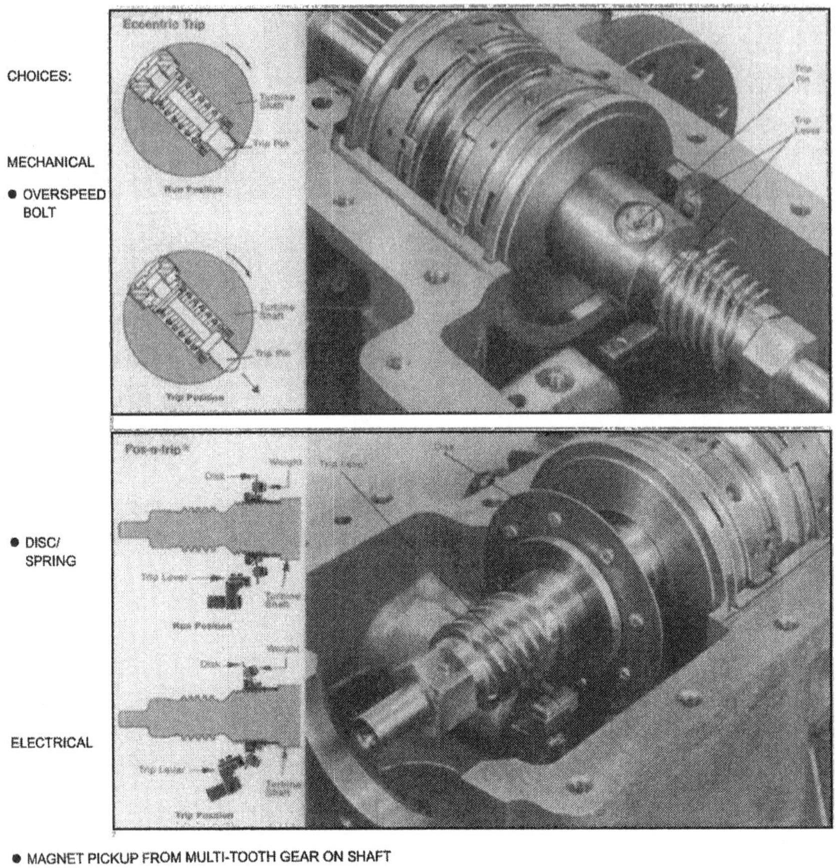

Figure 24.18 Overspeed detection (Courtesy of Elliott Co.)

Internal protection

- Loss of control oil pressure
  Spring force automatically overcomes oil force holding valve open (approximate set point 50–65% of normal control oil pressure)
- Manual trip (panic button)
  Manually dumps control oil on command
- Optional

Turbine excessive axial movement

Figure 24.19 Internal protection

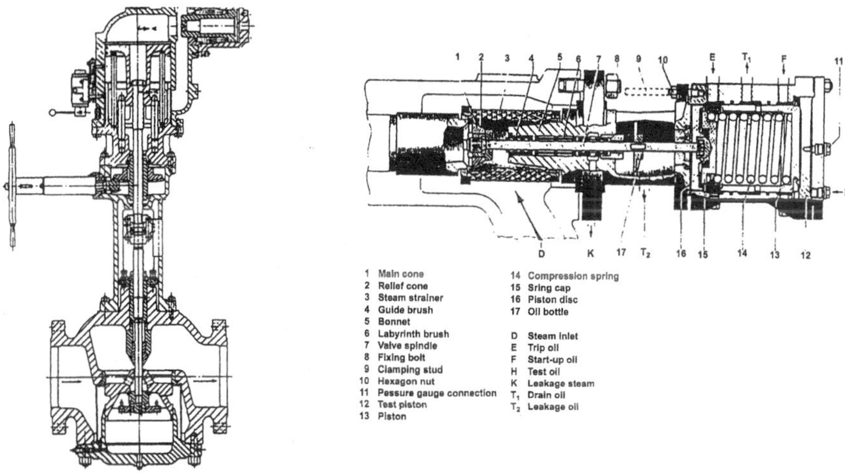

- FUNCTION RAPIDLY TO CUT OFF STEAM TO TURBINE ON OVERSPEED OR ANY DESIGNATED UPSET. STRONG SPRING FORCE RAPIDLY (ONE (1) SECOND) CLOSES VALVE.

Figure 24.20 Steam turbine shut-off valves. Left: Trip and throttle (Courtesy of Gimple Corp.). Right: Trip (Courtesy of Siemens)

Two popular types of steam turbine shutoff valves are shown in Figure 24.20.

Both types use a high spring force, opposed by control oil pressure during normal operation, to close the valve rapidly on loss of control oil pressure.

It is very important to note that the trip valve will only close if the spring has sufficient force to overcome valve stem friction. Steam system solid build up, which increases with system pressure (when steam systems are not properly maintained) can prevent the trip valve from closing.

To assure the trip valve stem is free to move, all trip valves should be manually exercised on line. The recommended frequency is once per month.

All turbine trip valves should be provided with manual exercisers to allow this feature. Figure 24.21 presents facts concerning manually exercising a turbine while on line.

On line manual exercise of trip valve

- Trip valve is only as reliable as valve to move
- Should periodically (minimum one per month) exercise valve to assure movement
- **Exercisers will not trip turbine**
- If valve does not move, must be remedied immediately

Figure 24.21 On line manual exercise of trip valve

Protection system philosophies have tended to vary geographically with steam turbine vendors. Figure 24.22 presents these facts.

Protection system philosophies

- Most domestic vendors rely only on trip valve to shut off steam supply. (Throttle valves remain open)
- European vendors close both trip and automatic throttle valve on trip signal

Figure 24.22 Protection system philosophies

25 Steam turbine operation

- Introduction
- Safety and reliability considerations
- Pre start-up considerations
- Steady state operation
- Shutdown and post shutdown
- Single stage turbine guidelines

Introduction

At this point, we have completed our study concerning the steam turbine design and have also explained the functions of the auxiliaries associated with the turbine. In this section, we will conclude the subject of steam turbines by examining steam turbine operation.

The section will begin with the discussion of safety and reliability considerations necessary for the operation of any steam turbine. Practical information will be presented based on actual field experience.

The next subject of discussion will involve pre start-up requirements. A step-by-step procedure will be reviewed and will be applicable to all types of turbines by design and by vendor. In this section we will also discuss an important stationary emergency shutdown system test that is carried out without the turbine in operation.

We will then proceed to discuss an actual start-up sequence and again cover all specific areas involved. During this discussion, we will detail specific areas of concern, namely proper use of turning gear, what to do if high vibration (rotor bow) is present during start-up and how to accelerate safely through a critical speed. Once the minimum governor speed is attained we will discuss at speed operation. It should be noted that considerations in soloing a turbine (operating uncoupled to check the overspeed trip) will also be discussed in this section.

Steady state operation will be discussed in the context of condition monitoring, baseline conditions, trending and predictive maintenance. We will define each term and highlight those performance and mechanical items that must be monitored during turbine operation. We will also cover in this section typical parameter limits (vibration, bearing temperature, steam seal operation, etc.).

Finally, shutdown procedure will be covered in the same manner as start-up procedures. Normal shutdown and emergency shutdown procedures will be discussed. Also contained in this section will be post shutdown requirements and recommendations.

Safety and reliability considerations

Steam turbine operation involves high steam pressures and temperatures and high shaft speeds. Following are some basic facts that are essential to steam turbine safety and reliability.

Failure to completely drain condensate from steam turbines can result in catastrophic failure. Figure 25.1 presents important considerations and problem areas.

Steam turbine operational safety and reliability

Complete condensate draining

- Hot steam creates condensate
- Sudden expansion damage – 'Water Hammer'
- Mass of liquid can cause severe damage (slugs)
- All external lines to turbine block valves must be drained until superheated steam appears

Problem areas

- Extraction lines
- Up-exhaust nozzle turbines
  - Proper operation of:
    - Automatic steam traps
    - Vacuum pots

Figure 25.1 Steam turbine operational safety and reliability

The writer personally has experienced turbine failure resulting from the failure to continually drain condensate from condensing turbines with up exhaust connections. These turbines usually have fan cooled condensers. Provisions should be made for low flow operation where the steam velocity is not great enough to carry saturated steam up to the condenser. Since the exhaust is under vacuum, vacuum pots which permit removal of condensate from the exhaust must be spared and have effective preventive maintenance to assure their continued operation. It is recommended that vacuum pot internals be inspected and cleaned if necessary during every turnaround. It has been the writers' experience that steam deposits 'harden up' during down periods (turnarounds) and render these devices inoperable during start-up.

Maintaining proper turbine steam conditions is essential to maintaining optimum turbine efficiency and reliability. Figure 25.2 presents these facts.

Steam turbines operational safety and reliability

Proper steam conditions

- Inlet pressure
- Inlet temperature +/- 30°C (50°F) of specified
- Exhaust pressure
  - Proper vacuum
  - Within limits
- Exhaust temperature
  - +/- 30°C (50°F) of specified (back pressure)

Problem areas

- High exhaust temperature (start-up)
- Low inlet temperature (condensing turbines)
- Extraction turbines

Figure 25.2 Steam turbines operational safety and reliability

During low load start-up conditions, attention must be paid to the exhaust temperature of condensing turbines. Usually maximum allowable exhaust temperatures are 200–250°F. This limit is based on exhaust end vertical thermal expansion which can cause shaft misalignment. Figure 25.3 presents the relationship for determining turbine case growth as a function of exhaust temperature change.

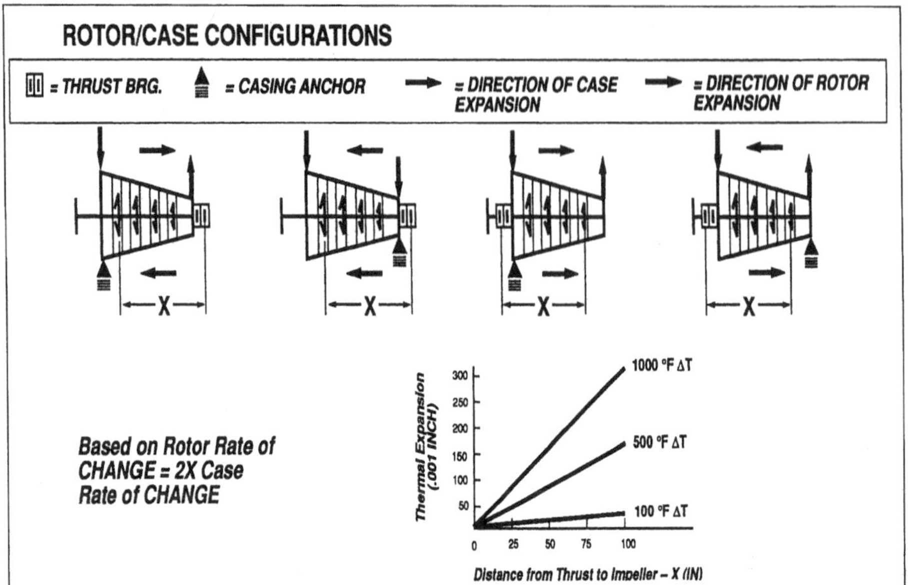

Since the rotor is in direct contact with the gas and has significantly less mass than the case, a rapid gas temperature increase can yield potentially damaging effects to the rotor and stationary compressor parts.

Both the rotor and case change dimensions by the relationship

$\Delta L\ (IN) = L\ (IN) \times 0.0000065\ (IN/IN\text{-}°F) \times \Delta T°F$

But the rotor rate of change will be greater than the casing (stationary parts).

Figure 25.3 Rotor/case thermal movement

The rated load condensing turbine exhaust temperature under maximum vacuum conditions is normally 120°F. In addition to shaft alignment concerns, high exhaust temperature can cause condensing turbine expansion joint failures. Unless specified, most expansion joints are limited to approximately 250°F operating temperature. Low load operation (start-up solo operation) can result in exhaust temperatures in excess of this value.

Low inlet steam temperatures (reduction of 50° or more from rated inlet condition) can lead to extensive exhaust end blade erosion (nozzles and rotor blades) in short periods of time. Steam turbines should never be continuously operated lower than 50°F from the rated steam temperature.

Referring to Figure 25.3, imagine what would happen if the top half of a turbine rotor were hotter than the bottom half. The top half would expand a greater amount resulting in a slight rotor deflection or bend.

In the case of heavy multistage steam turbine rotors, a deflection of 0.001" could cause a very high rotor unbalance resulting in high vibration and bearing failure. This phenomena is known as rotor bow. Facts concerning rotor bow and its solution are presented in Figure 25.4.

Steam turbine operational safety and reliability

Uniform rotor heating

- Rotor bow results from non-uniform heating or cooling of the rotor. If one area exists at a higher temperature a bow in the rotor can form:
- The bow is usually in the first critical mode shape shown and will cause high vibration (1 x frequency)
- Solution is to prevent bow by:
  - Rotation – turning gear
  - Proper cool down (slow-roll) procedure
  - Proper steam seal operational procedures (where applicable)

Figure 25.4 Steam turbine operational safety and reliability

For rotors sensitive to rotor bow a turning gear is the best solution. Facts concerning a turning gear are shown in Figure 25.5.

There are no specific industry guidelines for when to require a turning gear. Figure 25.6 presents the important considerations. The best advice is to require the vendor to provide references for similar operating turbines and follow up to determine if a turning gear is required based on field operating experience.

Pre start-up considerations

Figure 25.7 presents important items to check during the pre start-up. Readers are strongly cautioned to thoroughly check and use vendor procedures contained in the appropriate instruction books.

FUNCTION: TO ALLOW THE ROTOR TO ADAPT TO TEMPERATURE CHANGES BY ROTATING THE ROTOR

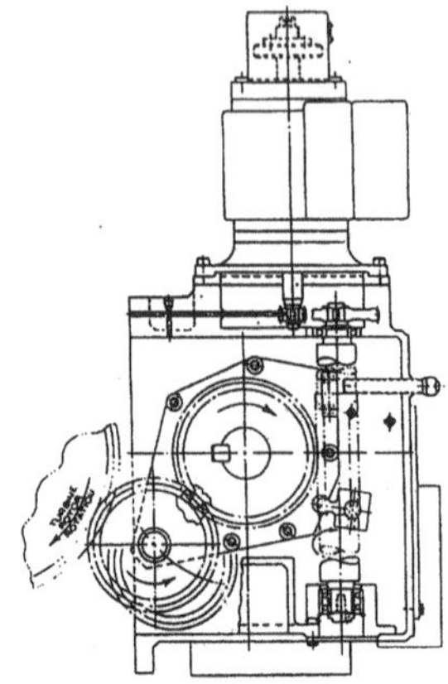

TURNING GEAR FACTS:

- MANUAL - RATCHET TYPE
- AUTOMATIC - USUALLY MOTOR DRIVEN (ON U.P.S. SYSTEM)
- CAN BE DESIGNED TO AUTO START
- AUTO DISENGAGE
- SPEED RANGES
    - LOW (AUTOMATIC RATCHET TYPE) = 2-5 RPM
    - MEDIUM = APPROXIMATELY 20 RPM
    - HIGH = 60-80 RPM

Figure 25.5 Turning gear

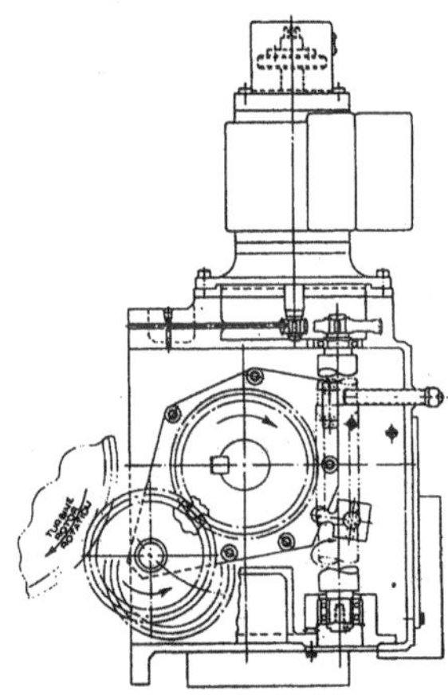

- NO SPECIFIC GUIDELINES
- CONSIDERATIONS:
    - ROTOR STIFFNESS
      L/D > APPROXIMATELY 10
    - OPERATING TEMPERATURE RANGE
    - ARE START-UP TIMES LIMITED?
- AUTOMATIC OR MANUAL TYPE?

Figure 25.6 Turning gear requirements

Steam turbine operation pre start-up

- Auxiliary systems – ready to run

■ Lube oil	
■ Control oil	■ fault free
■ Steam seal	
■ Cooling water	■ all utilities lined up
■ Monitoring	
■ Governor	■ properly cleaned
■ Protection	(initial start-up)

- All steam lines to and from turbine properly warmed and drained
  - Minimum 50°F (30°C) above saturation
  - Drain until total steam (vapor)

Note: confirm proper facilities are installed

- Stationary ESD test (turbine not operating)
  - Start lube/control oil system
  - Confirm proper system pressures
  - Fully open trip valve(s)
  - Reset trip system
  - Depress trip button
  - *Confirm* trip valve(s) close immediately (<1 second)
  - If above conditions are not met take appropriate action before operating turbine

Figure 25.7 Steam turbine operation pre start-up

Figure 25.8 is schematic of a typical condensing steam turbine piping system showing warm-up and drain lines.

Figures 25.9 and 25.10 present start-up guidelines. Readers must refer to appropriate instruction books for vendors' specific start-up guidelines.

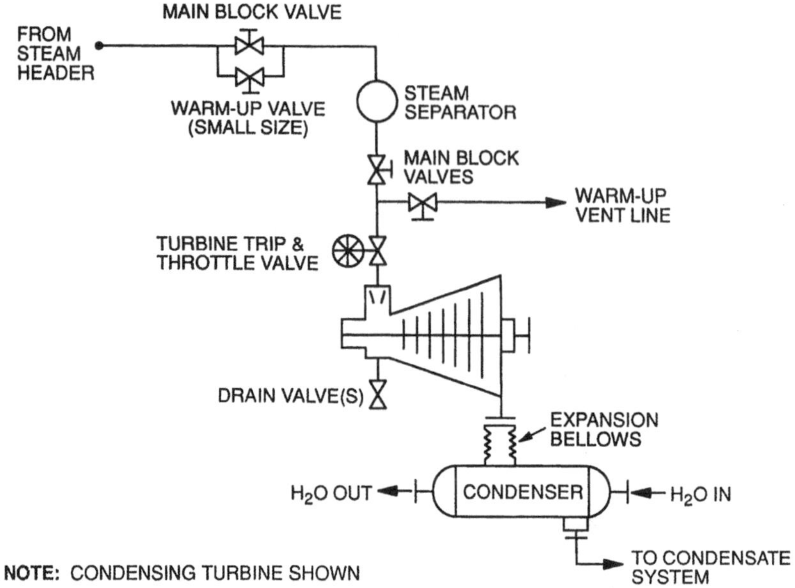

Figure 25.8 Typical steam turbine piping schematic

Steam turbine operation start-up sequence

- All applicable auxiliary systems operating fault free
  - Lube oil
  - Control oil
  - Steam seal
  - Cooling water
  - Monitoring
- Turning gear operating (if applicable)
- All safety systems (trips) armed
  - All auxiliary pumps in auto mode
  - Accumulator(s) properly pre-charged

Figure 25.9 Steam turbine operation start-up sequence

Steam turbine operation start-up sequence

- Starting the turbine
  - slowly open trip and throttle valve
    Observe 'Break Away'
  - Slow roll for prescribed period
    (refer to vendor's instruction book)
  - Proceed to increase speed as noted. Observe the following:
  - Vibration (< 60 MM (2.5 mils))
    - Unusual noise
    - Any alarm signals

Note: Any abnormal valves/alarms require immediate action

Figure 25.10 Steam turbine operation start-up sequence

Since the temperature of the rotor and turbine casing change during start-up and shutdown all steam turbine vendors require certain warm up and cool down procedures. Since the mass of the turbine rotor is significantly less than that of the case, the rotor will expand or contract at a much faster rate than the casing. Therefore, starting up a turbine too quickly will cause rotor rubs and significant turbine damage.

Figure 25.11 presents a typical steam turbine start-up procedure. Note how rapid acceleration is required through the critical speed range. Note also that a warm and cold start-up map are presented.

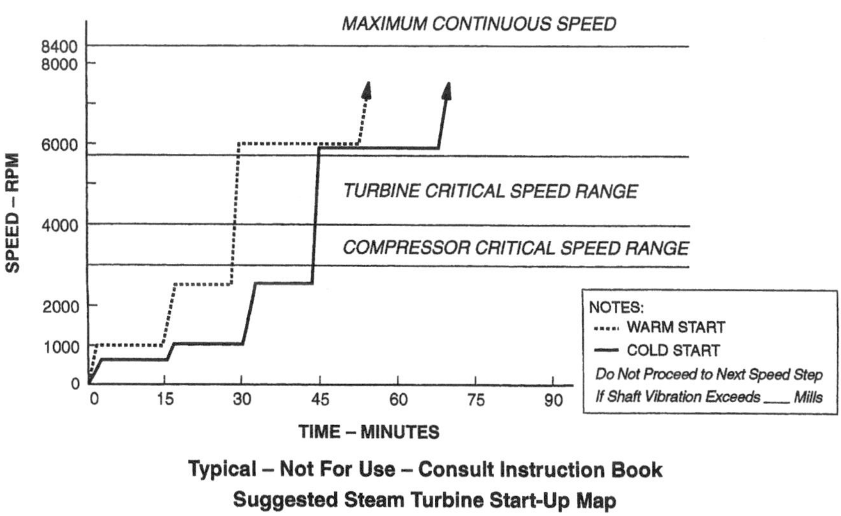

Figure 25.11 Steam turbine start-up map

It has been the writer's experience that many steam turbine start-ups are extended unnecessarily. Usually, as the turbine speed approaches the critical speed range, vibration will increase significantly. This is especially true if a turbine rotor bow exists. Rotor bows (unless significant rotor damage has been experienced) are temporary and are removed by rotor even heating and centrifugal force.

Unfortunately centrifugal force is a function of rotor speed because increased rotor speed in a 'bowed' condition will result in increased vibration. It has been my personal experience that when this condition exists, many a good operator will reduce the rotor to the 'slow roll' (lowest speed), wait 15–30 minutes and try again. This usually results in a long period (2–3 hours of delay) and high vibration being experienced through the critical speed range. Please try the following:

- Refer to the instruction book and/or contact vendor to establish a 'Go/No Go' radial vibration limit at a speed approximately 10% below the start of the critical speed range.
- If the vibration limit is exceeded at this speed or a lower speed, REDUCE SPEED UNTIL THE VIBRATION LIMIT IS EXACTLY MET.
- CONTINUE TO INCREASE SPEED AS SOON AS VIBRATION LIMIT REDUCES BUT DO NOT ALLOW LIMIT TO BE EXCEEDED.

In this procedure, maximum energy is continually put into the rotor to eliminate the bow as soon as possible. The 'rotor bow' will usually disappear suddenly with a dramatic reduction of vibration. The vibration limit mentioned above is usually in the 2–3 mil range (double amplitude, unfiltered) but ... consult the vendor to be sure.

Figures 25.12, 25.13 and 25.14 present important facts concerning turbine speed acceleration rates and vibration.

Steam turbine operation high vibration on start-up

- Immediately reduce speed until vibration is safe level (< 50mm 2.0 mil)
- Hold at above speed
- Increase speed as soon as vibration level falls – do not exceed 50 mm vibration

Note: do not attempt to proceed through a turbine critical speed if vibration exceeds 50 mm – 2.0 mil

Figure 25.12 Steam turbine operation high vibration on start-up

Steam turbine operation passing through steam turbine natural frequencies (critical speeds)

- It is required to quickly and smoothly accelerate through these areas because:
  - If a natural frequency is continuously excited, high vibrational energy will be transmitted to bearings case and foundation. This can cause significant, long term damage.
  - The primary excitation force is turbine rotor speed. Hence the name

'Critical Speed'

Figure 25.13 Steam turbine operation passing through steam turbine natural frequencies (critical speeds)

Steam turbine operation

- Attaining minimum governor speed
  - Speed control is transferred to the control (governor) system
  - Throttle valves close (units with trip and throttle valve only)
- Proceed to increase speed to desired value

Note: care must be exercised when 'soloing' turbine (uncoupled) to check overspeed trip.

In this case load is very low. Consequently, a small change in steam flow will produce rapid speed changes.

Figure 25.14 Steam turbine operation

Steady state operation

Once the turbine is under control of the governor, steady state operation is attained. Figure 25.15 defines the monitoring procedure for all critical parameters during steady state or 'at speed' operation.

Steam turbine 'at speed' (steady state) operation

During steady state operation, *condition monitoring* initial *baseline* conditions and changes *trends* provide the information necessary for predictive maintenance.

Figure 25.15 Steam turbine 'at speed' (steady state) operation

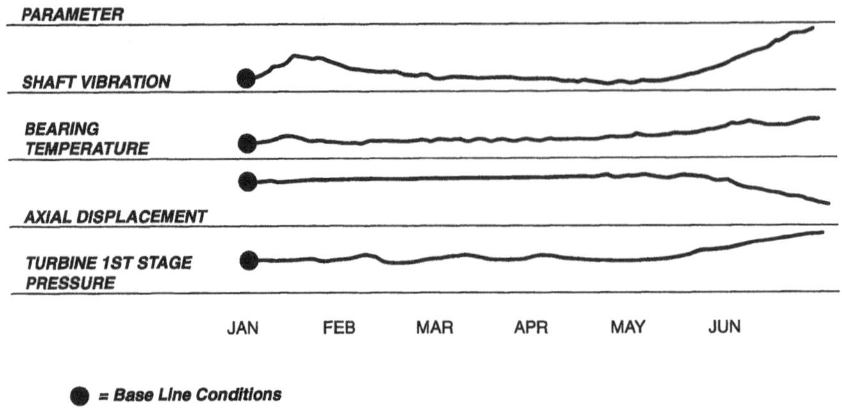

Figure 25.16 Parameter trending

An example of steam turbine condition monitoring using base data and trends is shown in Figure 25.16.

The definition of condition monitoring is stated in Figure 25.17.

Condition monitoring

Condition monitoring is the action of recording *baseline* and *operating* data to determine trends of critical factors.

Figure 25.17 Condition monitoring

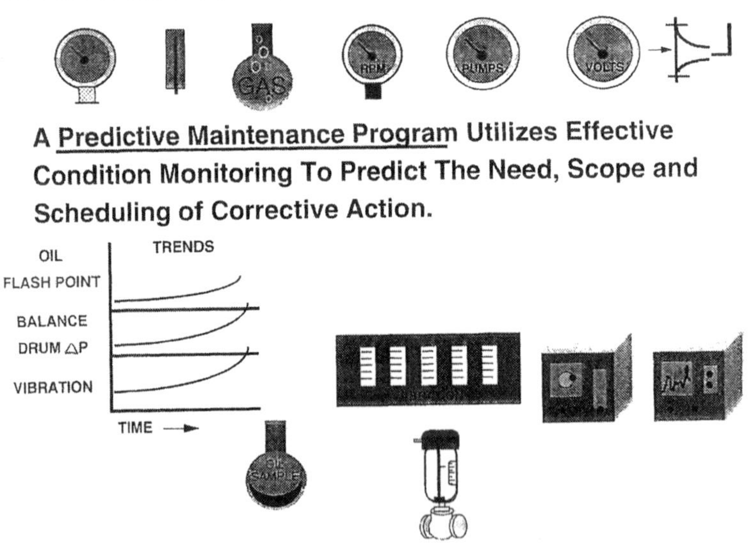

Figure 25.18 The basis of a predictive maintenance program (Courtesy of M.E. Crane Consultant)

Figure 25.18 defines the basis of a predictive maintenance program. Effective use of all steam turbine instrumentation will significantly increase reliability and reduce preventive maintenance.

A suggested list of performance condition parameters to monitor for expansion turbines is noted in Figure 25.19.

Performance condition monitoring for expansion turbines

The following data should be monitored

- Pressures (inlet, exhaust)
- Temperatures (inlet, exhaust)
- First stage pressure (steam turbines)
- Vapor flow
- Gas analysis (turbo-expander)
- Speed
- Throttle valve position (steam turbine)
- Nozzle position (turbo-expander)
- Generator amps (turbo-expander)

Figure 25.19 Performance condition monitoring for expansion turbines

Figure 25.20 presents a listing of mechanical condition parameters to be monitored.

Mechanical condition monitoring for expansion turbines

The following components should be monitored:

- Journal bearings
- Thrust bearings
- End seal leakage
- Lube oil system
- Control oil system (steam turbine)
- Gas seal system (turbo-expander)

Figure 25.20 Mechanical condition monitoring for expansion turbines

Typical condition limits for critical mechanical condition parameters are defined in Figure 25.21. Readers are advised to consult specific instruction books for vendors' recommended limits.

Typical condition limits

- Bearing pad temperature - 240°F
- Lube oil temperature rise - 50°F ΔT
- Axial displacement - ± 0.020
- Shaft seal leakage - visible steam
- Shaft vibration - 2.0 mills (unfiltered, peak - peak)

Figure 25.21 Typical condition limits

Shutdown and post shutdown

Shutdown of any type of rotating equipment can be divided into two (2) categories:

- Planned or normal shutdown
- Emergency shutdown

Suggested normal shutdown procedure guidelines are shown in Figure 25.22. Always consult the appropriate instruction book for the vendor's recommended procedure.

Steam turbine operation

Normal shutdown

- Advise boiler room (if applicable)
- Reduce speed to minimum governor speed setting
- Continue to reduce speed with manual throttle valve (or starting device if manual throttle valve is not supplied)
- Reduce to slow roll speed and operate for specified period of time
- Put on 'turning gear' or manually turn (if applicable) for specified period of time
- Open all steam drains

Figure 25.22 Steam turbine operation

However, planned shutdowns are a great opportunity to prove some important things under a controlled, safe environment. Please refer to Figure 25.23.

Steam turbine operation

Planned shutdowns (before turnarounds) are an excellent opportunity to check or prove the following:

- Manual exerciser
- Main/auxiliary lube pump transfer
- Operation of the circuits

Note: always alert operations and have operation personnel ready in event of malfunction.

Figure 25.23 Steam turbine operation

Emergency shutdown guidelines are presented in Figure 25.24.

Steam turbine operation emergency shutdowns (ESD)

- Turbine trips due to any ESD actuation or by manual signal
- Inspect entire train and control panel for any indication of damage
- If trip did not relate to train or cause immediate damage:
  - Slow roll and listen for rubs
  - Put on turning gear (if applicable)

Figure 25.24 Steam turbine operation emergency shutdowns (ESD)

Single stage turbine guidelines

The five (5) common problems with single stage turbines are noted in Figure 25.25.

Single stage steam turbines common reliability problems

- Bearing bracket oil contamination (inadequate carbon ring steam seal design)
- Slow governor system response (inadequate governor linkage maintenance and governor power)
- Hand valve(s) closed on critical services
- Bearing bracket oil viscosity reduction and bearing wear (high pressure service)
- Use of sentinal valves on turbine cases

Figure 25.25 Single stage steam turbines common reliability problems

We will now discuss each problem in detail. Please refer to Figure 25.26 which has each problem area circled.

Bearing bracket oil contamination

Please refer to Item 1 in Figure 25.26.

The most common reliability problem with single stage steam turbines is the contamination of the oil in the bearing housing with water. The root cause of the problem is the ineffectiveness of the floating carbon ring shaft seal system to stop.

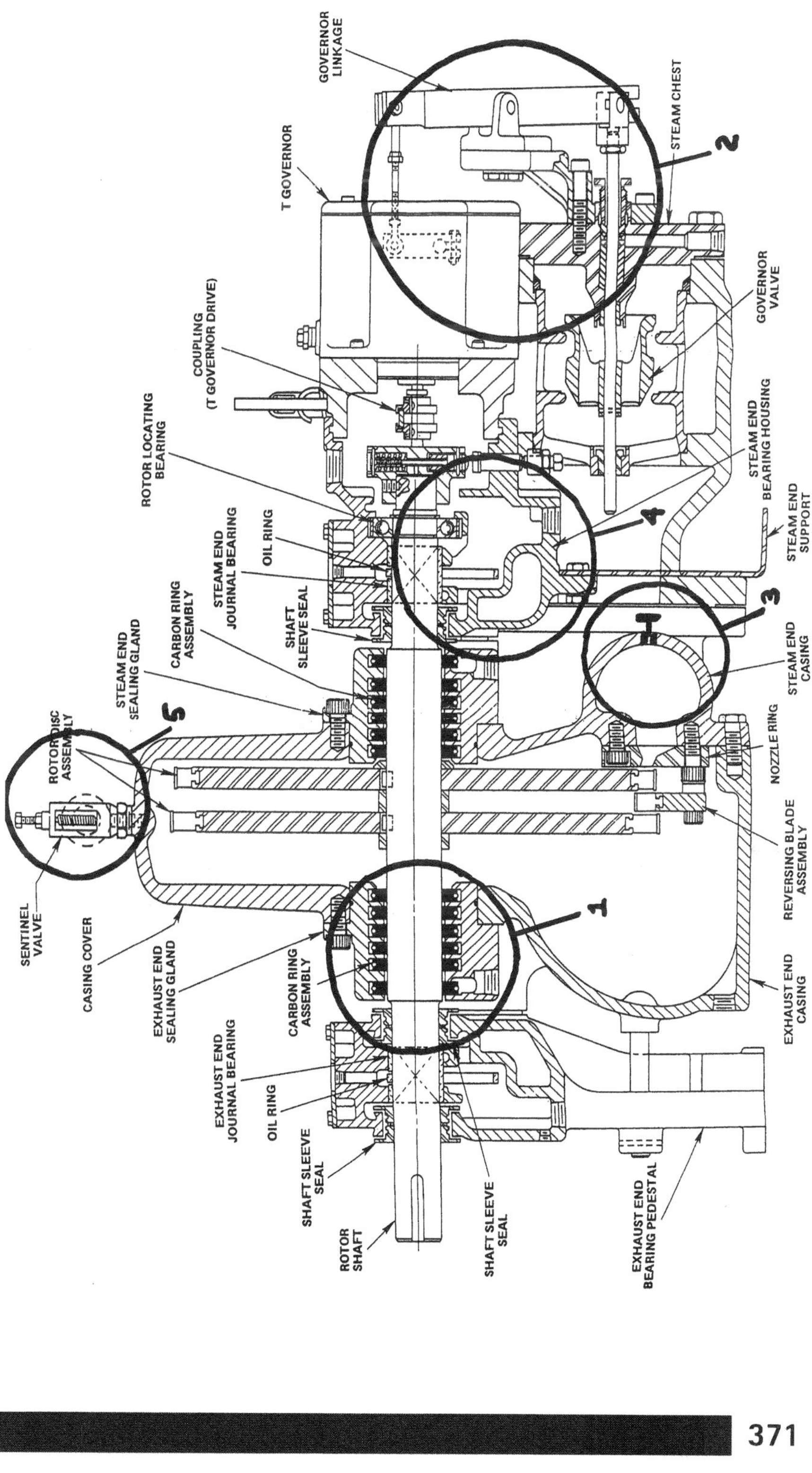

Figure 25.26 Single stage steam turbines common reliability problems

Bearing bracket oil contamination (root cause)

- Shaft carbon ring seal cannot positively prevent steam leakage

Figure 25.27 Bearing bracket oil contamination (root cause)

Unless present site systems are modified to eliminate the root cause, the best action plan is to minimize the effect of the contamination so a bearing failure will not occur. Such an action plan is presented in Figure 25.28.

Steam turbine bearing bracket oil contamination monitoring action plan

- Install oil condition site glasses in bearing bracket drain connection
- Inspect once per shift
- Drain water as required
- Sample oil monthly initially

Figure 25.28 Steam turbine bearing bracket oil contamination monitoring action plan

The action plan to eliminate the root cause of this reliability problem is presented in Figure 25.29.

How to correct carbon ring seal ineffectiveness

- Install steam eductor on each seal chamber leak off drain (between 4th and 5th carbon ring)
- Design eductor to pull 5–10" of H_2O vacuum at this point
- Alternative approach – install bearing housing isolation seal ('Impro' or equal)

Figure 25.29 How to correct carbon ring seal ineffectiveness

Slow governor system response

Please refer to Item 2 in Figure 25.26. Another very common reliability problem is the slow or non-movement of the governor system linkage during start-up and normal operation during steam condition changes. It will appear that the governor is not responding because speed will not be controlled when it should. Typical examples are:

- Speed will continue to increase when throttle valve is opened, turbine will trip on over speed
- Speed will increase or decrease when:
  - Steam conditions change
  - Driver equipment changes

These facts are presented in Figure 25.30.

Slow governor system response can cause:

1. Rapid speed change & trip on start-up
2. Speed increase or decrease on steam condition or load condition change
3. Governor instability (hunting) around set point

Note: #1 usually occurs on "solo", #2 occurs during steady state operation

Figure 25.30 Slow governor system response

Since most single stage steam turbines are not supplied with tachometers, it is difficult, if not impossible to condition monitor this problem. A condition monitoring action plan is provided in Figure 25.31.

Slow governor system response condition monitoring action plan

- Install tachometer on all single stage steam turbines
- Always test speed control on "solo run" (1)
- Monitor turbine speed once per shift. Take corrective action if speed varies +/- 5% (200 rpm)

Note: (1) since load is very low, test acceptance is the ability to stabilize speed & prevent overspeed trip when throttle valve is slowly opened.

Figure 25.31 Slow governor system response condition monitoring action plan

The usual root cause of the problem is that the friction in the mechanical linkage and/or valve stem packing exceeds the maximum torque force that the governor output lever can deliver. The governor designations TG-10, TG-13 & TG-17 simply mean "turbine governor with ... FT-LB torque. Therefore, if a TG-10 governor is installed and the torque required to move the valve steam exceeds the value of 10 FT-LBs, the governor system will not control speed. Taking the governor to the shop, will not solve the problem. Causes of excessive friction are shown in Figure 25.32.

Causes of excessive governor mechanical linkage system & valve friction

- Linkage bushings not lubricated with high temp. grease
- Valve steam packing too tight
- Steam deposits in valve &/or packing after extended shut down (turbine cold)
- Bent steam valve stem

Figure 25.32 Causes of excessive governor mechanical linkage system & valve friction

The action plan to eliminate this reliability problem is presented in Figure 25.33.

Slow governor system response condition monitoring action plan

- If problems occur (Fig. 25.30), disconnect linkage and confirm ease of valve movement
- Replace bushings and/or lubricate with "molycote" or equal
- Clean deposits from valve and packing as required
- If above action does not correct problem, replace governor (inspection and/or adjustment of governor droop is required)

Figure 25.33 Slow governor system response condition monitoring action plan

Hand valve(s) closed on critical services

Most single stage steam turbines are supplied with one (1) or more hand valves in the steam chest. Refer to Figure 25.26, Item 3. The purpose of the hand valves is to allow more or less inlet steam nozzles to be used during operation. Optimizing the steam nozzles used, maintains turbine efficiency during load changes. **However, the efficiency of single stage steam turbines is only 35% at best!** Therefore, adjustment of hand valves, other than during start-up or during slow roll, should not be required. Figure 25.34 is a top view of a one (1) hand valve.

We have witnessed many unscheduled shutdowns of critical (unspared) compressor units because the general purpose steam turbine that is the main lube oil pump driver, had the hand valves closed. An upset in the steam system reduced steam supply pressure and caused the turbine and lube pump to slow down. This was because, hand valves were closed and the throttle valve, even when full open, could not meet steam flow requirements. When the speed of the steam turbine decreased, the lube oil pressure dropped and guess what? ... The auxiliary pump did not start in time and the unit tripped.

Figure 25.35 presents the recommended action plan in the refinery for single stage steam turbine hand valves.

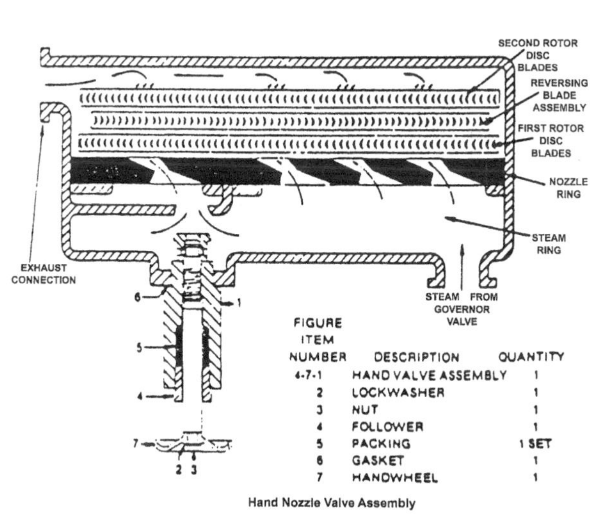

FEATURES:

- **NOZZLE BLOCK IN LOWER HALF OF CASE**
- **ONE OR MORE HAND VALVES FOR EFFICIENCY AND INCREASED LOAD**
- **HAND VALVES NOT MODULATING!**
  (FULL OPEN OR FULL CLOSED)

Figure 25.34 Single valve turbine admission path

Single stage steam turbine hand valve recommendations

- Never throttle hand valves
- Hand valves should be open on main oil pump and auto-start steam turbines

Figure 25.35 Single stage steam turbine hand valve recommendations

Bearing bracket oil viscosity reduction & bearing wear on high pressure single stage steam turbines.

Please refer to Figure 25.26, Item 4.

Observe the jacket in the bearing housings. The purpose of this jacket is to cool the oil in the bearing bracket. When the inlet steam pressure is high, the high temperature of the steam is transmitted to the steam end inlet bearing through the shaft. Although the jacket in the bearing housing does reduce the oil temperature in the bearing housing, it cannot effectively reduce the oil temperature at the shaft/bearing interface. Figure 25.36 presents these facts.

High pressure single stage steam turbine bearing problems and oil viscosity reduction

- Sleeve bearings (usually steam inlet end) wear out quickly
- Oil viscosity is reduced and difficult to maintain

Figure 25.36 High pressure single stage steam turbine bearing problems and oil viscosity reduction

This problem is a design issue. A small single stage turbine is not provided with an effective oil system to remove the heat between the shaft and bearing when the turbine is operating on high temperature (up to 750°F) steam. The solution is to require pressure lubrication for this application.

Naturally, it is difficult, and not cost effective to retrofit these turbines for pressure lubrication. The field proven solutions to this problem are presented in Figure 25.37.

Eliminate bearing wear & oil viscosity reduction (high pressure service) by:

- Assuring bearing housing jacket passages are open (flushed)
- Consulting with turbine vendor for bearing material change
- Using special high temperature service oil (synthetic based oil)

Figure 25.37 Eliminate bearing wear & oil viscosity reduction (high pressure service)

Continued use of sentinel valves on turbine cases

Please refer to Figure 25.26, Item 5. Sentinel valves were used, years ago, as alarm devices indicating that the steam turbine case (low pressure part) was under excessive pressure.

These devises are not pressure relief valves and will not protect the case from failure during over pressure events.

It is a known fact that the sentinel valves, wear, leak and require steam turbine shutdown for repair. Most large company specifications prevent the use of sentinel valves and require full relief valve protection on the inlet and exhaust of all single stage turbines. These facts are presented in Figure 25.38.

Prevent excessive sentinel valve maintenance by:

- Removing sentinel valves
- Assuring that inlet and exhaust casings are protected by properly sized and set pressure relief valves

Figure 25.38 Prevent excessive sentinel valve maintenance

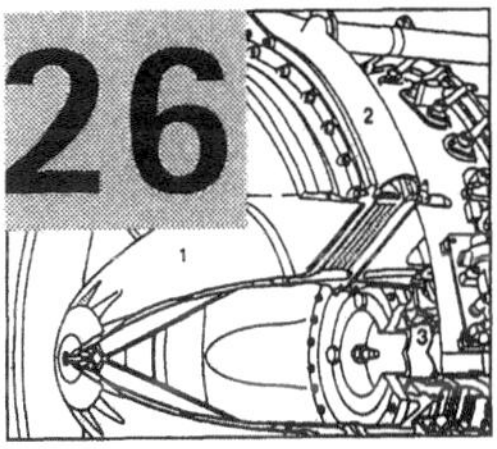

26 Gas turbine types and applications

- Introduction
- Comparison to a steam turbine
- Comparison to an automotive engine
- Building a gas turbine
- History of gas turbine development
- Gas turbine classifications
- Classification by design type
- The number of gas turbine shafts
- Gas turbine drive configurations
- Gas turbine cycles

Introduction

In this chapter, we will discuss gas turbine function and types. In my personal experience, the gas turbine is the most misunderstood rotating equipment item. Because of its many support systems and various configurations, the gas turbine is often approached with mystery and confusion. In order to thoroughly explain the gas turbine from a functional standpoint, we will build on prior knowledge. We will also compare the gas turbine to an automotive engine in terms of its combustion cycle. Having done this, we will then use a building block approach to explain the total configuration of a gas turbine and conclude this introduction by discussing a brief history of its evolution.

Gas turbine classifications will then be presented, specifically:

- Design type
- Number of shafts
- Drive and number of shafts
- Cycle
- Drive and location

We will discuss the major design difference between aero derivative and hybrid (aero derivative gas generator/industrial power turbine). Single and multiple shaft gas turbines will be discussed and reviewed. The three major application cycles for gas turbines: simple, regenerative, and combined will be presented and discussed.

Finally, we will present applications of different gas turbine types and provide the information concerning when the different types are used.

Comparison to a steam turbine

Figure 26.1 shows a typical condensing steam turbine and an industrial type gas turbine. The major difference is that a steam turbine is an external combustion engine, while a gas turbine is an internal combustion engine. That is, the motive fluid for a steam turbine is generated external (in the boiler) to the engine. In the case of a gas turbine, the motive fluid is generated internal to the engine (air compressor and combustor).

Figure 26.2 presents the comparison of the gas turbine and steam turbine cycles. The steam turbine cycle is known as the Rankine cycle. As shown, the hot vapor is generated in the boiler which is external to the steam turbine (expander). In the gas turbine cycle known as the

Figure 26.1 A gas turbine vs steam turbine. Left: Steam turbine (Courtesy of General Electric Co.). Right: Mars gas turbine (Courtesy of Solar Turbines, Inc.)

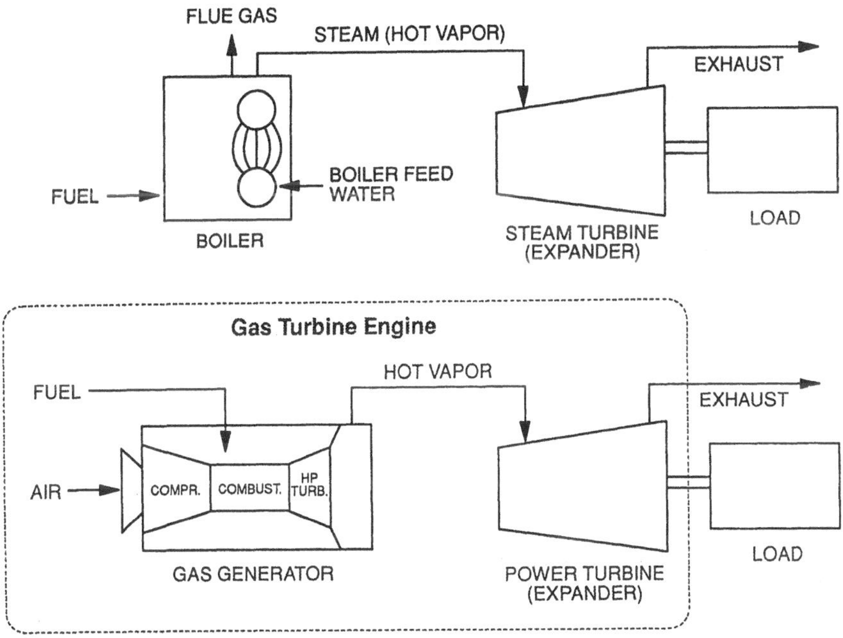

Figure 26.2 Comparison – gas turbine vs steam turbine cycles

Brayton cycle, air is brought into the engine by the axial compressor, combined with fuel and an ignition source in the combustor to produce a hot vapor which then is expanded through the HP (high pressure) turbine. The combination of the compressor, combustor and HP turbine is commonly known as the gas generator. This is because the function of the gas generator is to generate or produce a hot vapor from the combination of an air fuel mixture. Essentially, a gas generator can be considered to have the same function as a boiler – both produce a hot vapor. One can think of the gas generator then as a 'rotating boiler'. After the hot vapor is generated, it then is expanded additionally in the power turbine. The power turbine, therefore, serves exactly the same function as the steam turbine. That is, both components are hot gas expanders.

Comparison to an automotive engine

Figure 26.3 shows the similarities between an automotive engine and a gas turbine. If one considers that a gas turbine is only a dynamic internal combustion engine, the understanding of the gas turbine becomes significantly easier. As shown in the referenced figure, an automotive engine is a positive displacement internal combustion

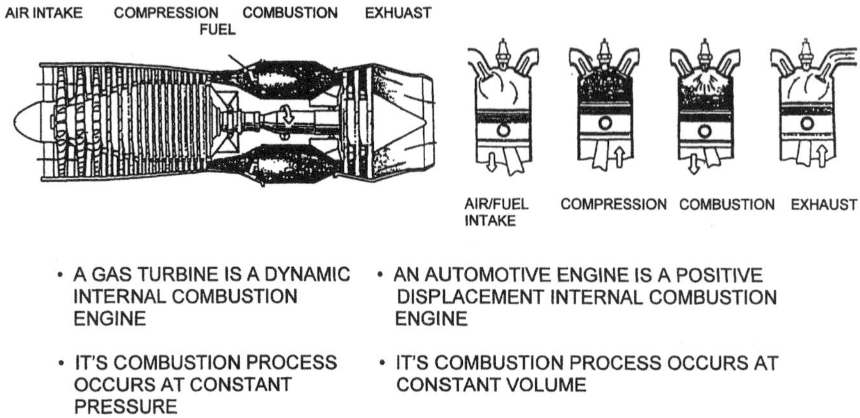

Figure 26.3 Gas turbine vs automotive engine (Courtesy of Dresser-Rand)

engine having an intake, compression, combustion and exhaust stroke. A gas turbine engine is a dynamic internal combustion engine. The process in this case is continuous and not intermittent, as is the case for the automotive engine. Both engines have compression, combustion and exhaust sections. When one considers the similarities of these engines in this manner, it can be seen that both require starters, ignition sources, inlet air filters, inlet fuel systems, cooling systems, and monitoring systems.

Building a gas turbine

We have already discussed turbo compressors and expansion turbines. A gas turbine is a combination of these components, plus a combustor that produces the hot gas for expansion. Figure 26.4 presents these facts.

Figure 26.5 shows a gas turbine configuration for a gas turbine with a regenerator. The function of the regenerator is to use gas generator exhaust vapors to preheat the air exiting the air compressor, thus reducing the amount of fuel required by the gas turbine. Figure 26.5 also shows the changes of gas temperature, pressure, energy, and the horsepower produced in the different sections of the gas turbine. Note that the power produced in a gas turbine is typically three times the output power. This is because the air compressor typically requires two-thirds of the produced power.

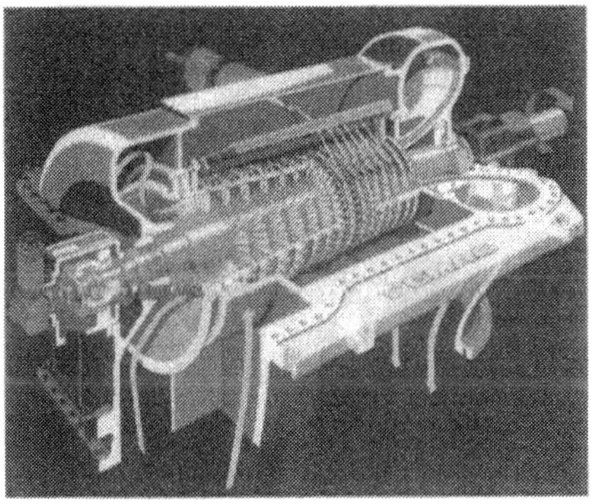

Figure 26.4 Building a gas turbine

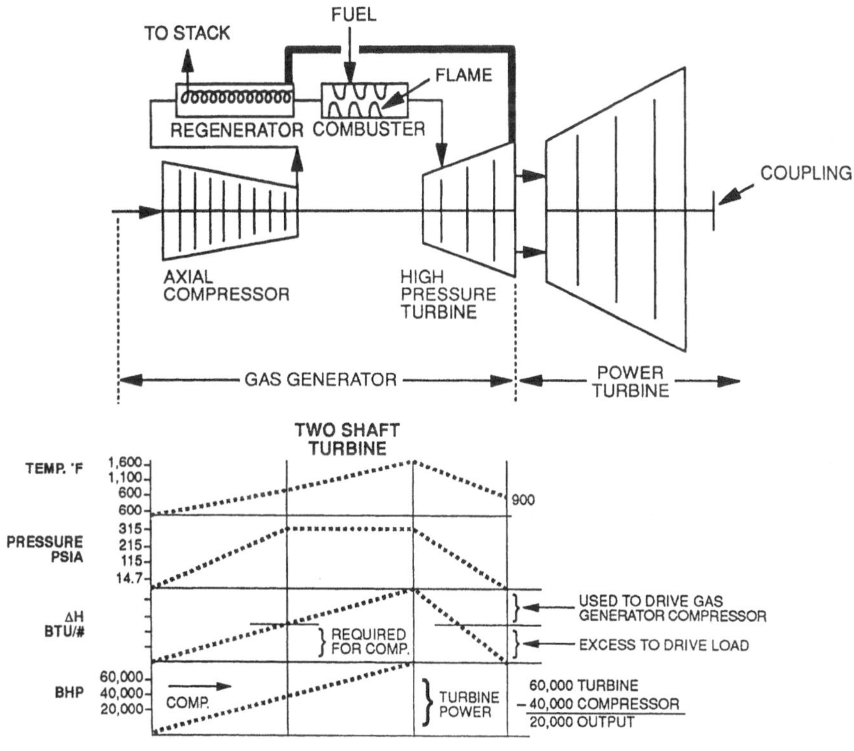

Figure 26.5 Gas turbine configuration

History of gas turbine development

A brief history of gas turbine development is presented in Figure 26.6. Gas turbines were initially used in the early 1940's for military purposes. In the 1950's, gas turbines first entered mechanical service applications. Since gas turbines are production type equipment and not custom designed, we usually refer to the generation of gas turbines. The first generation of gas turbines begins in the 1950's and progresses through the 1970's (second generation) 1980's (third generation) and present day efficiency improvements.

Gas turbines – history of development

Year	Milestone
1900–30	Various works – expansion turbines (Delaval, Parsons, etc.)
1930	Sir Frank Whittle granted gas turbine patent
1943	First successful gas turbine (jet engine)
1950s	Industrial gas generators and power turbines used in pipeline service (1st generation)
1960s	Improved efficiency through use of material and cooling improvements. Use in power generation and industrial plants
1970s	■ Development of 2nd generation – larger sizes, higher efficiency (higher firing temperatures) ■ First uses of aero derivative gas generators and power turbine for 'off shore' applications ■ Increased availability ■ Increased preventive maintenance cycle time
1980s	■ Extensive use of gas turbines in combined cycles for co-generation ■ Aero derivative types gain further acceptance ■ Continued efficiency increase (higher firing temperatures) – retrofits of 1st generation units ■ Development of 3rd generation gas turbines (use of advanced materials, processing, coating and cooling techniques
1990s	■ Further acceptance of gas turbine as an industrial prime mover ■ Simple cycle efficiencies approach 45%. Firing temperatures approach 2500°F

Figure 26.6 Gas turbines – history of development

Gas turbine classifications

In this section, we will discuss the different gas turbine classifications in terms of design type, number of shafts, drive location, and cycle.

Classification by design type

Figure 26.7 presents the industrial type of gas turbine. The two industrial divisions of gas turbines are shown. The older industrial type which was grass roots industrial; that is, never built to function as an aircraft engine. The modern approach to industrial type gas turbines (late 1960's) is the aero derivative influenced industrial gas turbine. This gas turbine design evolves from the aircraft industry and is a lighter weight type of industrial turbine. Maintenance is easier than the grass roots industrial since components are modulized and are changed out as opposed individual parts in the grass roots industrial. Both industrial types are differentiated from aero derivative types by the fact the radial bearings are always hydrodynamic. Facts concerning industrial type turbines – advantages and disadvantages are presented in Figure 26.8. In recent years, the high emphasis on maintainability has favored the aero derivative type gas turbine as opposed to either type of industrial gas turbine presented here. This is because the maintenance times for the aero derivative gas turbines in the field are significantly reduced over industrial type gas turbines because the aero derivative

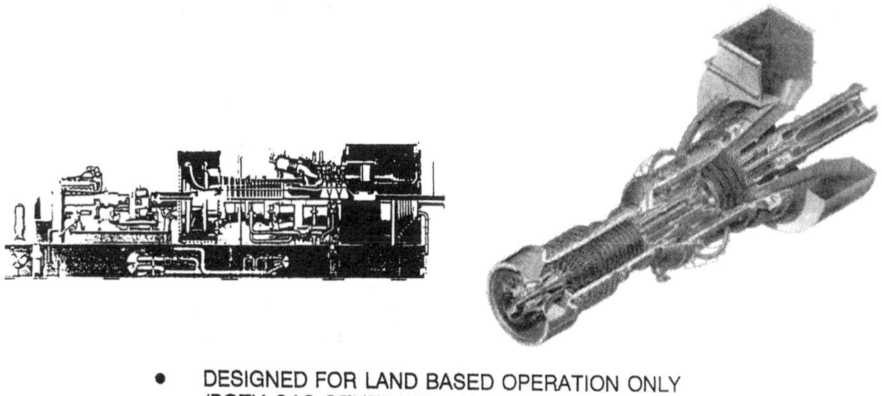

- DESIGNED FOR LAND BASED OPERATION ONLY (BOTH GAS GENERATOR AND POWER TURBINE)
- TWO VARIATIONS

Figure 26.7 Gas turbine classifications industrial type Left: Grass root's industrial (never built to fly) (Courtesy of General Electric Co.). Right: Aero-influenced industrial (lighter weight hydrodynamic bearings) (Courtesy of Solar Turbines, Inc.)

unit can be easily exchanged with a similar unit in the field. Therefore, the field maintenance time is significantly lower for the aero derivative gas turbine (typically 72 hours as opposed to 360 hours+).

Industrial type gas turbines

Advantages	Disadvantages
■ Longer cycle time between maintenance	■ Longer maintenance times
■ Longer bearing life (hydrodynamic bearings)	■ Large foot print
■ Greater tolerance to upsets	■ High specific weight
	■ Lower efficiency (1st and 2nd generation)
	■ Longer start sequences

Figure 26.8 Industrial type gas turbines

A single shaft grass roots industrial gas turbine – General Electric model 7000 (Frame 7) is shown in Figure 26.9. This turbine has been used for both generator drives and mechanical drives. Nominal ISO horsepower is in the 135,000 BHP range. Efficiency is approximately 35%.

Figure 26.10 shows an example of a two shaft aero derivative gas turbine – Solar Mars gas turbine used for mechanical drive applications (compressor and pump drives). Nominal ISO horsepower is in the 15,000 BHP range. Efficiency is approximately 35%

Figures 26.11, 26.12 and 26.13 are examples of various aero derivative gas turbines. In Figure 26.11, a General Electric LM 2500 gas turbine is shown in two applications, the first as a gas generator for a Dresser

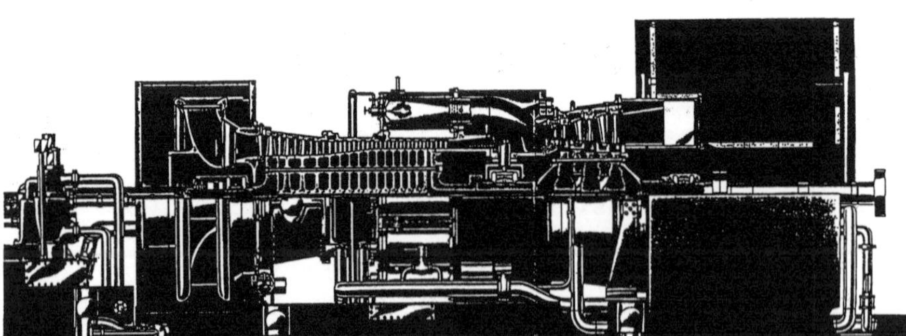

Figure 26.9 Single shaft industrial gas turbine (Courtesy of General Electric)

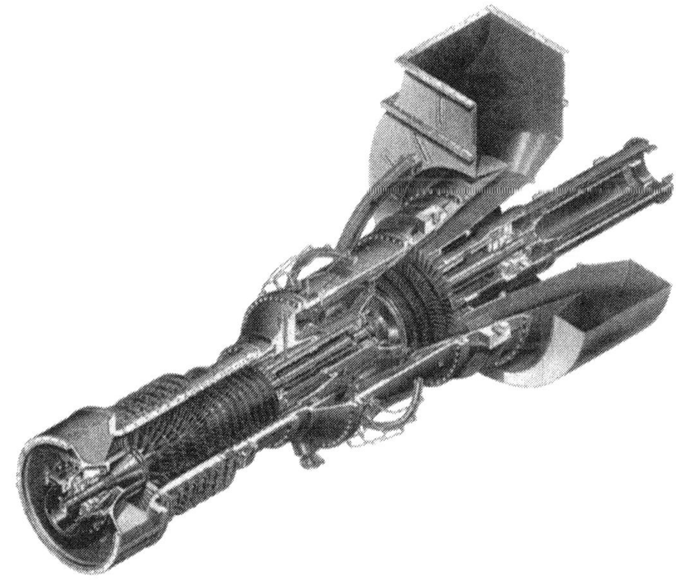

Figure 26.10 Mars gas turbine (Courtesy of Solar Turbines)

Rand DJ 270R power turbine. The other application uses the LM 2500s six-stage power turbine on a separate shaft. Nominal ISO horsepower is in the 25,000 BHP range. Efficiency is approximately 37%.

Figure 26.12 is a drawing of a Rolls Royce RB211 two shaft gas turbine. This gas turbine has intermediate and high pressure axial

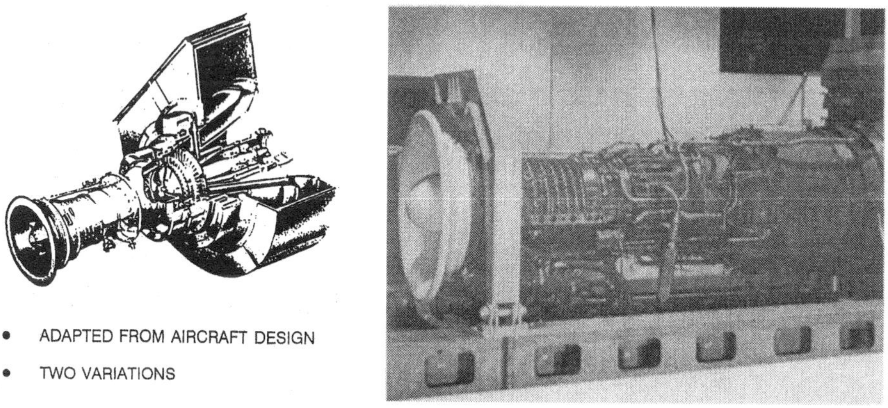

Figure 26.11 Gas turbine classifications aero derivative. Left: used as gas generator only (Courtesy of Dresser Rand). Right: entire engine adapted (gas generator and power turbine) Note: power turbine is either turbo-prop or bypasss fan drive in aero-engine version (Courtsey of General Electric)

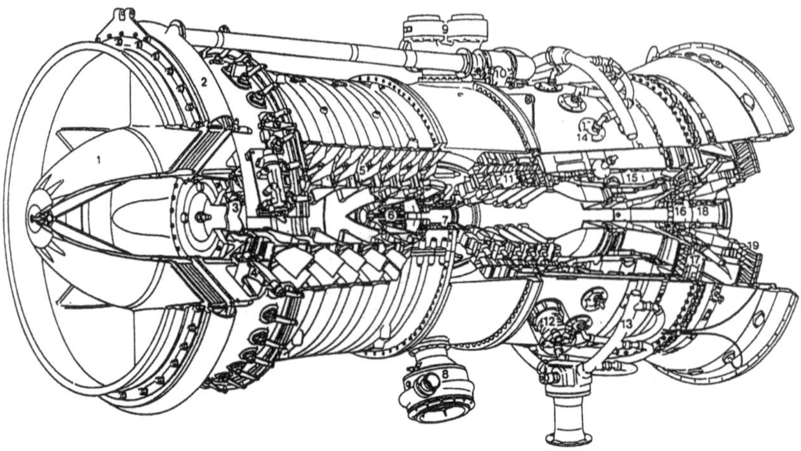

1 Hose fairing
2 Anti-locking duct
3 IP compressor front bearing'
4 Variable inlet guide vane
5 IP compressor
6 IP location bearings
7 HP location bearings
8 Air starter
9 IP handling bleed valves
10 Anti-locking hot air valves
11 HP compressor
12 starting bleed valve
13 Gas fuel manifold
14 Gas fuel burner
15 Combustion chamber
16 HP turbine bearing
17 HP turbine
18 IP turbine bearing
19 IP turbine

Figure 26.12 The industrial RB211 (Courtesy of Rolls Royce)

compressors mounted on separate shafts for increased gas turbine efficiency. Nominal ISO horsepower is in the 30,000 BHP range. Approximate efficiency is 38%.

The newest aero-derivative gas turbine used for generator and mechanical drive is the LM 6000 two shaft gas turbine shown in Figure 26.13. This gas turbine produces 60,000 ISO horsepower, has an efficiency of 43% and can drive a load on either or both ends.

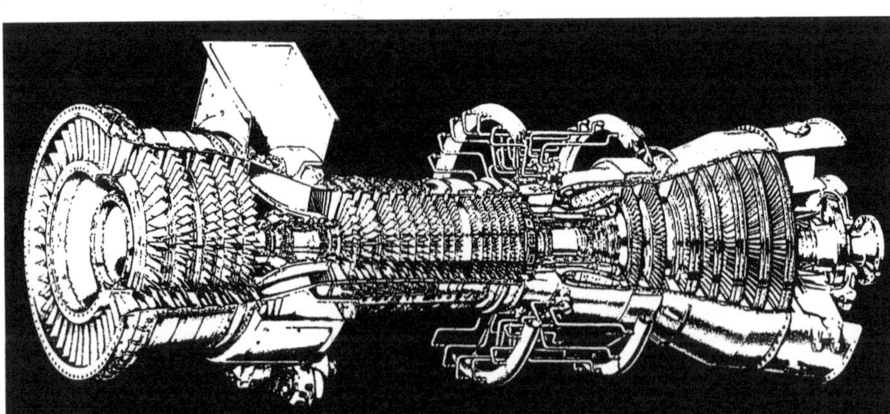

Figure 26.13 LM6000 gas turbine (Courtesy of General Electric)

The advantages and disadvantages of aero-derivative gas turbines are presented in Figure 26.14.

Aero-derivative type gas turbines

Advantages	Disadvantages
■ Shorter maintenance times	■ *Shorter cycle time between maintenance
■ Small foot print	■ Less tolerance to upsets
■ Low specific weight	■ Shorter bearing life (anti-friction bearings)
■ Higher efficiency	
■ Faster start sequence	

*Note: maintenance cycle time is increasing and approaching industrial types

Figure 26.14 Aero-derivative type gas turbines

Figure 26.15 shows a hybrid type gas turbine, which is a combination of an aero-derivative gas generator and an industrial power turbine. This design offers the advantage of maintainability on the 'hot section' of the gas turbine and high reliability in the power turbine. These facts are presented in Figure 26.16.

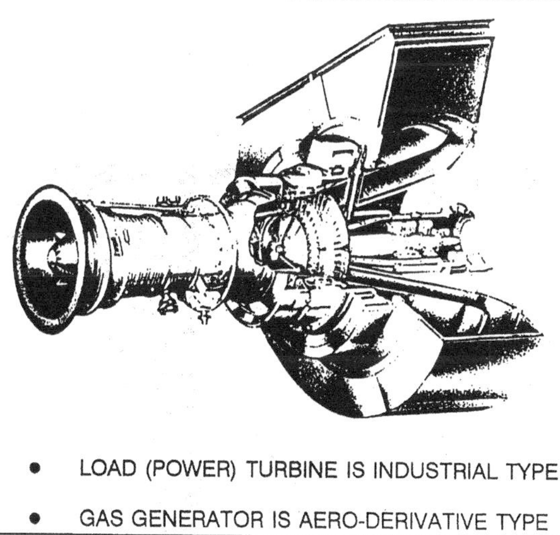

Figure 26.15 Gas turbine classifications hybrid type industrial (Courtesy of Dresser Rand)

Classification of industrial and aero-derivative gas turbines

- Usually based on power turbine type
- Depends on types of bearings
  - Anti-friction = aero
  - Hydrodynamic = industrial
- With time, both types will converge to a 'hybrid'
- Present 3rd generation designs are moving in this direction

Figure 26.16 Classification of industrial and aero-derivative gas turbines

Aero-derivative and industrial facts are discussed in Figure 26.17.

Aero-derivative vs industrial facts

Item	Aero-derivative	Industrial
Casing weight	Light	Very heavy
Casing material yield	3 times higher yield strength	–
Rotor weight	15–20 times lighter	–
Bearing type	Anti-friction	Hydrodynamic
Bearing life	50,000 hours	50,000–100,000 hours
Start-idle times	1–2 minutes	15–30 minutes
Boroscope locations	More than industrial	

Figure 26.17 Aero-derivative vs industrial facts

The number of gas turbine shafts

Gas turbines are configured as single, dual or triple shaft designs. The advantages and disadvantages of each type are presented in Figure 26.18. Most modern gas turbines are of the triple shaft design. Figure 26.19 shows a single shaft gas turbine where the gas generator and power turbine are mounted on the same shaft. This figure also shows a dual shaft gas turbine, where the gas generator and power turbine are mounted on different shafts. Single shaft gas turbines are usually

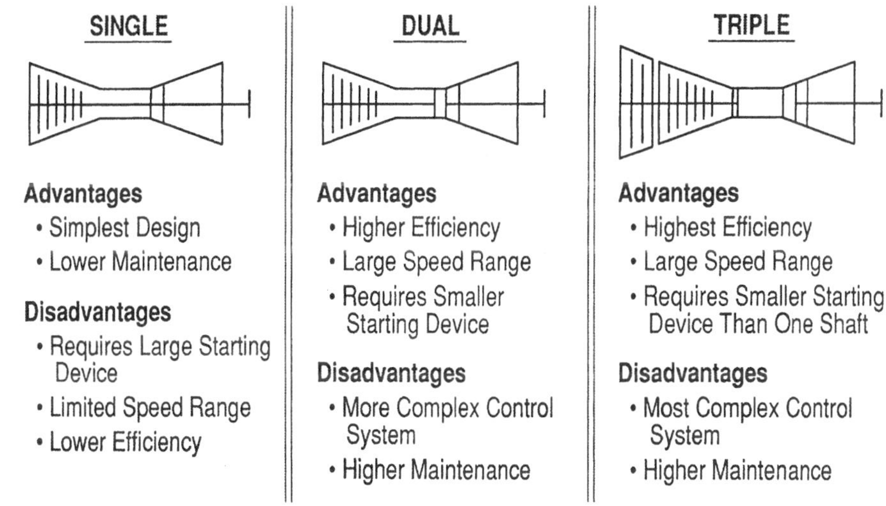

Figure 26.18 The number of gas turbine shafts advantages/disadvantages

limited to generator drive applications since the starting turbine load is significantly less for a generator application, because generator is started under no load. Dual shaft turbines are used for mechanical drive, pump and compressor applications.

- Single shaft – Gas generator and power turbine on one shaft (only industrial types) (left)

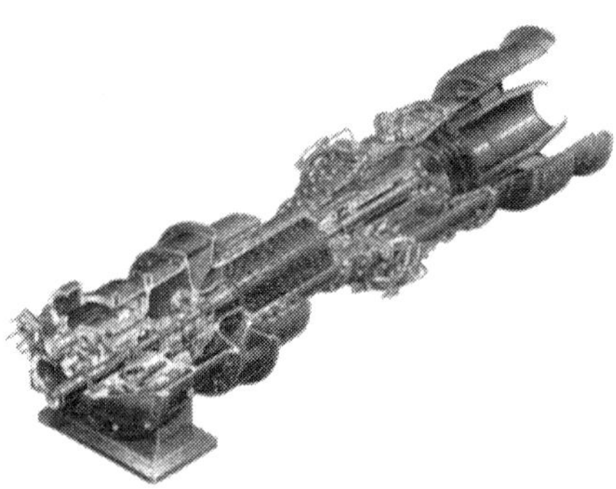

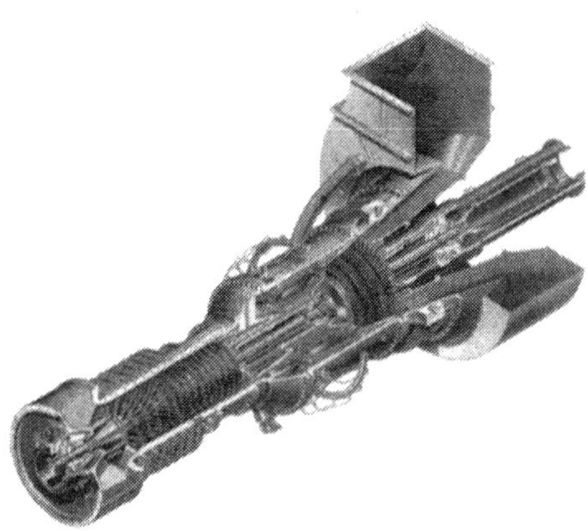

- Dual shaft – Gas generator and power turbine on individual shaft (industrial and aero types) (right)

Figure 26.19 The number of gas turbine shafts (Courtesy of Solar Turbines, Inc.)

Gas turbine drive configurations

Gas turbines can be designed as hot end drive, or cold end drive. Figure 26.20 presents these facts. The majority of first and second generation gas turbines were of a hot end drive. Most third generation gas turbines are of the cold and dry type. A cold end drive configuration is a more reliable approach, in the writer's opinion, since the coupling environment is significantly reduced in terms of temperature. This results in a much lower axial expansion of the drive coupling and subsequently increases the reliability of the gas turbine.

Gas turbine drive configurations	
Hot end drive (exhaust end)	**Cold end drive (inlet end)**
■ Majority of 1st, 2nd generation	■ Some 2nd generation ■ Most 3rd generation
Disadvantages	**Advantages**
■ Longer drive coupling spacer	■ Shorter drive coupling Spacer
■ Driver coupling in hot environment	■ Minimized thermal expansion effects

Figure 26.20 Gas turbine drive configurations

Gas turbine cycles

Gas turbine cycles are presented in Figure 26.21. There are essentially three types of gas turbine cycles. The simple cycle, where the gas is exhausted directly to atmosphere. The regenerative cycle, where the exhaust gas is used in an exchanger (regenerator) to preheat the compressor discharge air prior to the combustor and the combined cycle where the exhaust gas is used in a heat recovery steam generator (HRSG) to either generate steam for plant use or as an expansion fluid is a steam turbine. Typical efficiencies are as follows:

- Simple cycle 20% to 43%
- Regenerative cycle 30% to 45%
- Combined cycle 55% to 60%

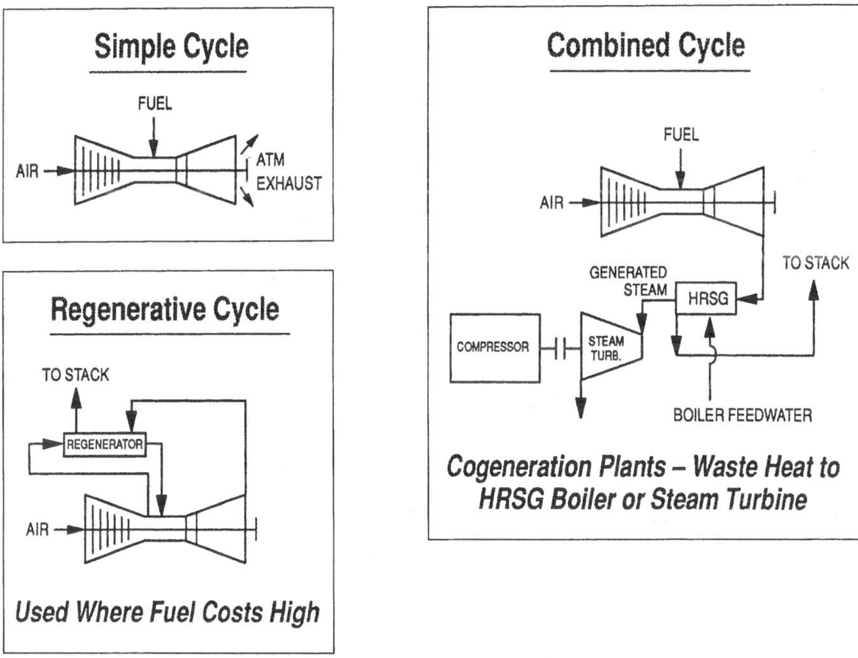

Figure 26.21 Gas turbine cycles

Gas turbine applications

Industrial single shaft

- Generator drive
- *Limited mechanical drive applications (Compressor – pump)

Industrial dual shaft

- Land based mechanical drive
- Pipeline
- Refinery, gas plant, petrochemical plant land based
- Generator drive

*Usually used only when a multi-shaft alternative in the designed power range is not available

Figure 26.22 Gas turbine applications

Gas turbine applications (continued)

Aero multi-shaft (dual or triple)

- Co-generation
- Off-shore
    - Generator drive
    - Compressor drive
- Pipeline
    - Pump drive
    - Compressor drive

Figure 22.23 Gas turbine applications (continued)

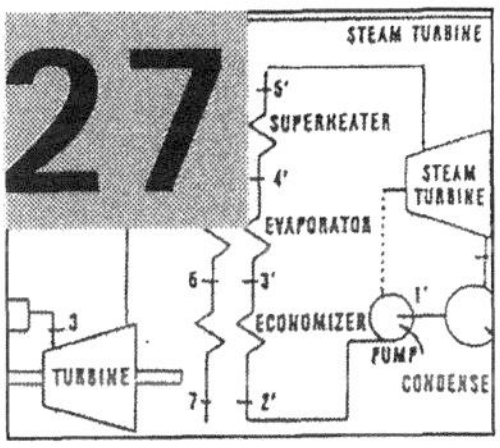

Gas turbine performance

- Introduction
- Gas turbine ISO conditions
- Site rating correction factors
- The effect of firing temperature on power and efficiency

Introduction

In this chapter we will consider the performance of a gas turbine. As previously discussed the gas turbine is a dynamic internal combustion engine. When we compare the performance of a gas turbine to a steam turbine, it becomes immediately evident that steam turbine performance is much easier to calculate since both the vapor and the vapor conditions are fixed. When we examine the performance of a gas turbine, we immediately see that the vapor condition is variable based on the type of fuel used and the atmospheric conditions. This is true since the inlet to the gas turbine engine is from atmosphere and any change of temperature, humidity or pressure will affect the mass flow and consequently the power produced by the gas turbine. The gas turbine cycle (Brayton) is open.

As a result, steam turbine performance can be expressed rather easily in terms of steam rate (pounds of steam per horsepower or kilowatt hour) and external efficiency. Since the gas turbine vapor conditions are variable however, gas turbine performance must be expressed in terms of heat rate, BTU's per horsepower or kilowatt hour, thermal efficiency and fuel rate. All of the above also must be expressed in standardized terms.

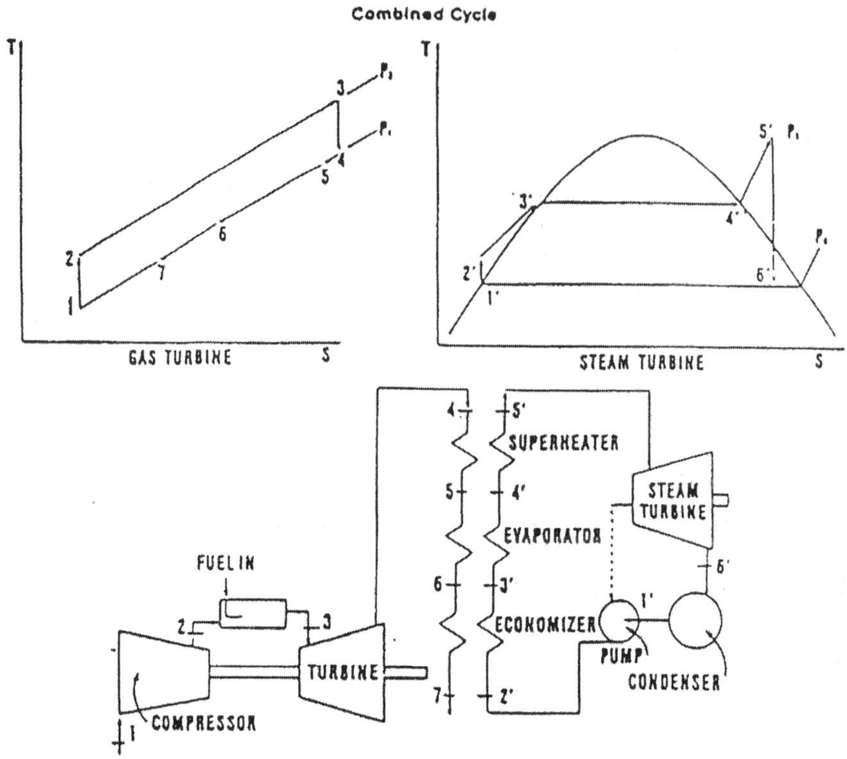

Figure 27.1 Gas turbine vs steam turbine performance (Reprinted with permission of Gas Producers Suppliers Association GPSA)

A set of standardized conditions has been established by ISO (International Standards Organization) to rate all gas turbines. We will discuss the various ISO standard requirements and how the site rating is obtained by using vendor ISO derating data for each turbine design. An actual gas turbine performance example will be presented and the effect of varying inlet conditions (temperature, pressure and humidity) on performance will also be presented.

Finally, the exhaust gas composition will be discussed and the emission products examined. In addition, various alternatives for meeting local emission requirements will be presented and discussed.

Figure 27.1 presents a comparison between gas turbine and steam turbine performance.

A gas turbine is an internal combustion engine in that the hot vapor is produced internal to the engine. The cycle is open since both inlet and exhaust conditions are "open" to the atmosphere and vary with atmospheric conditions.

The steam turbine is an external combustion engine since the hot vapor is produced external to the engine. The steam turbine cycle is closed in that both inlet and exhaust conditions are controlled by the steam generation system (boiler), therefore steam turbine conditions are constant and do not vary.

Figure 27.2 presents performance parameters for steam turbines. Since inlet and exhaust conditions are controlled and the steam turbine is an external combustion engine, steam rate and external efficiency can be used to express performances.

Gas turbine vs steam performance

- Since the steam turbine (Rankine) cycle is closed, the vapor and vapor conditions are constant!
- Therefore performance can be expressed in terms of:
- Steam rate = $\dfrac{\text{LBS. of steam}}{\text{BHP or KW - HR}}$
- External efficiency = $\dfrac{\text{actual work x loss factor}}{\text{ideal work}}$

Mechanical and leakage losses

Figure 27.2 Gas turbine vs steam performance

Since the gas turbine Brayton cycle is open, vapor conditions are variable and performance must be expressed as:

- Heat rate
- Thermal efficiency
- Fuel rate

These facts are shown in Figure 27.3.

Gas turbine vs steam turbine performance (continued)

- Since the gas turbine (Brayton) cycle is open, both the vapor and vapor conditions are variable.
- Therefore performance is expressed in terms of:
- heat rate (ISO) = $\frac{\text{BTU}}{\text{BHP or KW-HR}}$
- Thermal efficiency = $\frac{\text{BTU/HP-HR}}{\text{heat rate (ISO) BTU/HP-HR}}$
- fuel rate = $\frac{\text{(heat rate) x (horsepower)}}{\text{Fuel heating value (BTU/LB)}^{\dagger}}$

Note: * ISO conditions – standardized fuel, inlet conditions at design speed – no losses
† BTU/SCF (standard cubic foot) for gas fuel

Figure 27.3 Gas turbine vs steam turbine performance

Gas turbine ISO conditions

Since gas turbine performance varies as a function of fuel and inlet conditions, a set of standard conditions has been established by the International Standards Organization to define gas turbine performance. These facts are presented in Figure 27.4.

Gas turbine performance ISO conditions

- Since gas turbine performance varies as a function of fuel and inlet conditions, a set of standardized conditions has been established by ISO (International Standards Organization) to rate all gas turbines.
- ISO standard conditions are:
  - T inlet = 59°
  - P inlet = sea level
  - inlet and exhaust losses = 0"H_2O
  - Based on stated fuel heating value
  - relative humidity = 0%
  - design speed of rotors
  - power losses = 0
  - compressor bleed air = 0

Figure 27.4 Gas turbine performance ISO conditions

Dresser-Rand
Turbo Products Division Olean, New York-USA

MODEL	POWER RATING ISO Base Load Gas Fuel (hp)	HEAT RATE Lower Heating Value (LHV) (Btu/hp-hr)	POWER SHAFT SPEED (RPM)	PRESSURE RATIO	NUMBER OF COMBUSTORS	AT ISO RATING CONTINUOUS Turbine Inlet Temp. (°C)	Exhaust Flow (kg/sec)	Exhaust Temp (°C)
DR-22C	5,278	8,850	13,820	9.9	6	1,035	15.6	579
DR-990	5,900	8,350	7,200	12.2	1	1,082	20.0	482
DR-60G	18,750	6,840	7,000	21.5	1	1,216	45.6	482
DR-61G	31,200	6,777	3,600	18.8	1	1,235	69.0	523
DR-61	30,800	6,800	5,500	18.8	1	1,235	69.0	520
DR-63G	56,840	6,135	3,600	30.0	1	1,154	122.5	452

- ALL VENDORS PUBLISH ISO PERFORMANCE AND DE-RATING DATA SO THAT SITE PERFORMANCE (AT ACTUAL SITE CONDITIONS, FUEL AND LOSSES) CAN BE DETERMINED
- TYPICAL VENDOR'S ISO DATA

Figure 27.5 Gas turbine performance ISO and site performance (Reprinted with pemission of Turbomachinery International Handbook 1993 Vol. 34 No. 3)

Gas turbine vendors publish performance data in terms of ISO power rating and ISO heat rate. Typical vendor data is shown in Figure 27.5.

Site rating correction factors

Gas turbine site performance is directly effected by inlet air density and air environmental conditions as shown in Figures 27.6, 27.7 and 27.8 respectively.

The effect of inlet air density on produced power and heat rate

- A given engine design limits air volume flow capacity.
- Produced power is a function of actual energy extracted per pound of vapor and mass flow of vapor.
- For a given engine therefore, produced power varies directly with inlet air density.
- Produced power does become limited by low volume (stall and surge) flow.

Figure 27.6 The effect of inlet air density on produced power and heat rate

Gas turbine performance effect of inlet conditions on performance

- Care must be taken when selecting gas turbines to assure sufficient shaft power is available at:
  - High temperature conditions
  - Fouled inlet conditions
- Gas turbine applications tend to be "fully loaded" since gas turbines (unlike steam turbines) are not custom designed.

Figure 27.7 Gas turbine performance effect of inlet conditions on performance

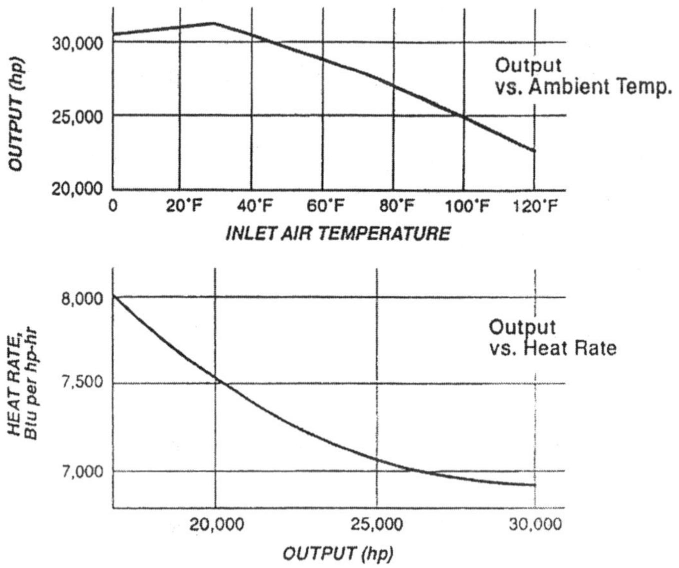

Figure 27.8 Typical gas turbine output power and heat rate vs ambient temperature

Since produced power and heat rate vary as a function of inlet temperature, pressure and inlet duct and exhaust duct pressure drop, vendors supply correction curves to convert ISO conditions to site conditions.

Figures 27.9A to 27.9F present an example of a typical gas turbine site rating exercise.

Typical gas turbine site rating exercise

1. Scope
   The purpose of this specification is to estimate the site shaft horsepower and heat rate for a given set of site conditions.

2. Applicable documents
   Figures 27.9D, 27.9E and 27.9F

3. Requirements
   3.1 The following site condition must be known:
   - A. Elevation (ft.)
   - B. Inlet temperature (°F)
   - C. Inlet duct pressure loss (inches of water)
   - D. Exhaust duct pressure loss (inches of water)

4. Procedure
   4.1 Read the shaft horsepower (SHP) and heat rate (HR) for the site inlet temperature (from Figure 27.9D)

   4.2 Read the elevation correction factor (δ) for the site elevation (from Figure 27.9E)

   4.3 Site shaft horsepower:
   - A. Read the inlet correction factor (Ki) for the site inlet duct pressure loss (from Figure 27.9F).
   - B. Read the exhaust correction factor (Ke) for the site exhaust duct pressure loss (from Figure 27.9F).
   - C. Calculate the site shaft horsepower:

     Site SHP = SHP (from Figure 27.9D) x δ x Ki x Ke

   4.4 Site heat rate:
   - A. No elevation correction factor (δ) is used for the heat rate.
   - B. Read the heat rate correction factor (Kh) from Figure 27.9F for the duct pressure loss (sum of site inlet and exhaust duct pressure losses).
   - C. Calculate the site heat rate:

     Site HR = HR (from Figure 27.9D) x Kh

5. Sample calculation
   5.1 Assume the following site conditions:

A. Elevation (ft.)	1000 ft
B. Inlet temperature (°F)	59°F
C. Inlet duct pressure loss (inches of water)	3.5 inches of water
D. Exhaust duct pressure loss (inches of water)	4.5 inches of water

Figure 27.9A Typical gas turbine site rating exercise

Typical gas turbine site rating exercise (continued)

5.2 Read the shaft horsepower (SHP) and heat rate (HR) with no inlet or exhaust duct pressure losses for the site inlet temperature (from Figure 27.9D).

59°F site inlet temperature

Shaft horsepower (SHP) 29,200

Heat rate (HR) 7,035 BTU/HP-HR

5.3 Read the elevation correction factor (δ) for the site elevation (from Figure 27.9E).

1000 ft. site elevation:

Elevation correction factor (δ) 0.964

5.4 Site shaft horsepower:

A. Read the inlet correction factor (Ki) for the site inlet duct pressure loss (from Figure 27.9F).

3.5 inches of water site inlet duct pressure loss:

Inlet correct factor (Ki) 0.9845

B. Read the exhaust correct factor (Ke) for the site exhaust duct pressure loss (from Figure 27.9F).

4.5 inches of water site exhaust duct pressure loss:

Exhaust correction factor (Ke) 0.991

C. Calculate the site shaft horsepower:

Site SHP = SHP (from Figure 27.9D) x δ x Ki x Ke

= 20.200 x 0.964 x 0.9845 x 0.991

Site SHP = 27,463

5.5 Site heat rate:

A. Read the heat rate correction factor (Kh) from Figure 27.9F for the duct pressure loss (sum of site inlet and exhaust duct pressure losses).

Duct pressure loss is the sum of 3.5 inches of water site inlet duct pressure loss and 4.5 inches of water site exhaust duct pressure loss, equaling 8.0 inches of water duct pressure loss:

Heat rate correction factor (Kh) 1.016

Figure 27.9B Typical gas turbine site rating exercise (continued)

Typical gas turbine site rating exercise (continued)

B. Calculate the site heat rate:

Site HR = HR (from Figure 27.9D) x Kh

= 7,035 x 1.016

Site HR = 7,148 BTU/HP-HR

Figure 27.9C Typical gas turbine site rating exercise (continued)

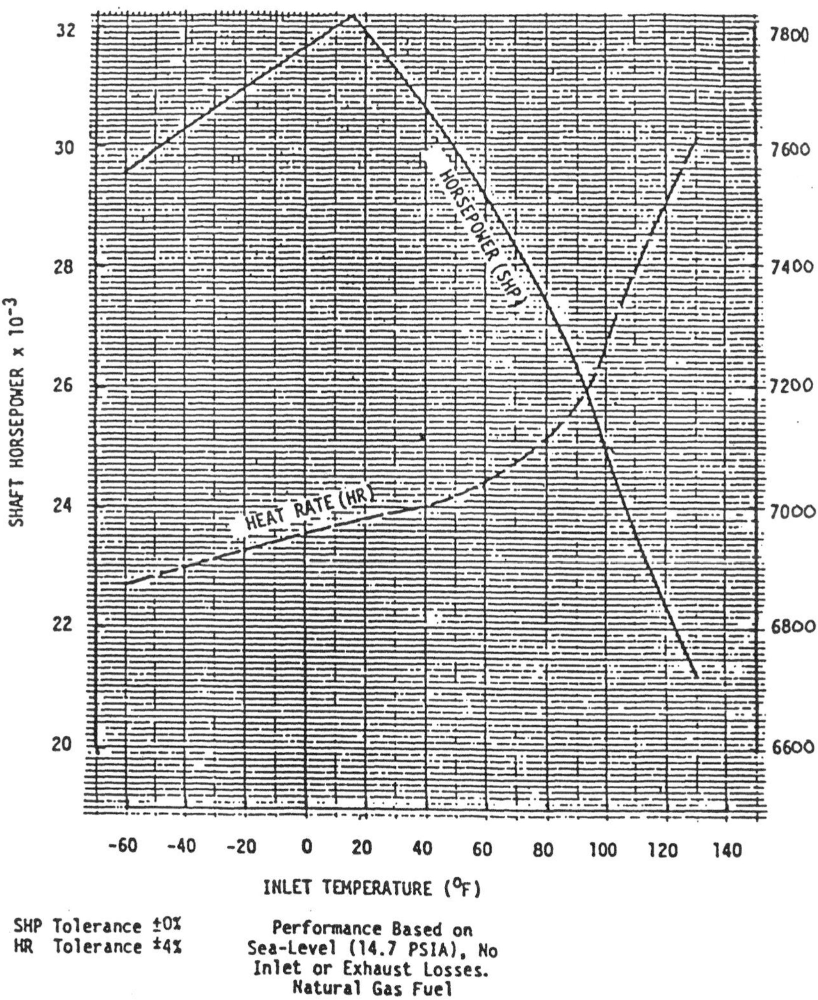

Figure 27.9D Figure for typical gas turbine site rating exercise (Courtesy of General Electric)

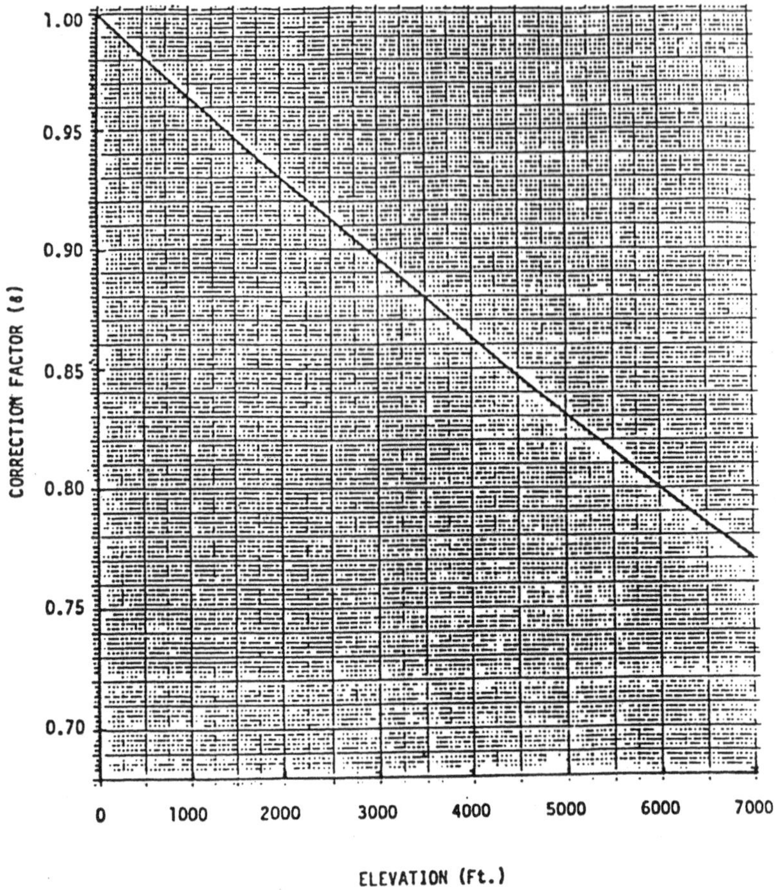

Figure 27.9E Figure for typical gas turbine site rating exercise (Courtesy of General Electric)

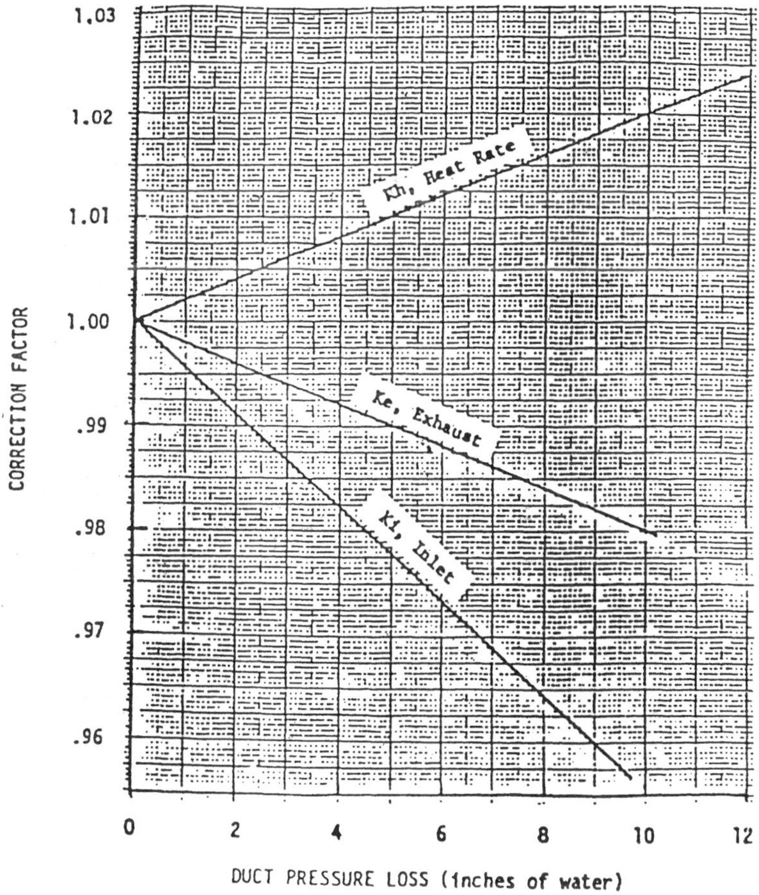

Figure 27.9F Figure for typical gas turbine site rating exercise (Courtesy of General Electric)

- THE HOTTER THE VAPOR (FIRING TEMPERATURE) THE GREATER AMOUNT OF ENERGY EXTRACTED FROM EACH LB OF VAPOR.

Mollier Diagram – Typical Fuel Vapor

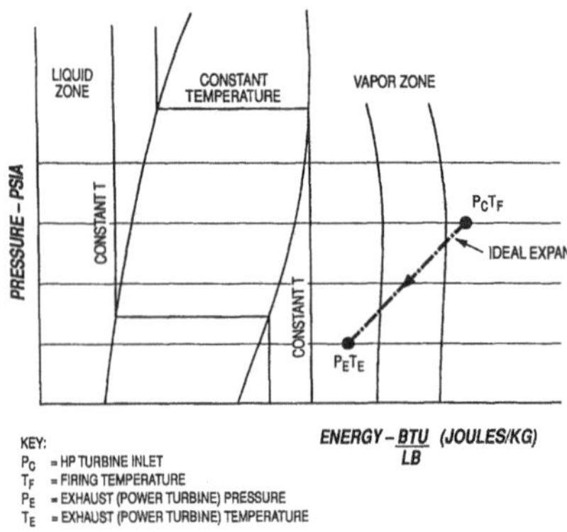

<u>TYPICAL VALUES</u>

100°F INCREASE IN FIRING TEMPERATURE
= APPROXIMATELY 10% INCREASE IN PRODUCED POWER

= APPROXIMATELY 3% INCREASE IN ENGINE EFFICIENCY

Figure 27.10 The effect of increased firing temperature on produced Power and engine efficiency

The effect of firing temperature on power and efficiency

A small increase in firing temperature has a significant effect on produced horsepower on engine efficiency. These facts are shown in Figure 27.10.

As will be shown in the next chapter (gas turbine mechanical component design), hot gas path material places constraints on firing temperature. Development of new material, cooling schemes and high temperature coatings will lead the way to increased turbine point output and efficiency.

Gas turbine mechanical design

- Introduction
- Major component assemblies
- Air compressor section
- Combustor design
- Power turbine design

Introduction

In this section we will discuss the function of the major components of a gas turbine. We will begin however, by discussing the design similarities to other types of equipment with which you are already familiar. Namely, turbo-compressors and steam turbines. We will then proceed to define the major parts of a gas turbine. The gas generator consists of the air compressor, combustor and high pressure turbine. We will then present the power turbine. The major components of both the gas generator and the power turbine will then be covered in detail.

As in the case of the steam turbine, we will focus on blade design, stage types and their applications. We will then proceed to discuss blade efficiency considerations and will conclude the section by discussing blade root types, blade natural frequencies and blade loading considerations.

Major component assemblies

Figure 28.1 presents an overview of the four (4) major component assemblies of any type of gas turbine.

Gas turbine major component assemblies

- Gas generator
  - Compressor
    - Low pressure
    - High pressure
  - Combustor
  - Turbine
    - High pressure
    - Low pressure
  - Power turbine

Figure 28.1 Gas turbine major component assemblies

As already discussed, regardless of design (industrial or aero-derivative) or configuration (number of shafts), the major parts remain the same. In a single shaft configuration, the gas generator turbine (HPT) will also function as the power turbine by delivering excess power to the load.

The function of turbo-compressors and expansion turbines are already known. Therefore, the compressor section (usually axial type) increases intake air energy by increasing the air velocity in the blade sections and reducing it in the stator vanes. Gas velocities off the blades exceed 600 mph. The stator vanes slow the air velocity to approximately 300 mph, thereby increasing air energy (head) and pressure and efficiently directing the air to the following stage.

The function of any type of expansion turbine is exactly the opposite of a turbo-compressor. The stationary nozzles contained in diaphragms or blade carriers increase the gas velocity to over 600 mph and inpart high values of kinetic energy to the blades to produce rotor rotation and work. The typical gas exit velocity off the turbine blades is between 300–350 mph depending on the types of blades used (impulse or reaction).

However, even though the function of the turbo-compressor and expansion turbine are exactly the same when employed in a gas turbine, significant high gas temperatures require component and assembly modifications. These facts are presented in Figure 28.2.

Required component design modifications

- The function of the component parts have been discussed for:
  - Turbo-compressors
  - Expansion turbines
- However, the high temperatures result in the following design enhancements:
  - Stacked rotor design
  - Internal rotor drive shafts (concentric drives)
  - Compressor air bleedoff's
  - Nozzle blade cooling
  - Nozzle blade coatings
  - Super alloy nozzle and blade material selection

Figure 28.2 Required component design modifications

Air compressor section

The major components of the air compressor are defined in Figure 28.3.

Air compressor assembly components

- Compressor casing
- Inlet air plenum
- Stator vanes (guide vanes – radial type)
- Rotor blade rows (impellers – radial type)
- Rotor
- Low and high pressure air compressor assemblies (aero-derivative type)

Figure 28.3 Air compressor assembly components

Each of the above components will now be reviewed in detail.

Air compressor casing

The function of the air compressor casing, similar to that of any other type of compressor, is presented in Figure 28.4, below.

Air compressor casing

Function: contain gas and support components

- Industrial – longer heat soak time
- Aero-derivative – short heat soak time

Figure 28.4 Air compressor casing

It should be noted that the shorter heat soak time for the aero-derivative casing, due to lower mass, was initially seen as a design disadvantage. However, the lower case mass results in relatively equal thermal expansions of the aero-derivative rotor and casing resulting in faster start-up times.

Compressor inlet plenum

Figure 28.5 shows the difference between an industrial and aero-derivative inlet plenum. Regardless of design approach, they both have the same function.

Compressor stator vanes

Figure 28.6 presents the function of fixed and variable stator vanes for axial compressors. Note: adjustable guide vanes are not throttle devices. They alter compressor performance curve shape by changing the entrance gas angle.

INDUSTRIAL AERODERIVATIVE

FUNCTION: DISTRIBUTE AIR TO FIRST ROW OF COMPRESSOR VANES

Figure 28.5 Compressor inlet plenum (Courtesy of General Electric)

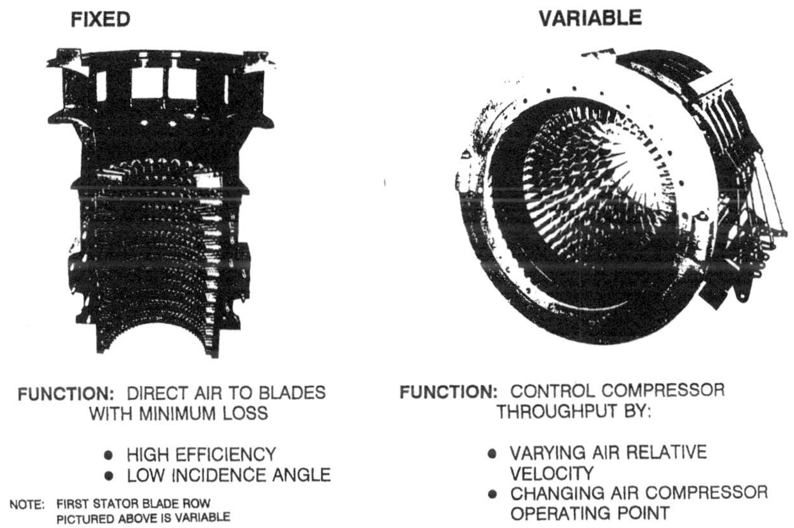

Figure 28.6 Compressor stator vanes (Courtesy of Ruston Gas Turbine Ltd)

Axial compressor blades

Axial air compressor blades, as opposed to radial impellers, are used in the majority of gas turbine designs. The reason for their use is the significantly higher attainable efficiency for high flow (over 100,000 cubic feet per minute) applications as compared to radial type impellers. Most gas turbines in the power range above 5.000 bhp use axial type compressor blades. The facts surrounding axial compressor blade design are presented in Figure 28.7.

Axial compressor blades

- General
  - High stage efficiencies (92%)
  - Compression ratios as high as 30.1 (lp and hp compressor)
  - Reaction type blading
- Aero-dynamic considerations
  - Instability areas (stall)
  - Effect of air bleed (stage mismatching)
  - Aero-dynamic excitation of natural frequencies
- mechanical considerations
  - blade natural frequencies
  - blade loading

Figure 28.7 Axial compressor blades

Radial air compressor impellers

Radial air compressor impellers are used in lower power (less than 10,000 H.P.) gas turbines. In this lower power range, the low required air flow significantly reduces the efficiency of axial compressor blades. Figure 28.8 presents radial compressor design criteria.

Radial compressor impellers

- General
  - used on smaller gas turbines (<10,000 hp)
  - lower efficiencies (80–85%)
  - lower compression ratios (10–12:1)
  - three dimensional semi-open impellers
- Aero-dynamic considerations
  - Instability area (stall)
  - Aero-dynamic excitation of natural frequencies
- mechanical considerations
  - blade natural frequencies
  - Impeller loading
  - Impeller/diaphragm clearance

Figure 28.8 radial compressor impellers

Rotor design

Gas turbine rotors have significantly more design consideration than other types of rotating machinery rotors. This is the result of higher transient and operating temperatures and blade stresses. Figure 28.9 illustrates these considerations.

Gas turbine rotor design consideration

- General requirements
  - Adaptable to rapid temperature changes (optimum tip clearances - no rubs)
  - Maintains balance (concentricity of parts)
  - Relatively lightweight

Figure 28.9 Gas turbine rotor design consideration

Figure 28.10 shows the differences between industrial and aero-derivative type rotor designs.

It should be noted that anti-friction type bearings must be used for all aero-derivative type gas turbines. This is necessary to assure close,

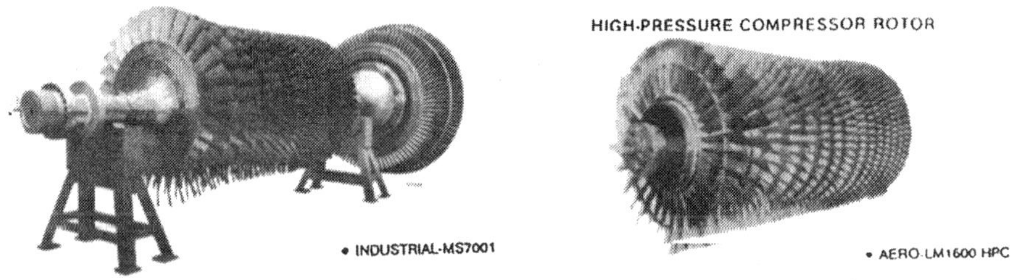

Figure 28.10 Industrial and aero-derivative rotor design differences (Courtesy of General Electric Co.)

accurate rotor position during any encountered flight air turbulence. Industrial type gas turbines always use hydrodynamic type bearings. Most manufacturers use sleeve or two-lobe (Lennon Bone) bearings. However, tilt-pad types are being used more and more and have the advantages of better rotor system stability.

Figure 28.10A presents facts concerning gas generator bearings.

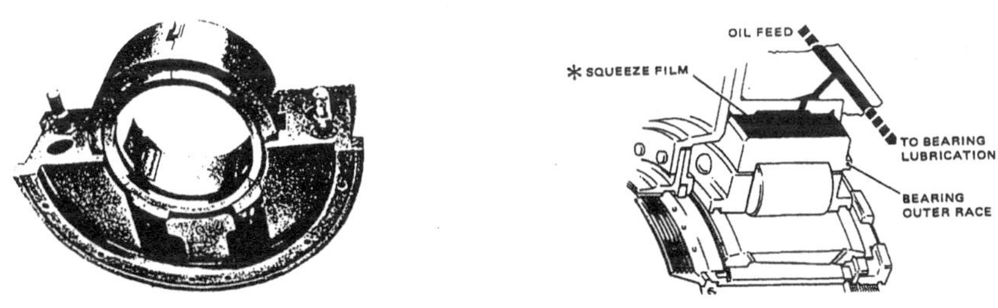

Figure 28.10A Gas generator bearings journal and thrust. Left: Industrial (Courtesy of General Electric Co.). Right: Aeroderivative (Courtesy of Rolls Royce PLC.)

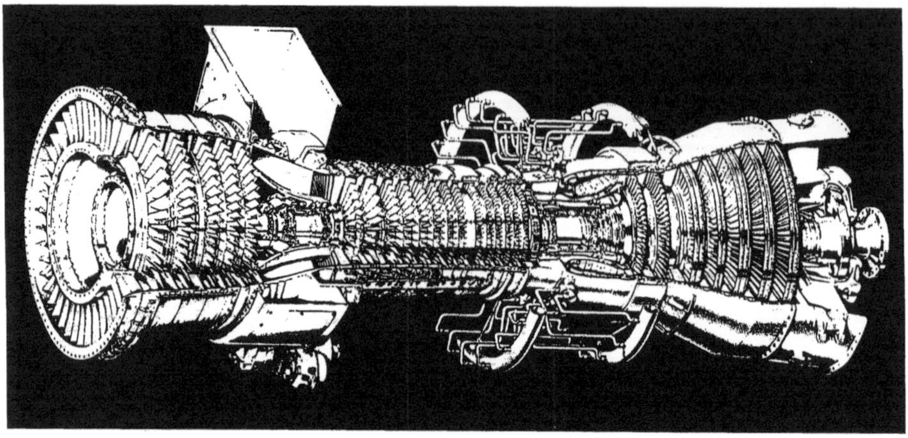

- PRESENTLY AVAILABLE IN AERO-DERIVATIVE GAS GENERATORS
- TWO COMPRESSORS IN SERIES
- DIFFERENT OPERATING SPEEDS

- INTER COMPRESSOR BLOW OFF (SURGE CONTROL) IS REQUIRED FOR:

  - START-UP, SHUTDOWN
  - OFF DESIGN OPERATION

Figure 28.11 Dual compressor rotors (Courtesy of General Electric)

The newest generation of aero-derivative gas turbines frequently employ a two-speed concentric shaft air compressor design to increase compressor efficiency by increasing the shaft speed (and therefore the specific speed) of the second (high pressure) rotor. A description of this design approach is shown in Figure 28.11 for a G.E. LM 6000 gas turbine.

Combuster design

As previously discussed, a gas turbine is an internal combustion engine. It is the dynamic alternative to the positive displacement engine used in cars, trucks, locomotives and airplanes. As such, it requires an internal source of producing a high energy vapor. A steam turbine, on the other hand, is an external combustion engine since the hot vapor is produced outside the turbine – in the steam generator (boiler).

Figure 28.12 defines the function of the combuster and its major parts.

Figure 28.12A defines the major components in a combuster and their function.

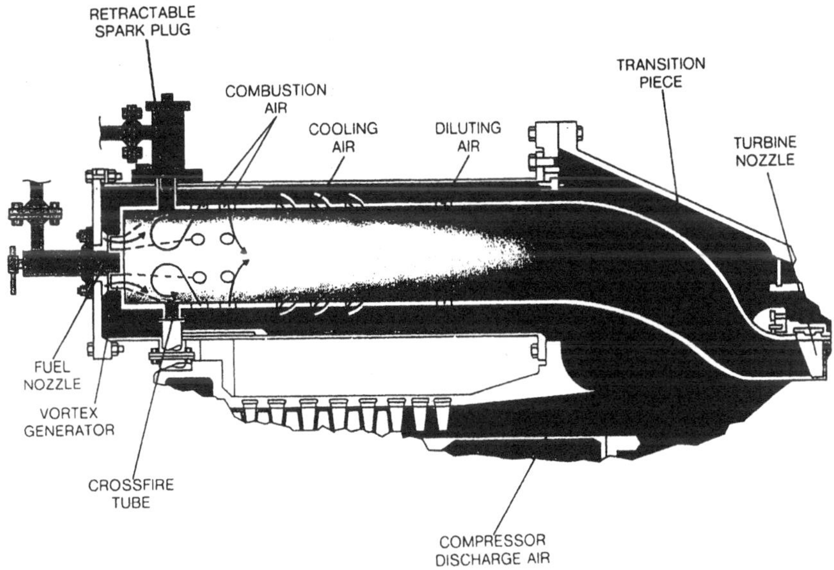

FUNCTION: A DEVICE FOR MIXING LARGE QUANTITY OF FUEL AND AIR THAT PRODUCES A HOT GAS

Figure 28.12 Combuster function and components (Courtesy of General Electric Co.)

Combuster major components function

- Liner – mixes fuel/air mixture, cools and provides constant flame and temperature profile
- Nozzle – gas, liquid or dual – atomizes fuel for efficient combustion
- Spark plugs – ignition source (not in every combuster location)
- Cross fire (distribution) tubes – distribute ignition instantaneously to all combusters
- Flame detectors – immediately trip turbine on loss of flame detection
- Transition piece(s) – evenly distribute hot gas to high pressure turbine

Figure 28.12A Combuster major components function

There are many possible configurations of combusters. Among these options are the following configurations:

- Single
- Multiple (can or annular)

- Staged
- Catalytic combuster

Figure 28.13 presents details concerning these combuster configurations.

Combuster configuration options

Single

- Less parts
- Simplest combuster
- Requires H_2O injection scr (catalytic converter) to meet emission requirements

42–25 ppm
NOx

Multiple (can or annular)

- Better temperature profile
- Smaller parts easier to handle
- Requires H_2O injection or scr to meet emission requirements

42–25 ppm
NOx

Staged (fuel and air)

- Can meet emission requirements – dry (without H_2O injection)
- Combuster control system becomes more complicated
- Varies temperature of flame and burn time

25–9 ppm
NOx

Catalytic combustion (future)

Figure 28.13

Figure 28.14 presents facts concerning gas turbine exhaust, gas components, methods for reduction of pollutants and established limits.

General facts concerning combuster design are shown in Figure 28.15.

Gas turbine exhaust gas facts

- Relative to other internal combustion engines, the gas turbine is a low emitter of pollutants
- However, local emission regulations are a major factor in combuster design, selection and operation.
- Emission products are:
  - Carbon monoxide (c o)
  - Unburned hydrocarbons (uhc)
    - Sulphur oxides (must be considered in process gas, coal gas and well head fuel applications)
    - Nitrogen oxides (limited to 25-42 ppm)
  - Particle matter

Usually very small acceptable amounts will present combuster design

- If exhaust emissions exceed local requirements, the available options are:
  - Use a motor (just kidding)
  - Water or steam injection
  - Modification of fuel/air ratio
  - Catalytic converters

Figure 28.14 Gas turbine exhaust gas facts

Combuster – general facts

- Ideal combustion - stoichiometric

$$\frac{\text{1 part air}}{\text{1 part fuel}} = \text{equivalent ratio} = 1$$

  - Highest combustion temperature
  - Complete combustion - minimum produced pollutants
- Not achieved in present combusters because:
  - Combuster material and design limitations
- Flame turbulence produces:
  - Uneven gas temperature distribution
  - Combuster thermal stresses
- Fuel/air turndown ratio (minimum load)
  - Very lean fuel mixtures (equivalent ratio = 0.1)
  - Possibility of flame out

Figure 28.15 Combuster - general facts

There are many acceptable gas and liquid fuels for use in gas turbines. Also, dual fuel designs are available that you can use either gas or liquid fuel (see Figure 28.16).

Gas turbine fuel alternatives

Fuel source:	gas	liquid	dual
	■ Natural gas	■ diesel	■ start, operate and change over to either gas or liquid fuel
	■ process gas*	■ heavy fuel oil	
	■ coal-gas*	■ kerosene	
		■ naptha	
		■ crude*	
		■ residual fuels*	

fuel system = simplest **flow divider system = most complex**

Note: fuel pressure to combusters must be > compressor discharge pressure.

*May require fuel treatment

Figure 28.16 Gas turbine fuel alternatives

Combuster component and material design must be compatible with the design firing temperature. Facts concerning firing temperature are noted in Figure 28.17.

Firing temperature facts

- The temperature of vapor at high pressure turbine nozzle.
- Limited by hot path materials, coating, cooling and clearances.
- Therefore, must be accurately controlled.
- Monitored by:
  - Actual firing temperature
  - Exhaust temperature
- Approximate hot path material temperature limit = 1500°F (800°C)

Figure 28.17 Firing temperature facts

Power turbine design

The high pressure turbine (HPT)

The high pressure turbine provides power to the gas generator and also functions as a power turbine in single shaft design. Figure 28.18 presents facts concerning HPT design.

H.P.T. design facts

- Experiences highest vapor temperature (nozzles & blades)
- Component profile exactly as steam turbine
  Either:
  - Impulse (usually industrial)
  - reaction (usually aero-derivative)
- Nozzle and blade stress safety factors are increased by:
  - material selection
  - component processing
  - coating
  - cooling

Figure 28.18 H.P.T. design facts

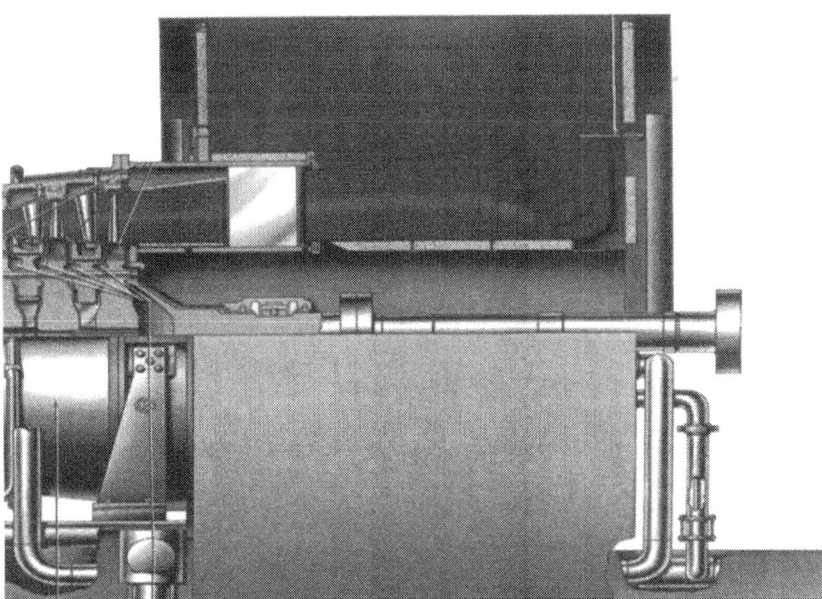

Figure 28.19 G.E. Frame 7 high pressure turbine

FUNCTION: MEET LOAD DEMANDS OF POWER TURBINE WITH MINIMUM EFFECT ON ENGINE PERFORMANCE

Figure 28.20 Adjustable inlet nozzle (Courtesy of General Electric Co.)

A drawing of a G.E. frame 7 high pressure turbine is shown in Figure 28.19.

Certain industrial gas turbine designs incorporate an adjustable inlet turbine nozzle. Figures 20 and 20A contain a drawing of this arrangement and facts concerning operation.

Adjustable inlet nozzle operation

Function	Action	Explanation
■ Power turbine load dump	wide open	immediately reduce gas generator load
■ Optimize heat rate	modulate as required	allow maximum gas generator energy extraction (largest Δp across H.P.T.)
■ Maximum output	modulate as required	allow gas generator to operate at maximum firing temperature set point
■ Lower starting power	wide open	reduces load requirement of gas generator (compressor)
■ Optimum combined cycle efficiency	modulate as required	provide optimum exhaust temperature to HRSG
■ Simultaneous with fuel stop valve closure		

Figure 28.20A Adjustable inlet nozzle operation

Mechanical design features for a high pressure turbine are defined in Figure 28.21.

High pressure turbine mechanical design features

- Long shank blading
  Purpose - remove root from direct contact with hot gas path
- Turbine rotor disc attachment
  Purpose - allows component differential expansion with a minimum of:
    - unbalance
    - component stress increase

Figure 28.21 High pressure turbine mechanical design features

It's a known fact that the "hot gas path" materials (blades, nozzles, combustion liners and transition pieces) require air cooling to meet design life requirements. The design and effectiveness of the air cooling system plays an important part in determining the reliability of a gas turbine. Figure 28.22 contains facts concerning air cooling systems.

Gas turbine component cooling air sources

- Cooling air is provided by the air compressor at various pressure levels to the components

Air pressure range	Cooled components
■ Low pressure (15–20 psig)	Bearing housing, power turbine disc face
■ Medium pressure (80–100 psig)	Power turbine nozzle cooling (impingement, convection, film)
■ High pressure (compressor discharge)	■ high pressure turbine nozzle cooling (impingement, convection, film) ■ high pressure turbine blade cooling (impingement, convection, film)
■ Cooling and thrust balance	Aero-derivative power turbines balance drum (reaction blading) Note: necessary because power turbine was originally designed to drive reaction bladed bypass fan.

Figure 28.22 Gas turbine component cooling air sources

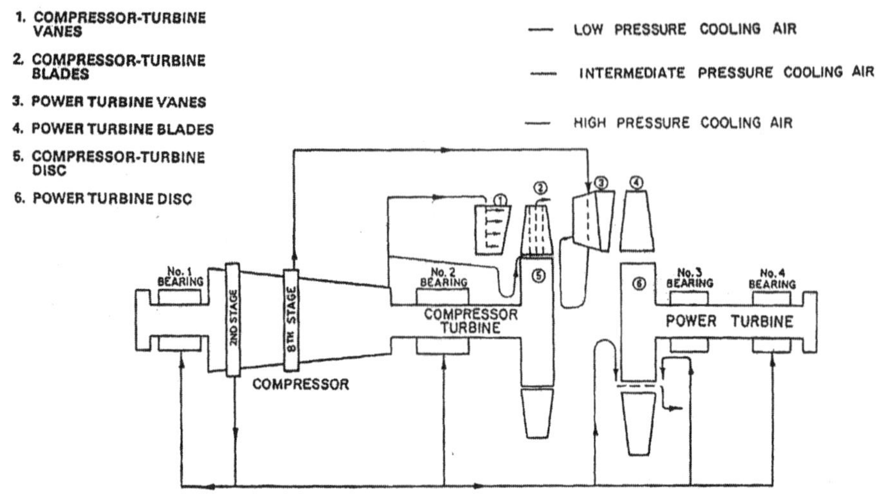

Figure 28.23 CW 352 gas turbine cooling air system (Courtesy of Westinghouse Canada)

A typical industrial gas turbine air cooling schematic is shown in Figure 28.23.

Figures 28.24 and 28.25 contain facts concerning gas turbine component cooling and examples of methods used.

Gas turbine component cooling facts

- Component cooling is necessary in the hot gas path to:
  - Keep component temperatures below limits
  - Prevent rubs
- Types of cooling are:
  - Impingement
  - Convection
  - Film

Figure 28.24 Gas turbine component cooling facts

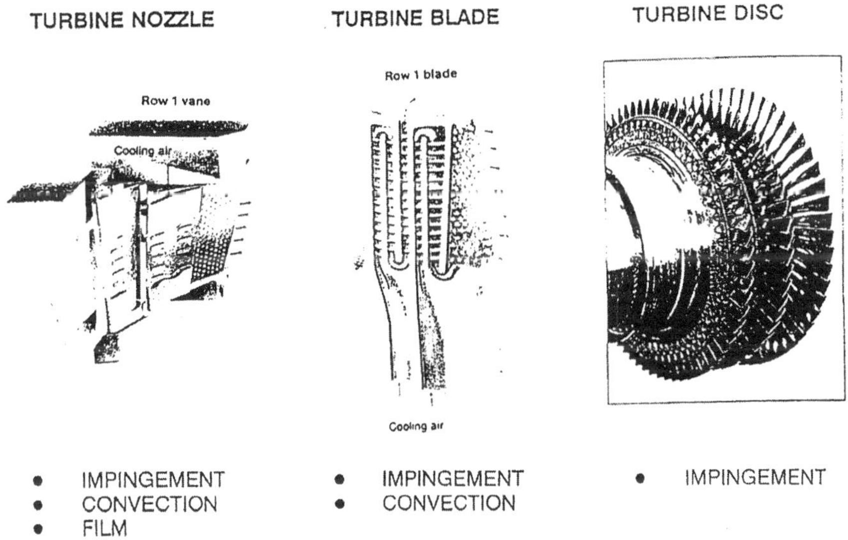

Figure 28.25 Examples of cooling methods (Courtesy of MHI Gas Turbines)

Power Turbine

The final major component assembly in the gas turbine is the power turbine. The power turbine may be integral with the H.P.T. in single shaft designs or separate in multiple shaft gas turbine designs.

Facts concerning power turbine design for industrial and aero-derivative type gas turbines are presented in Figure 28.26.

FUNCTION: EXTRACT VAPOR ENERGY REMAINING AFTER GAS GENERATOR TO PROVIDE USABLE POWER

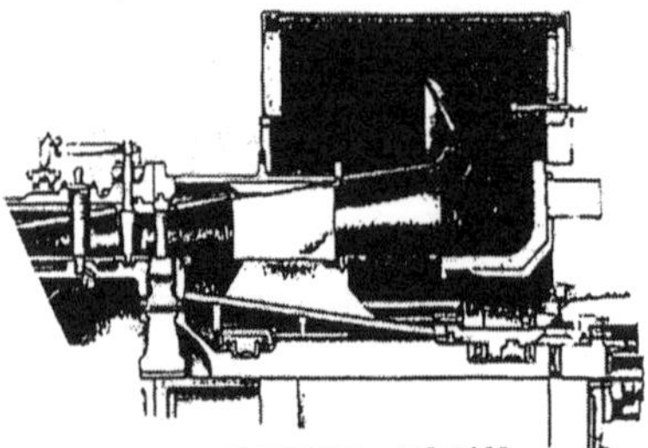

INDUSTRIAL

- IMPULSE BLADING
- LOW THRUST FORCE
- HYDRODYNAMIC BEARINGS
- COOLED NOZZLES AND DISCS
- BLADE SHROUDS-SEALING AND DAMPING

LM2500 Power Turbine Assembly

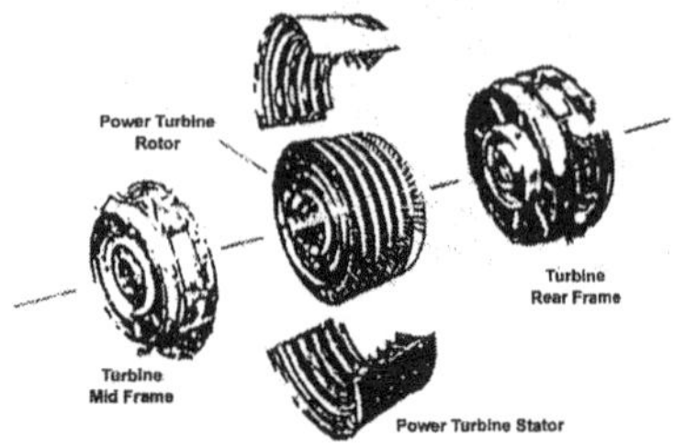

AERODERIVATIVE

• AERO - LM 2500

- REACTION BLADING
- HIGH THRUST FORCE
- ANTI-FRICTION BEARINGS
- COOLED NOZZLES AND ROTOR CORE
- BLADE SHROUDS - SEALING AND DAMPING

Figure 28.26 Power turbine facts (Courtesy of General Electric Co.)

29 Gas turbine support systems

- Introduction
- Types of support systems by classification
- Accessory gearbox

Introduction

In this chapter, we will present and discuss the various gas turbine support systems. The gas turbine, on first look, appears to be a very complicated piece of rotating equipment. Part of the reason for this perception is due to the complexity and number of the various support systems involved.

The objective of this chapter will be to classify each type of support system and present its particular function. It is felt that this approach will make the reader realize that the support systems used for a gas turbine are very similar, in most cases, to those used by turbo-compressors and steam turbines.

After each major type of support system is defined and discussed, the accessory gear box will be presented. The accessory gear box is a very critical piece of equipment since it provides power take-offs to the majority of support system pumps, starter and blowers. Accessory gear boxes will be discussed for both industrial and aero-derivative type gas turbines.

Figure 29.1 shows a sketch of the location and types of gas turbine support systems along with the definition of a support system.

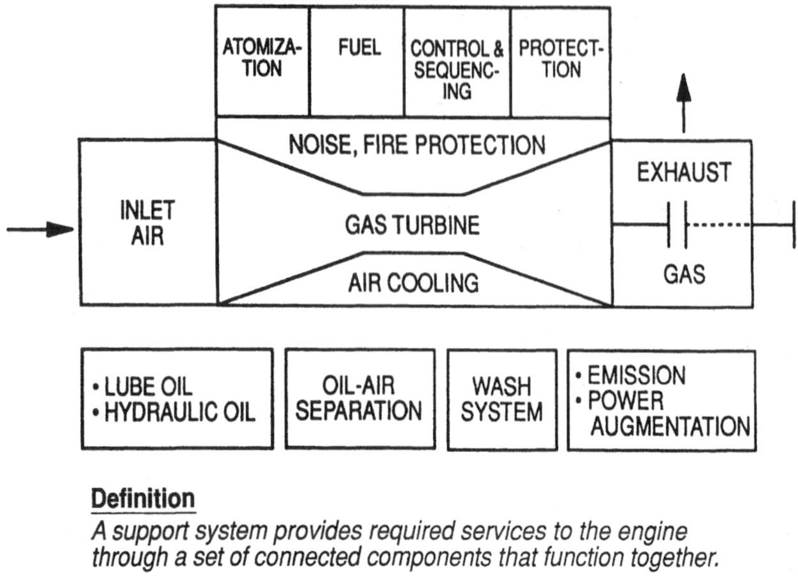

Figure 29.1 Gas turbine support systems

As can be seen in Figure 29.2, the availability of the gas turbine is a direct function of the support systems. Particular attention must be paid to the preventive maintenance (PM) and predictive maintenance (PDM) requirements of all support systems to achieve optimum gas turbine reliability.

Gas turbine support system facts

- Most unscheduled engine shutdowns are the result of a support system malfunction.
- The availability (reliability) of an engine is only as high as the support system component availability.
- There are typically 8–10 support systems per engine.
- There are approximately 1,000 support system components in a typical engine.

Figure 29.2 Gas turbine support system facts

All of the various types of gas turbine support systems are shown in Figure 29.3. The actual systems present in a specific gas turbine design are a function of vendor design preferences, customer requirements and local environmental requirements.

Types of gas turbine support systems

Air/gas handling

- Inlet
  - Filter
  - Silencer
  - Duct
- Exhaust
  - Duct
  - Silencer
  - Heat recovery (HRSG)

Lube/control oil

- Engine Lube
- Auxiliary Lube
- Driven Equipment Lube
- Engine control actuation
  - Guide vane
  - Stator vane
  - Trip
  - Control valve
- Air/Oil separation
- Scavenging

Control

- Engine start/stop sequencing
- Condition monitoring
- Governor System
- Protection System
- Fuel systems
  - Gas
  - Liquid
  - Dual
- Atomizing air

Cooling

- Engine internal
  - Industrial
  - Aero-derivative
- Engine external
  - Industrial
- Auxiliary cooling
  - Lube oil coolers (etc.)

Injection

- Emission control
  - H_2O
  - Steam
- Power augmentation
  - H_2O
  - Steam

Engine cleaning

Liquid

- Crank
- On-line

Solid

- On-line (industrial)

Engine enclosures

- Noise
- Pressurization
- Fire protection
  - Halon
  - CO_2

Figure 29.3 Types of gas turbine support systems

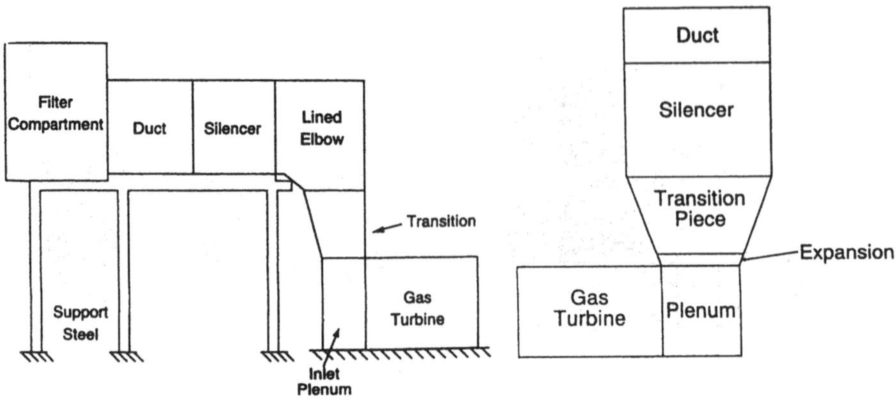

Figure 29.4 Gas turbine inlet (left) and exhaust (right) systems major components

Types of support systems by classification

Following are the function definition and details concerning each type of gas turbine support system.

Inlet and exhaust system

The inlet and exhaust systems provide the engine with an acceptable level of inlet air filtration, moisture removal and noise reduction. Figure 29.4 shows a typical arrangement for a simple cycle installation.

There are two basic types of gas turbine air filters in use today:

- Pulse air type
- Conventional staged type

Figure 29.5 presents the function definition of gas turbine inlet filters and a picture of each type.

Pulse type air filters have gained wide acceptance in regions of excessive dust (desert regions) and in regions of very low temperature conditions. They are highly efficient and can be changed on line. Figure 29.6 presents facts concerning this type of air filter.

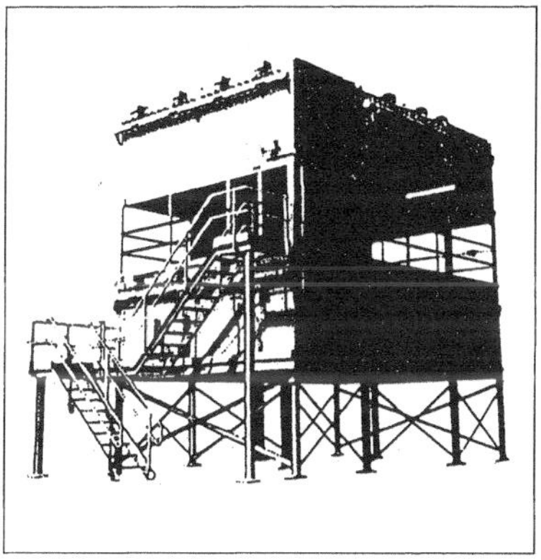

PULSE AIR TYPE

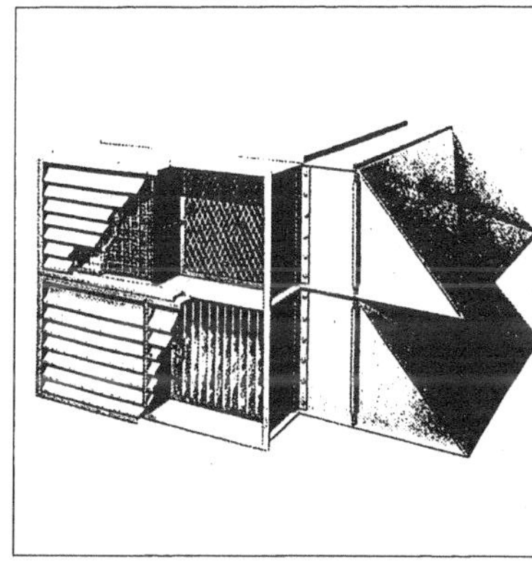

CONVENTIONAL STAGED

FUNCTION: REMOVES PARTICULATE MATTER FROM INLET AIR THAT CAN:

- REDUCE ENGINE POWER OUTPUT
- INCREASE ENGINE HEAT RATE
- CLOG AIR COOLING PASSAGES
- CAUSE DESTRUCTIVE VIBRATION
- ERODE OR CORRODE INTERNAL PARTS

Figure 29.5 Gas turbine inlet air filtration (Courtesy of American air filter)

CLEANING ACTION

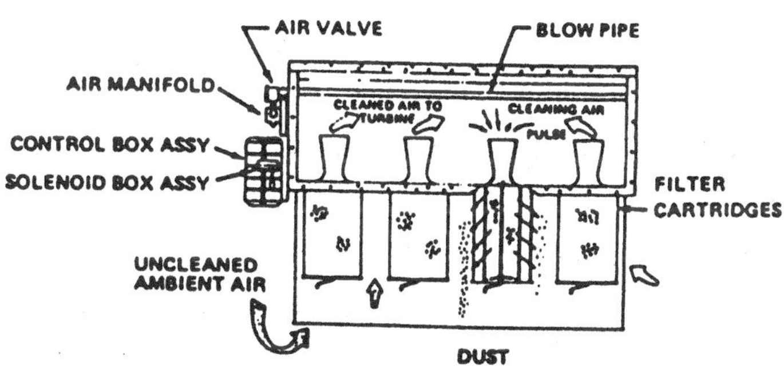

FEATURES:

- HIGH EFFICIENCY
- LOW PRESSURE DROP
- AUTOMATED SELF CLEANING
- ELEMENTS CHANGEABLE ON LINE

SINGLE STAGE MULTI-ELEMENT FILTER THAT UTILIZES REVERSE PULSE JET AIR TO CLEAN ON LINE

APPLICATIONS:

- DESERT REGIONS
- PROCESS OR BASE LOAD
- COLD CLIMATES

Figure 29.6 Pulse air filters

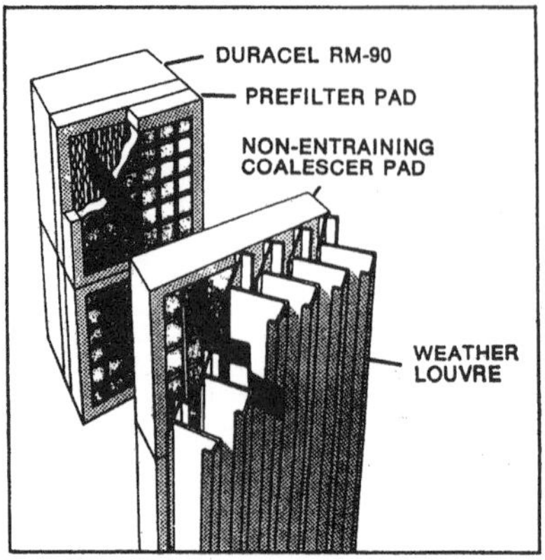

SPE-3 MULTI-STAGE LOW VELOCITY WET AND DRY SALT REMOVAL SYSTEM

TYPICAL FILTERS UTILIZED

- PRE FILTER
- INERTIAL FILTER
- HIGH EFFICIENCY
  - STATIONARY
  - MOVABLE
  (ROLL "O" MATIC)

APPLICATIONS:

- ALL

INSTALLED ON MOST FIRST GENERATION ENGINES - MULTI-ELEMENT FILTER NOT SELF CLEANING.

Figure 29.7 Multi-stage filters

Regardless of the type of air filter (pulse or conventional); filters are often 'staged' to meet local conditions. Figure 29.7 contains details concerning this application.

Engine noise abatement

The inherent result of energy input and extraction from a gas along with the gas velocities involved (in excess of 600 miles per hour!) result in a high engine noise level. A highly sophisticated noise abatement system along with an engine enclosure is required to reduce the generated noise to acceptable levels. Figure 29.8 describes the function, major components and design features of this system.

Engine noise abatement

Function: Reduce noise to specified level at specified distances from engine. Typical levels are:
- 90 DBA overall - 1 meter (3.3 ft)
- 55-65 DBA overall - 130 meters (400 ft)

Components:
- Inlet silencer
- Engine compartment
- Exhaust silencer

Comments:
- Rigid design
- Corrosion proof elements

Figure 29.8 Engine noise abatement

Control and protection

The control system is the heart of the gas turbine and is responsible for safe and reliable start-up, at speed operation, shutdown, monitoring and protection. Figure 29.9 defines the design objectives of the control system.

Gas turbine control systems

- Like steam turbines, gas turbines have similar control and protection systems
- In fact, third generation control systems are identical for steam and gas turbines
- However, due to high engine temperatures, gas turbine start-up (heat-up) and shutdown (cool down) time must be accurately controlled to prevent engine damage caused by:
  - Rubs
  - Rotor bow

Figure 29.9 Gas turbine control systems

The major functions of the control system are shown in Figure 29.10.

Gas turbine control

Sequencing system	Control system	Protection
■ Automated ■ Start cycle (starter) ■ Stop cycle	■ Governor ■ Gas generator ■ Power turbine	■ Overspeed protection ■ Power turbine ■ Gas generator
	■ Fuel control system ■ Liquid ■ Atomizing air ■ Gas ■ Dual fuel	■ Train parameter protection ■ Lube oil ■ Seal oil ■ Vibration ■ Engine temperature (etc.)

Figure 29.10 Gas turbine control

Lube and hydraulic systems

The lubrication and hydraulic systems continuously provide clean, cool lubrication and hydraulic fluid to the components at the proper pressure, temperature and flow. The lubrication system used for aero-derivative type gas turbines is different from the industrial gas turbine lube system in that a scavenge (vacuum) system is added to return lube oil to the sump under flight conditions. This system is retained on mechanical drive applications of gas turbines. Industrial gas turbines use gravity drain methods for lube oil return to the reservoir. Details concerning aero-derivative gas turbine lube systems are presented in Figure 29.11.

Aero-derivative gas turbine lube oil systems

In addition to the normal components, aero-lube systems utilize:
- Scavenge pumps
- Air/oil separators

These components are required because:
- Anti-friction bearings are used to minimize system flight weight
- Conventional atmospheric drains are not possible
- Oil/air mist from bearings must be:
  - Scavenged (drawn) back to reservoir
  - Separated prior to return to reservoir

Both scavenge pumps and separators are engine driven via the auxiliary gear box

Figure 29.11 Aero-derivative gas turbine lube oil systems

Figure 29.12 contains the function definition of the lube and hydraulic (control) system and differences between aero-derivative and industrial systems.

Gas turbine lubrication and control systems

- Function: identical to steam turbine/turbo-compressor – continuously provide cool, clean oil to bearings and control components at the proper pressure, temperature and flow rate.
- However, due to the high temperatures experienced in gas turbines and engine design features, there are differences.

Industrial types

- Main pump engine driven (thru accessory gearbox)
- Emergency cool down pump (D.C.)
- Compact design
- Guarded pipe (supply pipe within drain pipe)
- Possible use of:
  - Synthetic oil (high flash point)

Aero-derivative types

- Main pump engine driven (thru accessory gearbox)
- Smaller, very compact design
- Use of synthetic oil
- Requirement of:
  - Air/oil separator
  - Scavenge return oil pumps

Figure 29.12 Gas turbine lubrication and control systems

Cooling (engine external, internal and auxiliary systems)

Because of the high temperatures generated within the engine (in excess of 2000°F), cooling plays a very important role in engine reliability. Figure 29.13 presents the types of cooling required and the components serviced.

Gas turbine cooling systems

External engine cooling (air)	External engine cooling (H_2O or air)	Auxiliary cooling (H_2O)
■ Combustor	■ H.P. turbine case (jacket)	■ Lube oil coolers
■ H.P. turbine ■ Nozzles ■ Blades ■ Rotor/discs	■ Atomization air	
■ L.P. turbine ■ Nozzles ■ Blades ■ Discs ■ Rotors		
■ Bearing housings		
■ Struts		

Figure 29.13 Gas turbine cooling systems

Injection

Injection systems are required for pollution control and/or additional power (power augmentation). Figure 29.14 defines the function and features of these systems.

Gas turbine injection systems

- Function: Inject H_2O or steam into engine for the purpose of:
  - NOx emission reduction (25–42 PPM)
  - Increased engine power (power augmentation)
- Features:

NOx reduction	**Power augmentation**
■ Injected directly into combustor	■ Injected down stream of compressor
■ Self contained skids	■ Superheat requirements for steam (+50°F)
■ Stainless steel pipe	■ Maximum 3–5% by mass flow
■ High pressure centrifugal (sundyne or equal) for H_2O injection	
■ Amount approximately 2% mass	

- **Considerations:**
  - Hot path maintenance requirements

Figure 29.14 Gas turbine injection systems

Fire protection

High engine temperatures, the close proximity of fuel and potential ignition sources within an enclosure provide a potentially hazardous environment for the engine. As a result a fire protection system is required in the engine enclosure. Figure 29.15 presents the function and facts concerning the system features.

Gas turbine fire protection systems

Function: Quickly and effectively extinguish and confine engine fires.

Protection mediums:
- Carbon dioxide
- Halon (limited use in future – impact on ozone layer)

System features:
- Multiple fire detectors
- Medium control system
- Dispenses medium and trips engine on confirmation of fire

Figure 29.15 Gas turbine fire protection systems

Internal component cleaning

Available power can be significantly reduced by engine fouling (accumulation of dirt on air compressor blades and stators). Most gas turbines incorporate some type of crank and/or on-line cleaning system. Figure 29.16 shows the function and options for these systems.

Gas turbine internal component cleaning systems[1]

Function: To remove air compressor blade deposits and to restore power output and efficiency in a safe and reliable manner.

Options:
- Crank – performed at low (crank) speeds
  - H_2O/glycol/detergent
  - H_2O/glycol[2]
- On-line – performed at usually idle speed
  - H_2O/glycol[2]
  - H_2O/glycol/detergent
  - Solid particle (walnut shells, catalyst) heavy duty only

Notes: 1. Manufacturer must be consulted regarding acceptable cleaning procedures.
2. Must be used where operation below approximately 40° is possible.

Figure 29.16 Gas turbine internal component cleaning systems

Figure 29.17 contains details concerning the cleaning system.

Gas turbine cleaning systems on-line cleaning precautions*

- **On-line cleaning** should be performed **periodically** and **frequently** because:
  - All particles must be removed immediately in a uniform manner to avoid destructive vibration.
- **Cleaner injection rates** and **injection periods** should be **minimized** to prevent:
  - Erosion
  - Corrosion
- **Cleaner composition** must be **confirmed** to be satisfactory **by manufacturer** to assure:
  - Material and coating capability
  - Acceptable hot path component capability
    - Combustors
    - Nozzle, blade cooling passages

*Entire cleaning procedure must be reviewed with engine manufacturer.

Figure 29.17 Gas turbine cleaning systems on-line cleaning precautions

FUNCTION: PROVIDE VIBRATION FREE POWER TO ACCESSORIES AT REQUIRED SPEED AND MINIMUM TRANSMISSION LOSSES

Figure 29.18 The gas turbine accessory gearbox DR 990 drive gear module (Courtesy of Dresser Rand)

Accessory gearbox

A picture of an accessory gearbox used on a Dresser Rand DR 990 gas turbine is shown in Figure 29.18.

The typical accessory gearbox connections for both industrial and aero-derivative gas turbines are shown in Figure 29.19.

Typical accessory gearbox connections

Industrial	Aero-derivative
■ Engine starter	■ Engine starter
■ Main lube oil pump	■ Main lube oil pump
■ Hydraulic (control oil pump)	■ Hydraulic (control oil pump)
■ Atomization air compressor (liquid fuels)	■ Scavenge pump(s)
■ Main fuel pump (liquid fuel)	■ Air/oil separator
■ Cooling water pump (optional)	

Figure 29.19 Typical accessory gearbox connections

Remember, the engine reliability is a direct function of the reliability of each individual system component! Required PM and an effective PDM program is a must!

30 Gas turbine control and protection

- Introduction – total train control and protection objectives
- Start-up and shutdown sequencing
- Gas turbine control
- Protection systems

Introduction – total train control and protection objectives

At this point, we have covered the design and operation of a gas turbine. We will now discuss the gas turbine driven total train control and protection. The total train control and protection objectives for a gas turbine are identical to those of a steam turbine.

However, since a gas turbine operates at much higher temperatures than a steam turbine, an additional control function is required. This function is known as gas turbine sequencing. A sequencer automatically operates the gas turbine from a permissive to start up to the normal operating speed. In doing so, the sequencer will assure that all support functions, fuel purge functions, combustor firing functions and acceleration to normal speed is achieved in a safe and reliable manner. In addition, the sequence system also controls the deceleration of the gas turbine and operation of support systems during a normal shutdown.

As mentioned above, the basic control function is similar to a steam turbine. This is true however, since the fuel for a gas turbine is a combination of air/fuel mixture and since there are alternatives (liquid, gas or dual fuel) we will focus our attention on the various fuel system configurations and functions in this section.

Finally, the protection system will be discussed in detail with emphasis placed upon additional engine protection other than the over speed bolt which is similar to steam turbine operation. Namely, hot gas path temperature override function will be discussed as well as gas turbine air compressor protection. In addition we will also present a gas turbine control system brief history, which will detail the various types of control systems that are installed in the field.

Gas turbine control and protection objectives are similar to that of a steam turbine. However, due to the significantly higher operating temperature, certain additional features must be incorporated.

Figure 30.1 presents these features.

Gas turbine control and protection additional features

- In terms of control and protection, a gas turbine system is the same as a steam turbine system.
- Due to a gas turbine's characteristics, the following features must be added:
  - Automated start and stop (sequencing)
  - Hot gas path excessive termperature governor override
  - Air compressor operating point override

Figure 30.1 Gas turbine control and protection additional features

The design objective of the sequencing and temperature override systems are shown in Figure 30.2.

Gas turbine sequencing and temperature override

- Regardless of the type of gas turbine, start-up and shutdown must be sequenced (automated) to:
  - Assure all auxiliaries are operative
  - Gas path is urged
  - Differential thermal effects are minimized
  - Air compressor operates in the stable range
- In addition, the hot gas path temperature must be limited by continuously inputing the hot gas path average temperature to the governor

Figure 30.2 Gas turbine sequencing and temperature override

In addition, it is necessary to assure that the compressor operating point is optimized (as high an operating efficiency as possible). Figure 30.3 shows one such method – adjustable compressor stator valve control.

Air compressor operating point control

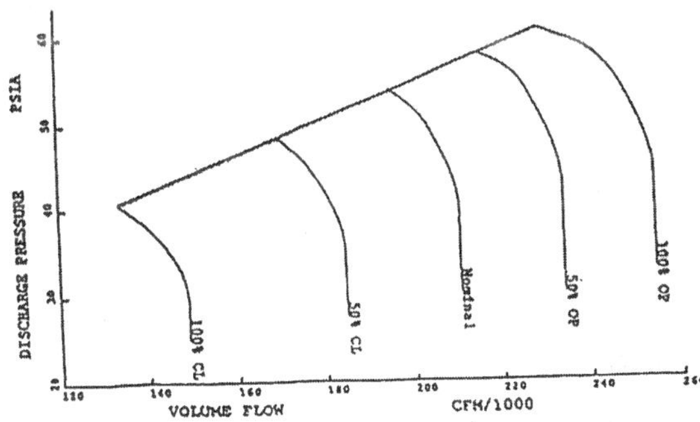

- Since air compressor inlet pressure and gas m.w. are constant, compressor discharge pressure (C.D.P.) is usually selected as the key parameter for determining the air compressor operating point.

Figure 30.3 Air compressor operating point control (Courtesy of Elliott Co.)

Gas turbine control system effectiveness and reliability has made great advances since the early years of gas turbine design. Since gas turbines and their control systems are production items, custom design control features have been minimized. Essentially, there are four (4) generations of control system design beginning in the 1950s. Figure 30.4 outlines a history of these systems.

Gas turbine control system history

Years	Sequencing	Control	Protection
1950–1970	relays	mechanical/hydraulic solid state components (analog type)	relays
1970–1980	solid state	integrated circuits (analog type)	solid state and relays
1980–1990	micro-processor redundancy	micro-processor (digital type)	micro-processor
1990–	micro-processor	micro-processor redundancy fault tolerance self diagnostics	independent micro-processor dual voting

Figure 30.4 Gas turbine control system history

Figure 5 presents an overiew of gas turbine control system input, set point and output functions.

We will cover each of the following sequencing, control and protection functions in detail.

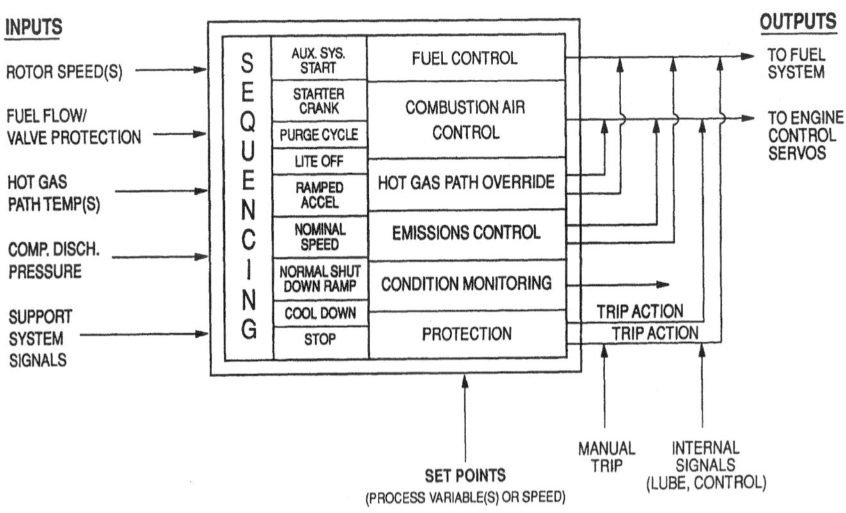

Figure 30.5 Gas turbine control system functions overview

Start-up and shutdown sequencing

The objectives of the gas turbine start and stop sequencing system are noted in Figure 30.6.

Gas turbine start-stop sequencing system

Sequenced (automated) start-up and shutdown are required to:

- Prevent engine fires or explosions
- Minimize thermal effects
- Eliminate damage caused by vibration when passing through natural frequencies (critical speeds)
- Eliminate damage caused by air compressor stall (surge)

Figure 30.6 Gas turbine start-stop sequencing system

A generic gas turbine starting sequence outline and a parameter travel are presented in Figures 30.7 and 30.8.

Gas turbine starting sequence[1]

Step	description	default parameter
1.	enable start	all permissives clear
2.	start acknowledged	any permissive default
3.	support systems started[2]	all support system parameters clear
4.	starter engagement/crank speed attained	starter operating/crank speed
5.	engine purge cycle	cycle timer set point
6.	igniter activation, fuel system activation lite off[3]	flame detectors combustion temp or speed change rate
7.	warm up idle, starter disengagement	warm up timer set point/starter disengagement
8.	ramped accelertion to normal speed (considers thermal effects, critical speeds)	vibration, exhaust temp, etc.
9.	warm up normal speed (no load) guide vanes, nozzles, etc. in auto position	all trip signals
10.	load engine	all trip signals

1. Generic procedure to demonstrate typical sequencer steps.
2. Some applications will have to have manual support system start.
3. All engine guide vanes, variable nozzles set at no load position.

Figure 30.7 Gas turbine starting sequence

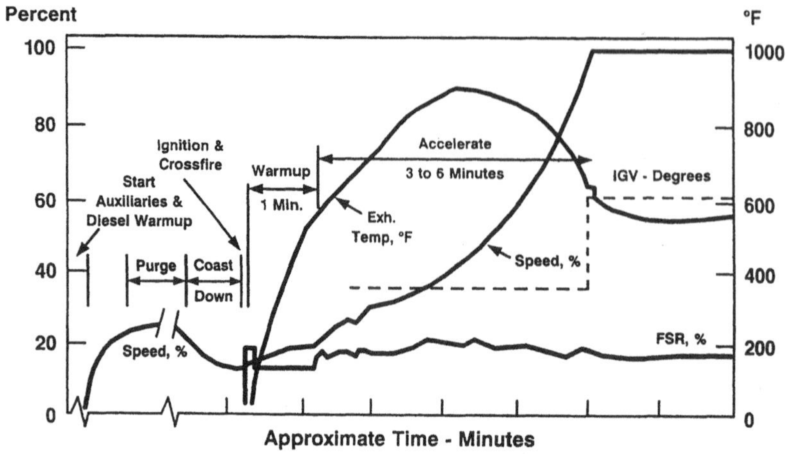

Figure 30.8 Typical gas turbine start characteristics – start-up unloaded (Courtesy of General Electric Company)

Facts concerning start sequences for industrial and aero-derivative type gas turbines are presented in Figure 30.9.

Start sequence facts

- The start-up sequence times vary with gas turbine type and manufacturer – typical values
  - Heavy duty – 10–20 minutes
  - Aeroderivative – 2–5 minutes
- Normal shutdowns are sequenced to allow uniform cool down of components
- Cool down is accomplished by low speed operation

Figure 30.9 Start sequence facts

Gas turbine control

Like steam turbines, the speed control (governor system) is the heart of the control system. It performs the identical function of 'cruise control' in your vehicle. Refer to Figure 30.10 for gas turbine control system objectives.

Gas turbine control objectives

- Meet load requirements
- Maintain optimum heat rate (firing temperature and efficiency)

Figure 30.10 Gas turbine control objectives

Input parameters are noted in Figure 30.11.

Gas turbine control system inputs

In order to meet control objectives, inputs to control system are:

- Power turbine rotor speed
- Air compressor
  - Rotor(s) speed[1]
  - Compressor discharge pressure (CDP)
  - Hot gas path temperatures

[1]if multi-shaft turbine

Figure 30.11 Gas turbine control system inputs

A typical control schematic for a gas turbine drive is shown in Figure 30.12.

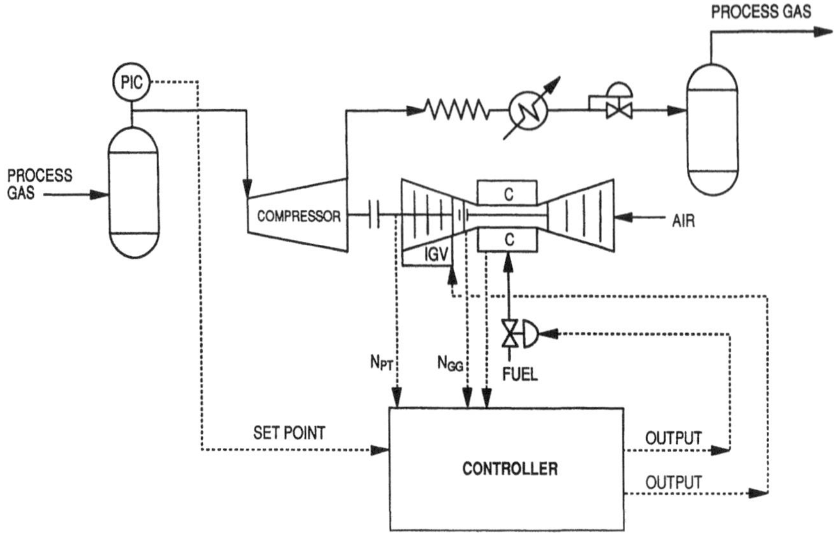

Figure 30.12 A gas turbine driven turbo-compressor (Courtesy of M E Crane Consultant)

Control system output for this system is discussed in Figure 30.13.

Gas turbine control system outputs

In order to control gas turbine output power,

- Vapor energy/mass – ΔH
- Mass rate – M

Must be controlled by:

- Fuel rate control (ΔH)
- Air flow rate (if furnished) (M)
    - Gas generator speed
    - Adjustable stator vanes
- Power turbine variable nozzles (if furnished) (M)

Figure 30.13 Gas turbine control system outputs

As discussed in the previous chapter, there are many available fuel options. Figure 30.14 presents fuel options and facts concerning fuel specific energy and specifications.

Gas turbine fuel systems facts

- Available fuel options:
  - Gas (most frequently used)
  - Liquid
  - Dual (gas or liquid)
- The lower the fuel heating value, btu/scf the greater required fuel system capacity (valve size, line size, etc.)
- Fuel characteristics must be accurately specified in terms of:
  - Type
  - Composition
  - Moisture content
  - Cleanliness
  - Viscosity (liquid fuels)
- All fuels must be treated as hazardous materials and require:
  - Proper electrical area classifications
  - Purging of all fuel areas prior to firing
  - Draining of liquids to safe locations
  - Tight shutoff valves isolating the fuel system from the engine
- A one quart slug of a liquid in a gas fuel system will generate a 10 m.w. load spike if ingested in one second!

Therefore:

- Appropriate gas fuel superheats must be continuously maintained (minimum of 50°F superheat)
- Possible heat tracing of lines and valves required in gas fuel systems
- Liquid knock out facilities

Figure 30.14 Gas turbine fuel systems facts

Refer to Figure 30.15 for typical gas turbine fuel alternatives.

Typical gas turbine fuels

Types of liquid fuels

- Conventional liquid fuels
- Distillate
- Crude oils
- Residuals

- Less conventional liquid fuels
  - Jet fuels
  - Kerosene

- Unconventional liquid fuels
  - Naptha
  - Natural gas liquids
  - Natural gasolines
  - Process residuals

Types of fuel gases

- natural gas
- lpg's
- propane
- butane

- refinery gases
  - high hydrogen content
  - coal-delivered gases

Figure 30.15 Typical gas turbine fuels (Courtesy of General Electric Company)

The remainder of this section concerning gas turbine control will cover the design requirements for the fuel control system. Figure 30.16 defines the system design requirements.

Fuel system general design requirements

- The fuel system must be designed for the maximum fuel pressure requirement
- Maximum fuel pressure requirement is a direct function of air compressor discharge pressure

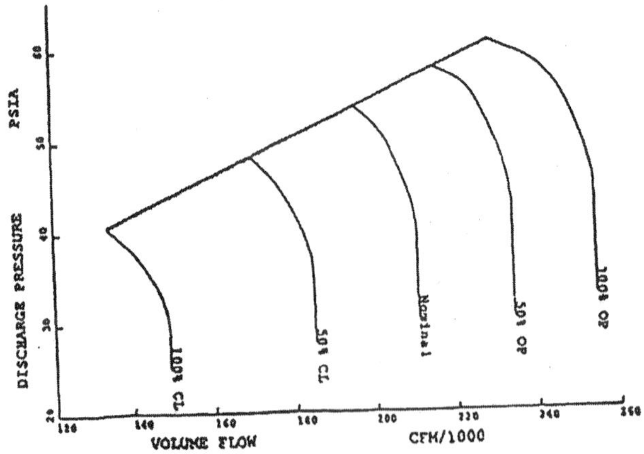

- Since compressor discharge pressure will vary significantly
  - Air compressor operating point must be referenced to fuel systems
  - Fuel flow requirement will vary significantly
    - Start-up
    - No load operation
    - Full load
  - Fuel heating values can vary
  - Fuel viscosity (liquid fuels) can vary
  - A different start-up fuel may be required

All of the preceding requirements impact

- Fuel control valve sizing and arrangement
- Fuel orifice sizing
  Pump and flow divider sizing (liquid systems)
- Fuel skid minimum and maximum supply pressures

Figure 30.16 Fuel system general design requirements

Specific fuel system requirements concern the selection of the stop (shut-off) and control valve. Guidelines are presented in Figure 30.17.

Gas turbine fuel system specific requirements

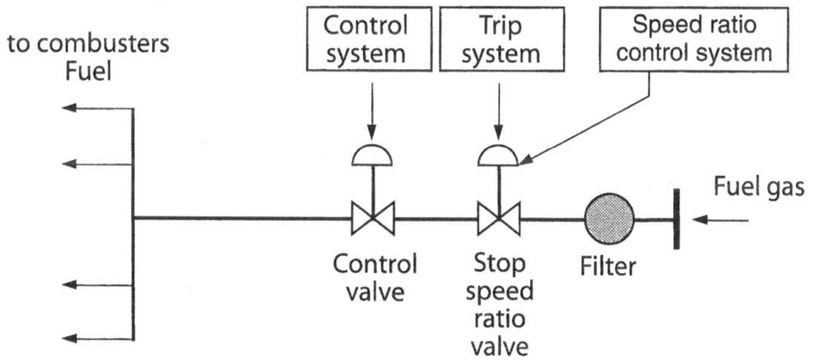

- Positive fast acting stop valve
- Large range of accurate control

Options:
- Split range control valves
- Two stage – pressure control and flow control

- Minimum fuel superheat requirement (50°F)

Figure 30.17 Gas turbine fuel system specific requirements

Figure 30.18 shows a typical gas fuel control system.

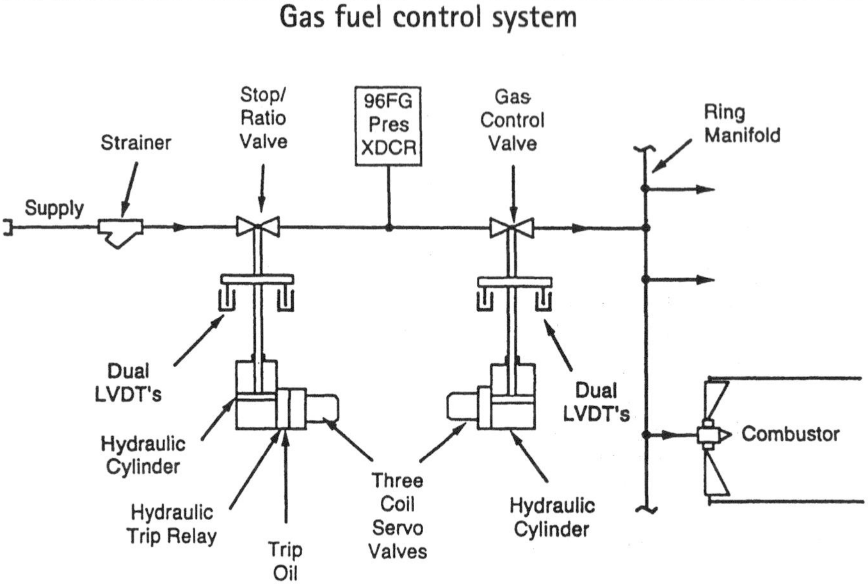

Operation: Control system regulates ratio valve and gas control valve to keep both valves in desired operation range, during start-up, normal operation and transient load changes.

Figure 30.18 Gas fuel control system (Reprinted from paper GER 3648A with permission of General Electric)

Liquid fuel system requirements are defined in Figure 30.19 below.

Liquid fuel system requirements

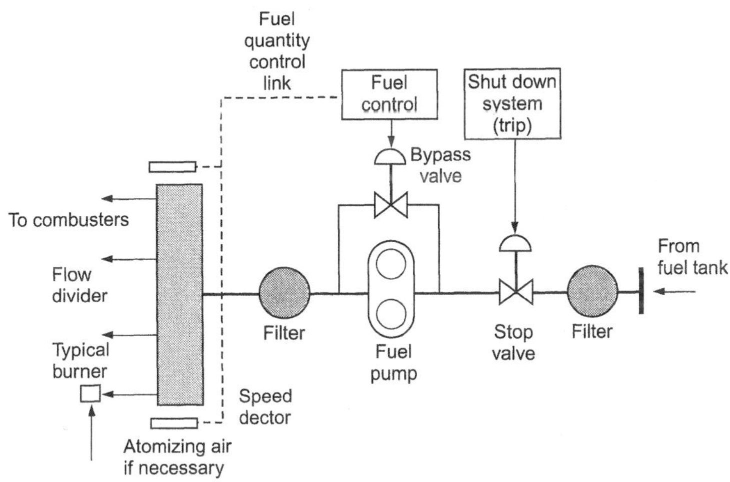

- Positive fast acting stop valve
- Minimum fuel skid supply pressure (prevent pump cavitation)
- Pump type compatible with fuel viscosity (centrifugal pumps for low viscosity)
- Large range of accurate controls (options)
  - Use of bypass valve
  - Variable speed flow divider (individual P.D. pump to each nozzle)

Figure 30.19 Liquid fuel system requirements (Courtesy of General Electric Company)

Figures 30.20 and 30.21 contain typical fuel system schematics for normal and low viscosity liquid fuels.

Normal viscosity fuel schematic (diesel, etc.)

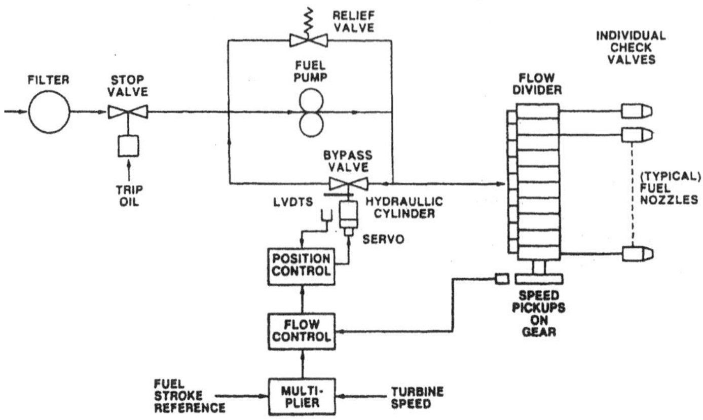

Operation: Control system regulates fuel flow through positive displacement fuel dividers to each nozzle. speed pickup on flow divider positions bypass valve to regulate required flow to nozzles (flow divider)

Figure 30.20 Normal viscosity fuel schematic (diesel, etc.) (Reprinted from article GER 3648A with permission of General Electric Company)

Low viscosity fuel system schematics

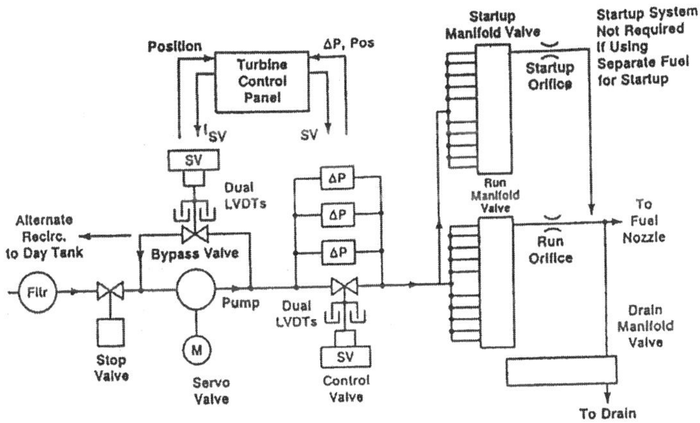

Operation: Since liquid fuel is low viscosity, positive displacement (gear pumps) cannot be used. A high speed centrifugal pump (sundyne or equal) and individual nozzle flow orifices are used. An in-line control and pump bypass valve are used to supply the required pressure to the nozzle orifices.

Figure 30.21 Low viscosity fuel system schematics (Reprinted from article GER 3648A with permission of General Electric Company)

There are also dual fuel control systems. Facts concerning dual fuel systems are shown in Figure 30.22.

Gas turbine control dual fuel systems

- Provide option of operation on liquid or gas
- Liquids usually back up fuel
- Can transfer or line without trip

Figure 30.22 Gas turbine control dual fuel systems

Protection systems

The design of the gas turbine protection system incorporates the features of steam turbine protection systems plus hot gas path over temperature control. These facts are presented in Figure 30.23.

Gas turbine protection systems

In addition to overspeed protection of each rotor, a gas turbine protection system also shuts the fuel valve on:

- Hot gas path overtemperature
- High compressor discharge pressure
- Train external trips
  - Lube oil low pressure
  - Control oil low pressure
  - Vibration
  - Manual trip
  - Etc.

Figure 30.23 Gas turbine protection systems

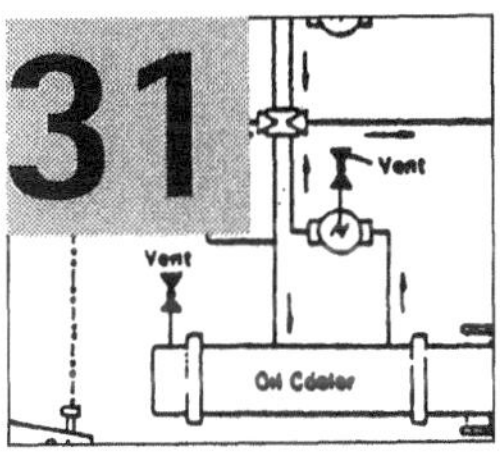

31 Lubrication system overview and types

- Introduction
- Types of lubrication systems
- Arrangement options

Introduction

In this chapter, we will confine our attention to lubrication system types and arrangement options only. A discussion of system operations of the more common types of systems used for critical equipment will be pursued. Refer to the definition of auxiliary system as specified in Chapter 2 in this book. For critical equipment systems, the function of a lubrication system is to continually supply cool, clean, lubricating fluid to each specified point at the required pressure, temperature and flow rate. All of the systems to be covered in this section follow the same function definition.

Types of lubrication systems

The classical lube system arrangement consists of a steam turbine driven main pump and motor drive auxiliary pump. In this section, we will devote our attention to other system variations.

Shaft driven positive displacement main pump, motor driven auxiliary pump

Refer to Figure 31.1. In this system, a shaft driven main pump is supplied. The function of having a shaft driven pump is to continuously provide some amount of lubrication to the system bearings while the

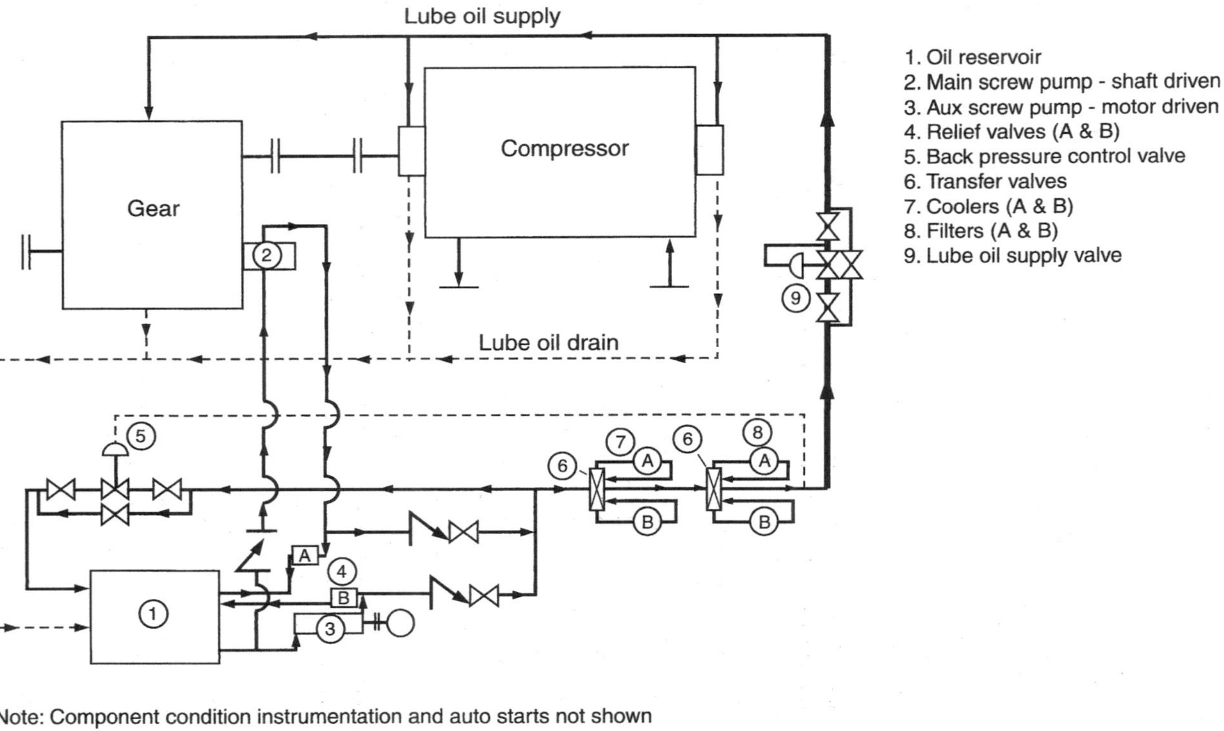

Figure 31.1 Lube oil system main pump shaft driven

critical equipment is in operation. The idea being, that on coastdown, or start-up, the system would never be without lubrication oil supply, even in the event of failure of the stand-by pump. Such a system is useful in areas where an alternate form of energy is not available for main and spare pumps, as on platforms, or in chemical plants where a utility steam system is not installed.

However, the system must still be designed to effectively lubricate all bearings prior to the main pump attaining full flow rate (speed). To facilitate this, the auxiliary pump is designed to start before the main critical equipment driver is started (permissive arrangement) and either automatically or manually shut off once the shaft driven pump attains sufficient speed. On shutdown, the reverse occurs. The auxiliary pump will start on deceleration of main pump speed. The signal to start the auxiliary pump can be critical equipment speed or system pressure. If a specific system is not designed for automatic shutoff of the auxiliary pump, care must be taken to assure that the system is designed for continuous two-pump operation. It is recommended that auxiliary pumps be shutoff during critical equipment main pump operation.

A particular concern with this system design is to assure priming of the main pump. Frequently, the main pump is at a significant height above the fluid reservoir. In this case, care must be taken to assure proper priming of main pump prior to start-up. Many systems incorporate a priming line from the auxiliary pump to fill the main pump suction line. This requires a check valve (foot valve) in the main pump suction line and a self-venting device to assure air is not entrained in the suction line of the main pump. Frequently, systems of this type do not incorporate a vent line and are susceptible to main pump cavitation on start-up.

Referring to Figure 31.1, let's examine the bypass valve response in the event the auxiliary pump does not shut off after the shaft driven pump has attained full flow (speed). Upon initial auxiliary pump start-up, flow is furnished to the system and the bypass valve opens to control system pressure to a pre-set value, thereby bypassing excessive auxiliary pump (stand-by pump) flow. Upon starting of the critical equipment unit, the bypass valve will react to increasing main pump flow and will gradually open to the point that it would be at maximum stroke if the stand-by pump remained in operation. When the stand-by pump is shutdown, the response of the bypass valve must be equal to the decrease of flow from the auxiliary stand-by pump such that system pressure does not drop below trip setting.

Refer back to the concepts of an equivalent vessel and orifice discussed in previous sections. In this case, the supply to the equivalent vessel (from the auxiliary pump) instantaneously drops while the demand from the critical equipment is constant thus causing an instantaneous drop in equipment vessel pressure. Prior to shutdown of the auxiliary

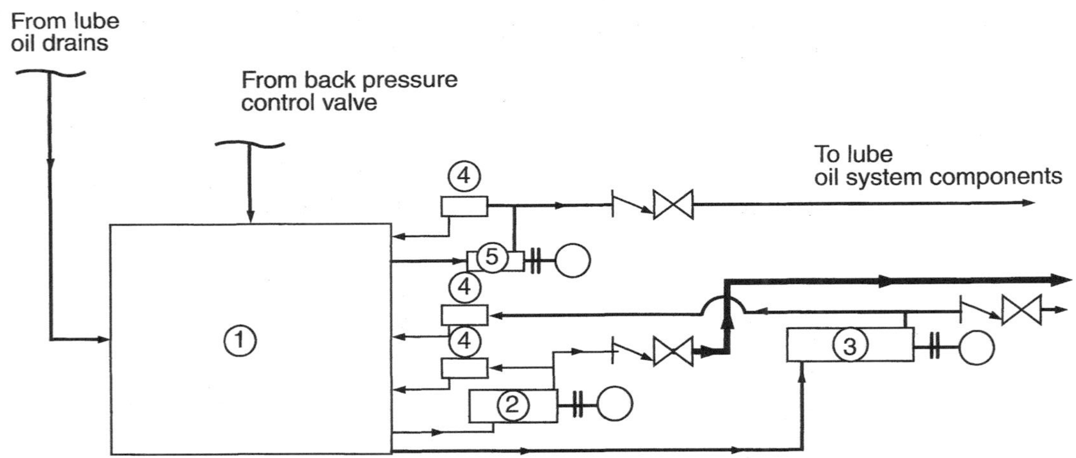

1.Oil reservoir
2. main oil pump - motor driven
3. Aux oil pump - motor driven
4. Relief valves
5. Emergency oil pump - DC motor driven

Notes: Other major components not shown
Component condition instrumentation and auto start of aux and emergency pumps not shown

Figure 31.2 Lube oil system main pump shaft driven

pump, excessive supply was recirculated by the bypass valve. Upon rapid decrease of supply flow, the bypass valve must decrease demand at the same rate to assure pressure in the system (equivalent vessel) is maintained at a constant value. Failure to do this can result in critical equipment shutdown.

Main and auxiliary A.C. motor driven pumps, D.C. motor driven emergency pump

Refer to Figure 31.2. This application would be used in facilities where steam systems are not available, such as chemical plants or in operations at remote locations or on offshore platforms. In this case, both pumps, main and stand-by pump, would shut down in the event of a power failure. If the critical equipment unit is also motor driven, it too will cease operation.

The design objective of this auxiliary system is to provide a sufficient flow, via an emergency pump driven by a DC source (direct current), for a sufficient time to promote the safe coastdown of the critical equipment unit. The DC power source can be obtained from a UPS (Uninterrupted Power Supply) system or other sources.

Coastdown times of the driven equipment vary between 30 seconds to in excess of 4 minutes. Sufficient emergency pump energy must be available. The emergency pump would be activated in this case on auxiliary system pressure drop. Note that the emergency pump is not a full capacity pump. It is sized to only provide sufficient flow during equipment coastdown to prevent auxiliary system component damage.

A.C. motor driven main pump, steam turbine driven auxiliary pump

Refer to Figure 31.3 for a diagram of this system. A system of this type would be used in a plant where dual pump driver power is available, but steam is at a premium. Therefore, the turbine driver is not desired to be used for continuous operation. This system requires continuous attention in terms of steam turbine steam conditions. The turbine must be capable of accelerating rapidly since a system accumulator is not included. Therefore, steam conditions at the turbine flange must be maintained as specified. That is, steam must be dry. A steam trap is recommended to be installed to continuously drain condensate from the steam. To assure rapid turbine acceleration, the turbine inlet valve must be a snap open type (less than 1 second full open time) to minimize auxiliary pump start-up time. Experience has shown that this type of system is not reliable in terms of auxiliary pump starts. If used, this type of system should employ a large accumulator to provide sufficient flow to the system for approximately ten seconds as a minimum since the start-up time of a steam turbine is much slower than that of a motor driven pump.

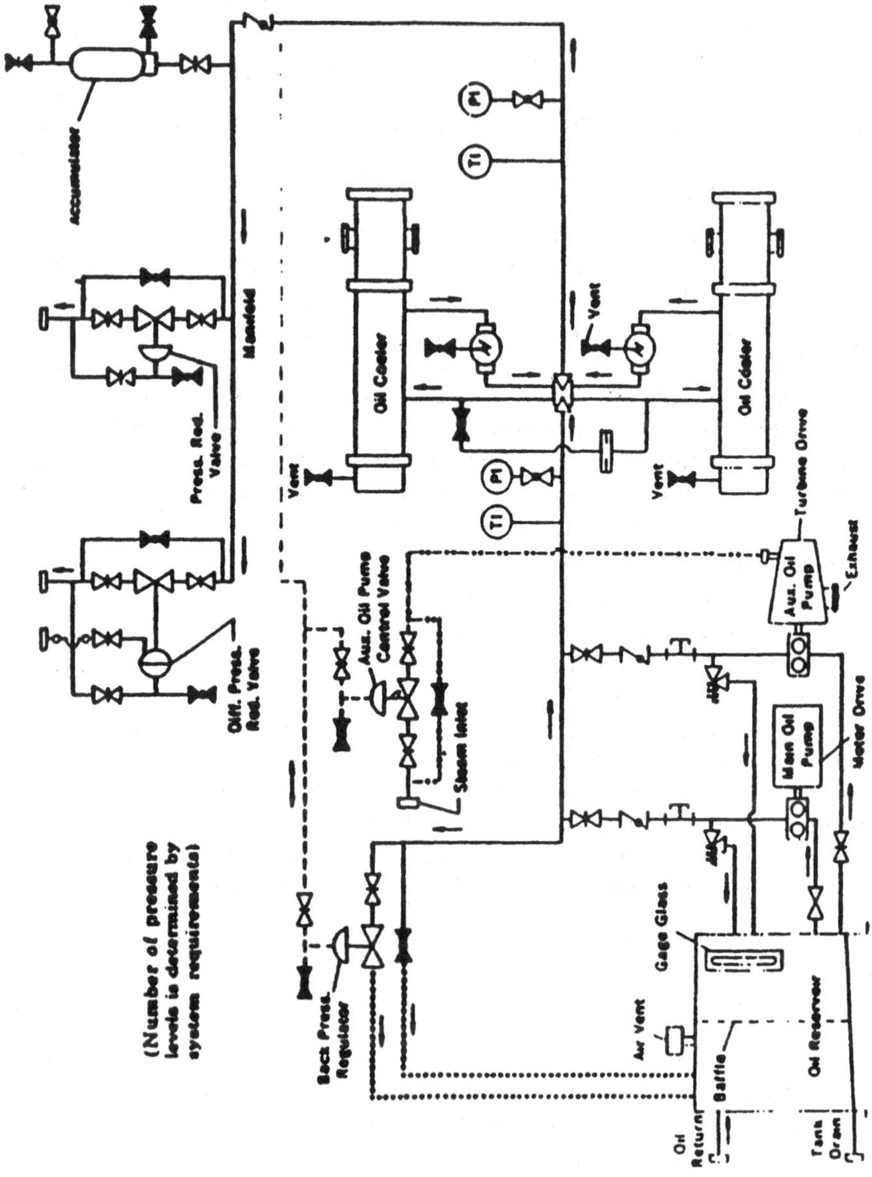

Figure 31.3 A.C. motor driven main pump, steam turbine auxiliary pump drive (Courtesy of Elliott Co)

Centrifugal pump system

Figure 31.4 depicts a dual centrifugal pump system. The main pump is steam turbine driven and the stand-by pump is motor driven. Such a system can be used in areas of the world where large ambient temperature fluctuations are not present, Middle East, etc. Since the pumps are dynamic type, note that a bypass valve is not used in this system. This is because the flow rate of dynamic pumps are determined by system resistance. Therefore, the control valve in this application senses pressure in the system and adjusts system resistance at the pump discharge to supply the required flow.

Referring back to the concept of an equivalent vessel, let us examine the case of increased system demand as in the case of bearing wear. Bearing wear will gradually increase system demand and reduce system equivalent vessel pressure. Consider in this case the system to be the equivalent vessel. Since the control valve senses equivalent vessel pressure, a reduction in the equivalent vessel pressure will open the valve to provide greater pressure at the sense point. This action in turn will reduce pump discharge pressure. Referring to the characteristic curve of a centrifugal pump, reduced pump discharge pressure results in increased pump output flow, thereby compensating for the increased demand requirement.

In the event of a sudden system flow increase, a sudden change in the equivalent vessel pressure would be experienced. This would result in a sudden drop of equivalent vessel pressure, resulting in a sudden opening of the control valve. However, in this case, the sudden pressure drop would initiate starting of the stand-by pump. As soon as the stand-by pump were to start, supply flow, to the equivalent vessel would quickly increase causing a corresponding increase in pressure. However, the control valve sense point would note the pressure increase and rapidly shut the control valve, thus increasing the system resistance to both pumps, and forcing pumps to a lower flow rate.

An important consideration in this system is that both main and auxiliary pumps operate at essentially the same speed and have essentially the same characteristic curves. (This would not be the case if one pump had excessively worn internals). Therefore, the total output flow of the pumps would be a result of the combined curve of the pump output. Refer to Figure 31.5 for a view of the combined effect of both pumps operating in parallel. From this example, the importance of assuring correct similar characteristics of both main and auxiliary pumps in systems incorporating dynamic pumps can be seen. If one pump is steam turbine driven, it is important to periodically check speed and correct the turbine governor setting if necessary. The motor and turbine speeds should be the same.

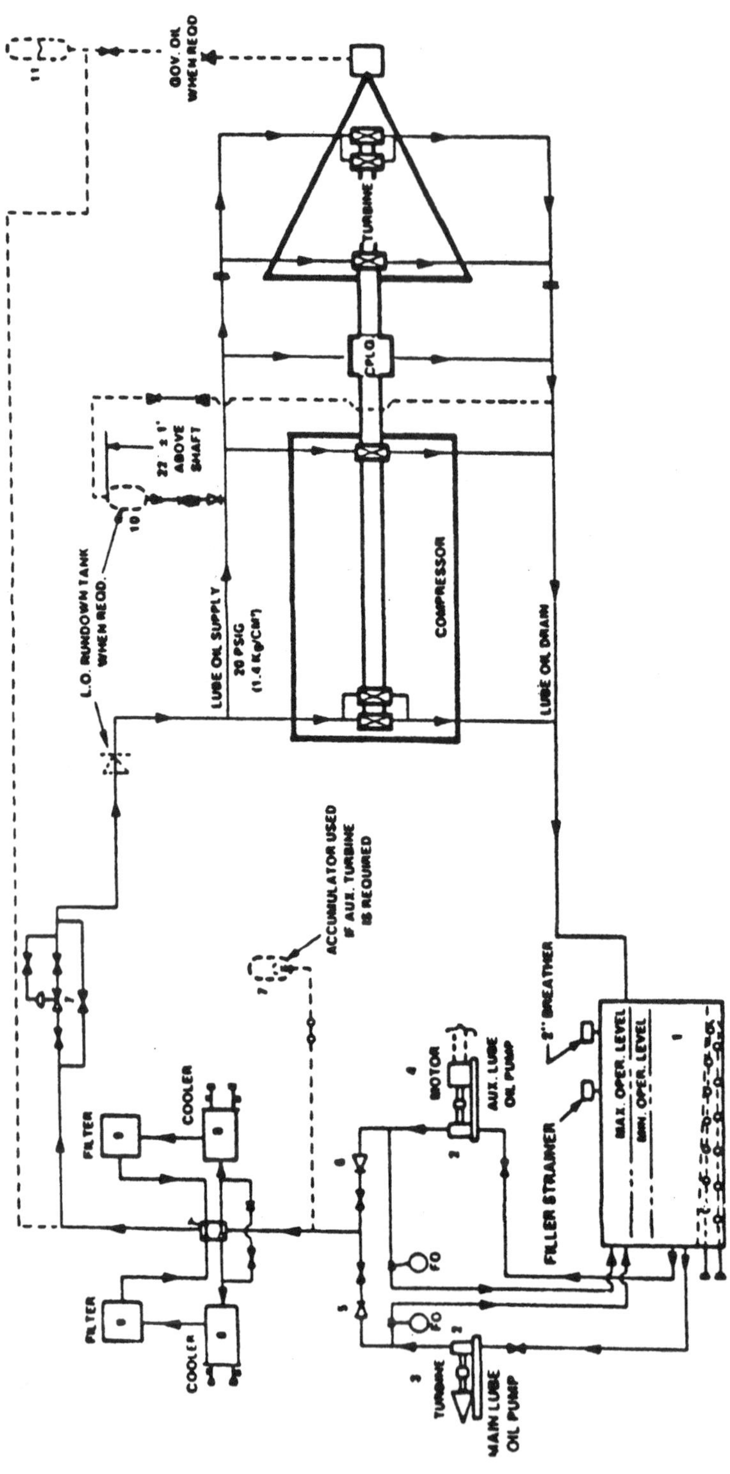

Figure 31.4 Lubrication system – centrifugal pumps (Courtesy of Dresser-Rand)

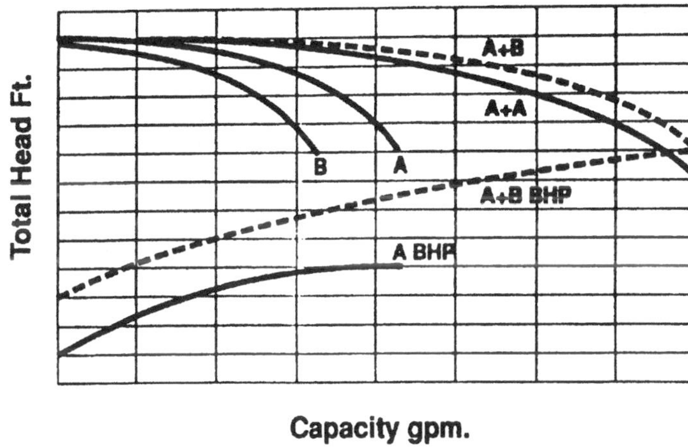

Figure 31.5 Non identical pumps with stable head curves

Dual motor driven positive displacement pump system with a rundown tank

This system is similar to that shown in Figure 31.2 with the exception of the absence of an emergency pump. In this case, the user has elected to have a rundown tank in the system that will supply sufficient lube oil for rundown. (Refer to Figure 31.6) The function of the tank is to

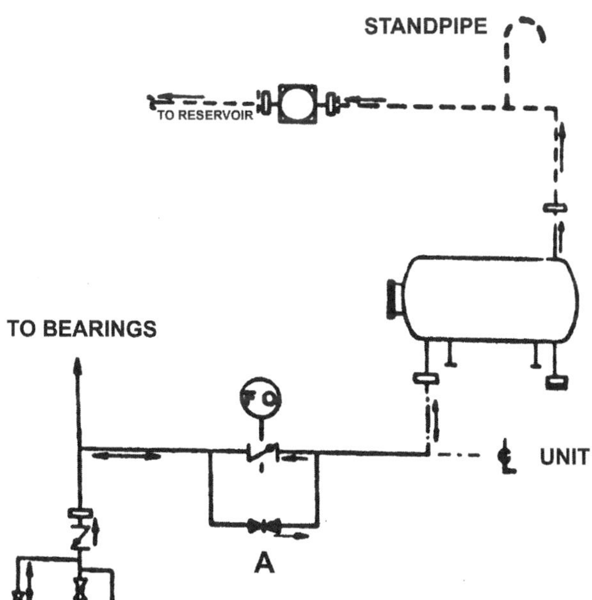

Figure 31.6 Lube oil rundown tank system with continuous overflow (Courtesy of Elliott Co)

provide oil to all bearings at a lower pressure than normal, but still sufficient to preclude bearing damage during emergency shutdown. In this case, the tank will drain when both main and auxiliary pumps cease to function. An important consideration is the material of the rundown tank since it represents a large vessel downstream of the filter in the system. Any rust scale or corrosion present in this tank will go directly into the bearings on shutdown and could cause a significant problem. Rundown tanks should also be sized to provide sufficient flow for the entire coastdown period. The calculation of the equipment coastdown time must include process system information. The lower the system pressure (resistance) during the coastdown period, the longer the equipment will continue to turn. The vendor and user must coordinate closely regarding the rundown tank sizing. It is recommended that a small amount of flow be continuously circulated through the rundown tank to prevent sediment accumulation and maintain operating viscosity (in cold regions). This is accomplished by installing a return line from the tank to the oil drain line that incorporates an orifice to regulate flow (approximately 2 GPM).

One final comment concerning lube systems with rundown tanks or emergency pumps. It is strongly recommended that in the case of equipment shutdown, even though equipment is furnished with rundown tanks and emergency pumps, bearings should be checked for wear upon coastdown. Failure to do so could result in catastrophic damage to the equipment if bearings were failed at shutdown. Additionally, in systems employing gears, gear box inspection covers should be removed and the gear mesh checked since spray elements, if clogged, may not distribute sufficient lubricating flow to the gear mesh in emergency conditions.

Centrifugal shaft driven main pump, motor driven positive displacement auxiliary system pump

Refer to Figure 31.7. This system combines two types of pumps, dynamic and positive displacement. An important consideration is that the maximum pressure of the auxiliary (positive displacement) pump be limited below the maximum pressure of the shaft driven (dynamic) pump. If this is not done, the shaft driven pump output pressure, when operating with the stand-by pump, will be less and its output flow will be reduced to zero. Continued operation in this mode will result in overheating of the shaft driven pump and failure. The stand-by pump discharge pressure setting is regulated by a bypass valve or a relief valve in this case. The setting of this valve must be checked to be sure it is less than the dynamic pumps discharge pressure at the dynamic pump minimum flow point.

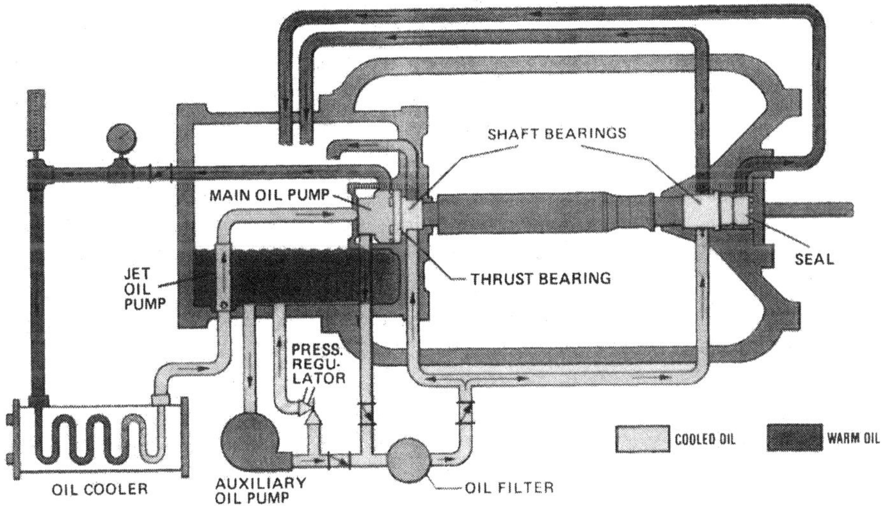

Figure 31.7 Shaft driven centrifugal pump motor driven auxiliary pump (Courtesy of York International Corp.)

Arrangement options

In this section we will briefly discuss a few arrangement options available for lubrication systems. As mentioned, the arrangement of auxiliary systems directly determines system reliability since arrangement determines accessibility to component parts that must be serviced and calibrated while critical equipment is operating. Attention must be drawn to particular applications and the need to maximize component accessibility. It is recognized that certain applications contain minimal space for auxiliary equipment and that equipment must be arranged for the available spaces.

Integral auxiliary systems

Refer to Figure 31.8. Such a system incorporating the lube oil system in the baseplate of the critical equipment is used in remote applications and frequently on platforms since space is at a premium. This system should be reviewed thoroughly in the design phase to optimize accessibility. Note that even though space is minimal, major components can be arranged such that accessibility is possible.

Horizontal console (no components on the reservoir)

Please refer to Figure 31.9. This console arrangement is fairly typical for a critical equipment lubrication system. Note the positions of

Figure 31.8 Modularized oil console arrangement (Courtesy of Fluid Systems, Inc)

Figure 31.9 Horizontal oil console arrangement (Courtesy of G.J. Oliver, Inc)

components affording ample space for maintenance and equipment calibration. In addition, note the placement of interconnecting piping, thus allowing for maximum mobility on the console.

Reservoir integral with component

Refer to Figure 31.10. This arrangement is frequently used in restricted space locations or could be the result of an attempt by the original equipment vendor to minimize cost. Careful scrutiny of arrangement design early during the project can avoid problems. Note from Figure 31.10 that even though all components are mounted on the reservoir, accessibility is maximized to components. In this case, the reservoir was mounted approximately three feet below grade, thus allowing all components to be within easy maintenance reach.

This concludes this chapter on auxiliary lubrication systems for critical equipment. Only the major types of critical equipment lubrication systems have been covered in this chapter. Various other alternatives in design are available.

Figure 31.10 Modularized oil console arrangement (Courtesy of Fluid Systems, Inc)

32 Monitoring reliability and component condition

- Introduction
- The major machinery components
- Specific machinery components and system monitoring parameters and their limits
- The rotor
- Radial bearings
- Thrust bearings
- Seals
- Auxiliary systems
- Predictive maintenance (PDM) techniques

Introduction

The effect of the process on machinery reliability is often neglected as a root cause of machinery failure. It is a fact that process condition changes can cause damage and/or failure to every major machinery component. For this discussion, the most common type of driven equipment – pumps will be used.

There are two (2) major classifications of pumps, positive displacement and kinetic, centrifugal types being the most common. A positive displacement pump is shown in Figure 32.1. A centrifugal pump is shown in Figure 32.2.

Figure 32.1 Positive displacement plunger pump (Courtesy of Union Pump Company)

It is most important to remember that all driven equipment (pumps, compressors, fans, etc.) react to the process system requirements. They do only what the process requires. This fact is noted in Figure 32.3 for pumps.

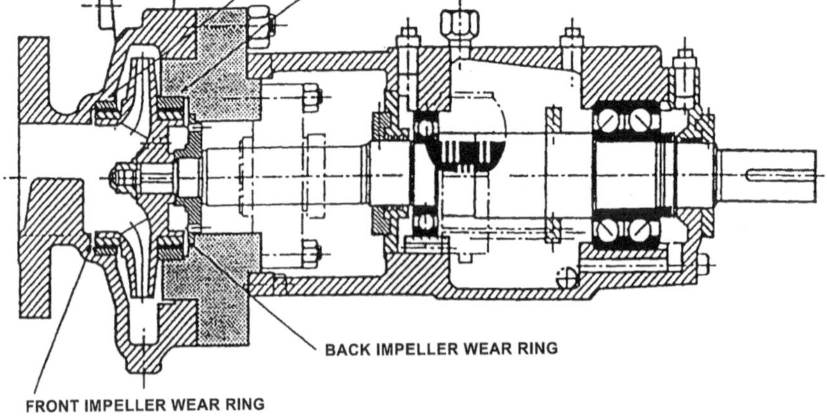

Figure 32.2 Centrifugal pump

Pump performance

- Pumps produce the pressure required by the process
- The flow rate for the required pressure is dependent on the pump's characteristics

Figure 32.3 Pump performance

Centrifugal (kinetic) pumps and their drivers

Centrifugal pumps increase the pressure of the liquid by using rotating blades to increase the velocity of a liquid and then reduce the velocity of the liquid in the volute. Refer again to Figure 32.2.

A good analogy to this procedure is a football (soccer) game. When the ball (liquid molecule) is kicked, the leg (vane) increases its velocity. When the goal tender (volute), hopefully, catches the ball, its velocity is significantly reduced and the pressure in the ball (molecule) is increased. If an instant replay 'freeze shot' picture is taken of the ball at this instant, the volume of the ball is reduced and the pressure is increased.

The characteristics of any centrifugal pump then are significantly different from positive displacement pumps and are noted in Figure 32.4.

Centrifugal pump characteristics

- Variable flow
- Fixed differential pressure produced *for a specific flow**
- Does not require a pressure limiting device
- Flow varies with differential pressure (P_1–P_2) and/or specific gravity

*assuring specific gravity is constant

Figure 32.4 Centrifugal pump characteristics

Refer again to Figure 32.3 and note that all pumps react to the process requirements.

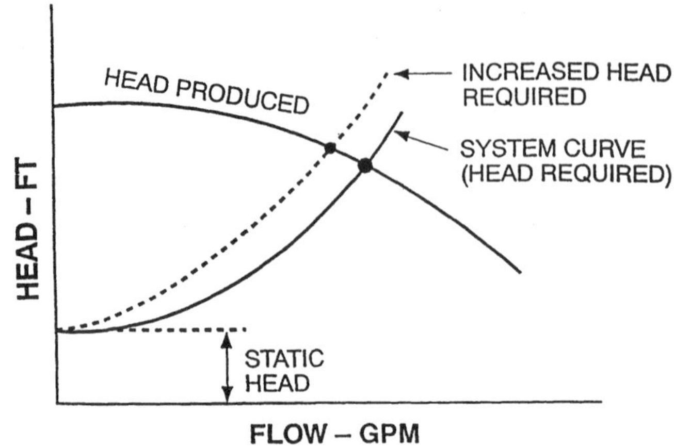

Figure 32.5 A centrifugal pump in a process system

Based on the characteristics of centrifugal pumps noted in Figure 32.4, the flow rate of all types of centrifugal pumps is affected by the Process System. This fact is shown in Figure 32.5.

Therefore, the flow rate of any centrifugal pump is affected by the process system. A typical process system with a centrifugal pump installed, is shown in Figure 32.6.

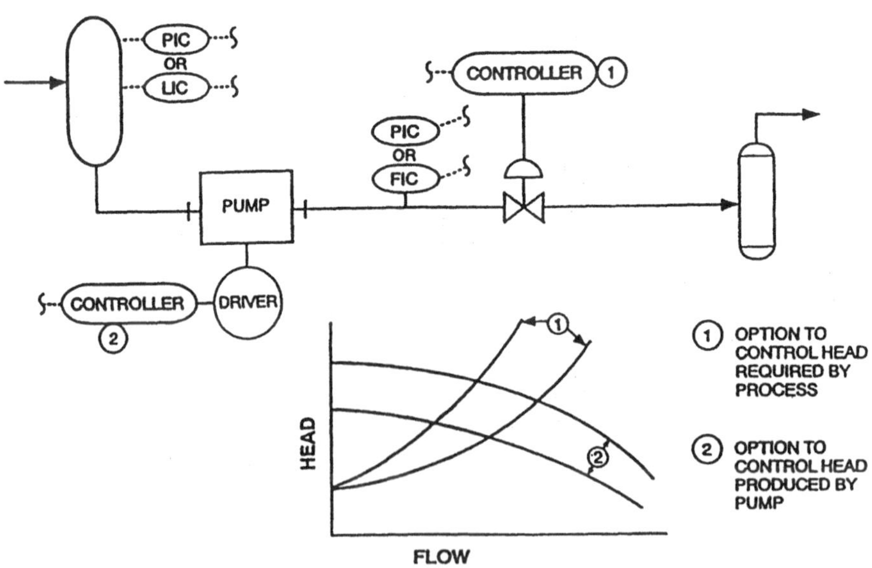

Figure 32.6 Centrifugal pump control options

The differential pressure required (proportional to head) by any process system is the result of the pressure and liquid level in the suction and discharge vessel and the system resistance (pressure drop) in the suction and discharge piping.

Therefore, the differential pressure required by the process can be changed by adjusting a control valve in the discharge line. Any of the following process variables (P.V.) shown in Figure 32.6, can be controlled:

- Level
- Pressure
- Flow

As shown in Figure 32.5, changing the head required by the process (differential pressure divided by specific gravity), will change the flow rate of any centrifugal pump!

Refer to Figure 32.7 and it can be observed that all types of mechanical failures can occur based on *where the pump is operating based on the process requirements.*

Since greater than 95% of the pumps used in this refinery are centrifugal, their operating flow will be affected by the process. Please refer to Figure 32.8 which shows centrifugal pump reliability and flow rate is affected by process system changes.

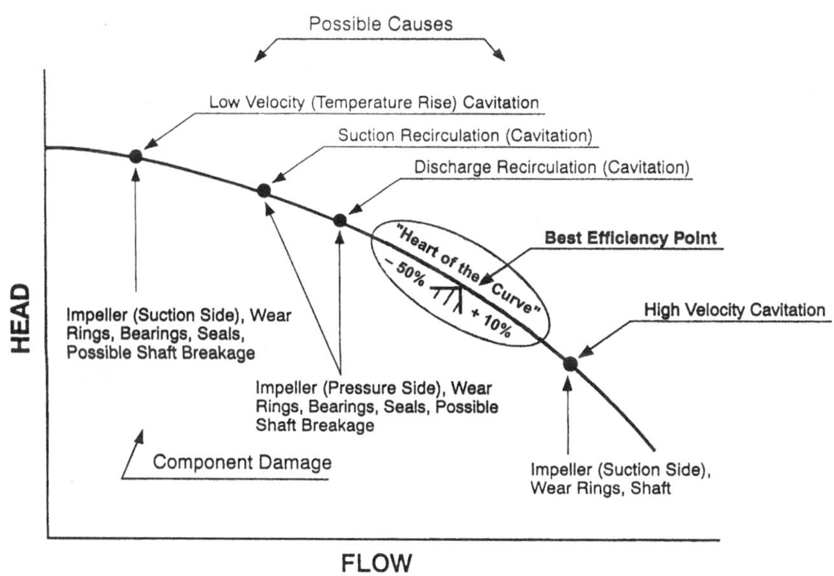

Figure 32.7 Centrifugal pump component damage and causes as a function of operating point

Centrifugal pump reliability

- Is affected by process system changes (system resistance and S.G.)
- It is not affected by the operators!
- Increased differential pressure (P_2-P_1) means reduced flow rate
- Decreased differential pressure (P_2-P_1) means increased flow rate

Figure 32.8 Centrifugal pump reliability

At this point it should be easy to see how we can condition monitor the centrifugal pump operating point. Refer to Figure 32.9.

Centrifugal pump practical condition monitoring

- Monitor flow and check with reliability unit (RERU) for significant changes
- Flow can also be monitored by:
  - Control valve position
  - Motor amps
  - Steam turbine valve position

Figure 32.9 Centrifugal pump practical condition monitoring

Driver reliability (motors, steam turbine and diesel engines) can also be affected by the process when centrifugal driven equipment (pumps, compressor and fans) are used.

Refer to Figure 32.10 and observe a typical centrifugal pump curve.

Since the flow rate will be determined by the process requirements, the power (BHP) required by the driver will also be affected. What would occur if an 8½" diameter impeller were used and the head (differential pressure) required by the process was low? Answer: Since the pressure differential required is low, the flow rate will increase and for the 8½" diameter impeller, the power required by the drier (BHP) will increase.

Therefore, a motor can trip out on overload, a steam turbine's speed can reduce or a diesel engine can trip on high engine temperature. These facts are shown in Figure 32.11.

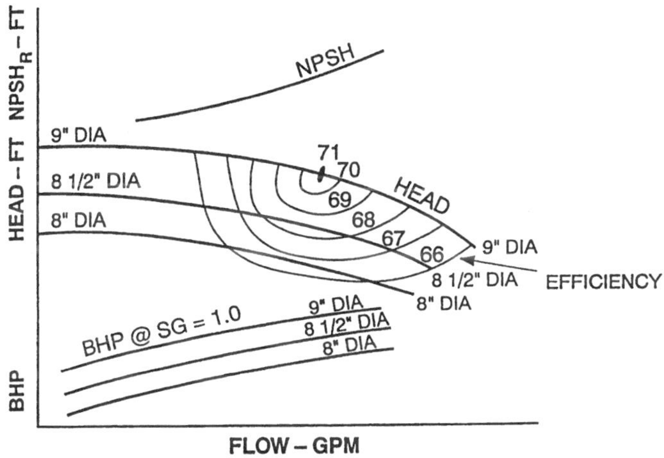

Figure 32.10 A typical centrifugal pump performance curve

Effect of the process on drivers

- Motors can trip on overload
- Steam turbines can reduce speed
- Diesel engines can trip on high engine temperature

Figure 32.11 Effect of the process on drivers

Auxiliary system reliability is also affected by process changes. Auxiliary systems support the equipment and their components by providing ... clean, cold fluid to the components at the correct differential pressure, temperature and flow rate.

Typical auxiliary systems are:

- Lube oil systems
- Seal flush system
- Seal steam quench system
- Cooling water system

The reliability of machinery components (bearings, seals, etc.) is directly related to the reliability of the auxiliary system. In many cases, the root cause of the component failure is found in the supporting auxiliary system.

As an example, changes in auxiliary system supply temperature, resulting from cooling water temperature or ambient air temperature changes, can be the root cause of component failure. Figure 32.12 presents these facts.

Component (bearing and seal) reliability

- Is directly related to auxiliary system reliability
- Auxiliary system reliability is affected by process condition changes.
- 'Root causes' of component failure are often found in the auxiliary system.

Figure 32.12 Component (bearing and seal) reliability

As a result, the condition of all the auxiliary systems supporting a piece of equipment must be monitored. Please refer to Figure 32.13.

Always 'think system'

- Monitor auxiliary system condition
- Inspect auxiliary system during component replacement

Figure 32.13 Always 'think system'

The major machinery components

Please refer again to Figure 32.7 which shows how process condition changes can cause damage and/or failure to any pump component.

Regardless of the type of machinery, the major component classifications are the same. The major machinery components and their systems are shown in Figure 32.14.

Major machinery components and systems

- Rotor
- Radial bearing
- Thrust bearing
- Seal
- Auxiliary systems

Figure 32.14 Major machinery components and systems

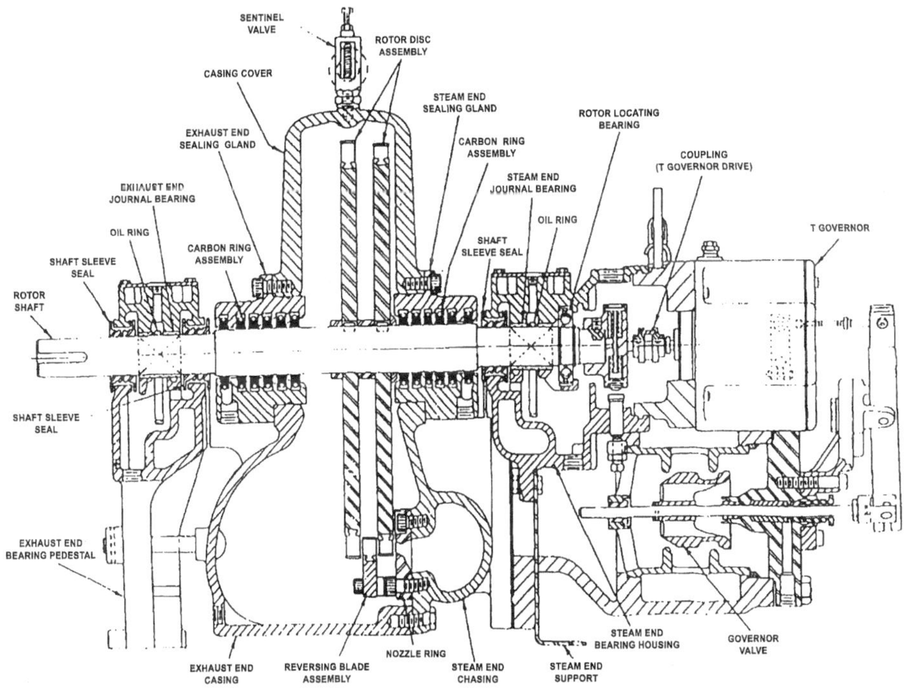

Figure 32.15 Typical general purpose (pump drive) turbine assembly (Courtesy of Elliott Company)

Regardless of the type of machinery, if we monitor the condition of each major component and its associated system, we will know the condition of the machine!

Refer to Figure 32.15 and define the major components and their associated systems.

The concept of component condition monitoring will be the focus of this chapter. If we know what to monitor to determine the condition of each component, we can eliminate unscheduled shut downs and failures. The component condition monitoring procedure is shown in Figure 32.16.

Component condition monitoring

- **Define** each major component
  - **List** condition monitoring parameters
    - **Obtain** baseline data
      - **Trend** data
        - **Establish** threshold limits

Figure 32.16 Component condition monitoring

Specific machinery components and system monitoring parameters and their limits

The following pages contain information concerning what parameters should be monitored for each major machinery component to determine its condition. In addition, typical limits are noted for each component.

These limits represent the approximate point at which action should be planned. They are not intended to define shutdown values.

The rotor

Rotor condition defines the performance condition (energy and efficiency) of the machine. Figure 32.17 presents this value for a pump.

Pump performance monitoring

1. Take value at minimum flow (shut off discharge valve)
2. Measure:
   - P_1
   - P_2
   - driver bhp
   - specific gravity
3. Calculate:

   A. head produced $\frac{ft\text{-}lb_f}{lbm} = \frac{\Delta p \times 2.311}{s.g.}$

   B. pump efficiency (%) = $\frac{*hd \times gpm \times sg}{3960 \times bhp}$
4. Compare to previous value if > –10% perform maintenance

*Where hd is defined in item A above.

Figure 32.17 Pump performance monitoring

Radial bearings

Figures 32.18 and 32.19 present the facts concerning anti-friction and hydrodynamic (sleeve) radial or journal bearing condition monitoring.

Condition monitoring parameters and their alarm limits

Journal bearing (anti-friction)

Parameter	Limits
1. Bearing housing vibration (peak)	.4 inch/sec (10 mm/sec)
2. Bearing housing temperature	185°F (85°C)
3. Lube oil viscosity	Off spec 50%
4. Lube oil particle size	
■ non metallic	25 microns
■ metallic	any magnetic particle in the sump
5. Lube oil water content	Below 200 ppm

Figure 32.18 Condition monitoring parameters and their alarm limits

Condition monitoring parameters and their alarm limits

Journal bearing (hydrodynamic)

Parameter	Limits
1. Radial vibration (peak to peak)	2.5 mils (60 microns)
2. Bearing pad temperature	220°F (108°C)
3. Radial shaft position*	>30° change and/or 30% position change
4. Lube oil supply temperature	140°F (60°C)
5. Lube oil drain temperature	190°F (90°C)
6. Lube oil viscosity	Off spec 50%
7. Lube oil particle size	>25 microns
8. Lube oil water content	Below 200 ppm

*Except for gearboxes where greater values are normal from unloaded to loaded

Figure 32.19 Condition monitoring parameters and their alarm limits

Thrust bearings

Figure 32.20 and 32.21 show condition parameters and their limits for anti-friction and hydrodynamic thrust bearings.

Condition monitoring parameters and their alarm limits

Journal bearing (anti-friction)

Parameter	Limits
1. Bearing housing vibration (peak)	
■ radial	.4 in/sec (10 mm/sec)
■ axial	.3 in/sec (1 mm/sec)
2. Bearing housing temperature	185°F (85°C)
3. Lube oil viscosity	Off spec 50%
4. Lube oil particle size	
■ non metallic	>25 microns
■ metallic	Any magnetic particles with sump
5. Lube oil water content	Below 200 ppm

Figure 32.20 Condition monitoring parameters and their alarm limits

Condition monitoring parameters and their alarm limits

Thrust bearing (hydrodynamic)

Parameter	Limits
1. Axial displacement*	>15–20 mils (0.4–0.5 mm)
2. Thrust pad temperature	220°F (105°C)
3. Lube oil supply temperature	140°F (60°C)
4. Lube oil drain temperature	190°F (90°C)
5. Lube oil viscosity	Off spec 50%
6. Lube oil particle size	> 25 microns
7. Lube oil water content	Below 200 ppm

**and* thrust pad temperatures >220°F (105°C)

Figure 32.21 Condition monitoring parameters and their alarm limits

Seals

Figure 32.22 presents condition parameters and their limits for a pump liquid mechanical seal.

Condition monitoring parameters and their alarm limits

Pump liquid mechanical seal

Parameter	Limits
1. Stuffing box pressure	< 25 psig (175 kpa) ** above suction pressure
2. Stuffing box temperature	Below boiling temperature for process liquid
3. Flush line temperature	+/-20°F (10°C) from pump case temp
4. *Primary seal vent pressure (before orifice)	>10 psi (70 kpag)

*On tandem seal arrangements only
** Typical limit - there are exceptions (sundyne pumps)

Figure 32.22 Condition monitoring parameters and their alarm limits

Auxiliary systems

Condition monitoring parameters and their alarm limits are defined in Figures 32.23 and 32.24 for lube and pump flush systems.

Condition monitoring parameters and their alarm limits

Lube oil systems

Parameter	Limits
1. Oil viscosity	Off spec 50%
2. Lube oil water content	Below 200 ppm
3. Auxiliary oil pump operating yes/no	Operating
4. Bypass valve position (p.d. pumps)	Change > 20%
5. Temperature control valve position	Closed, supply temperature >130°F (55°C)
6. Filter ΔP	>25 psid (170 kpag)
7. Lube oil supply valve position	Change +/- 20%

Figure 32.23 Condition monitoring parameters and their alarm limits

Condition monitoring parameters and their alarm limits

Pump seal flush
(single seal, flush from discharge)

Parameter	Limits
1. Flush line temperature	+/- 20°F (+/-10°C) of pump case temperature
2. Seal chamber pressure	< 25 psi (175 kpa) above suction pressure

Figure 32.24 Condition monitoring parameters and their alarm limits

Predictive maintenance (pdm) techniques

Now that the component condition monitoring parameters and their limits have been presented, predictive maintenance techniques must be used if typical condition limits are exceeded. This action will assure that we minimize site-troubleshooting exercises.

Figures 32.25, 32.26 and 32.27 present condition monitoring parameter for dynamic compressor performance, liquid seals and seal systems.

Compressor performance condition monitoring

1. Calibrated: pressure and temperature gauges and flow meter
2. Know gas analysis and calculate k, z, m.w.
3. Perform as close to rated speed and flow as possible
4. Relationships:

A. $$\frac{n-1}{n} = \frac{\ln\left[\frac{(T_2)}{(T_1)}\right]}{\ln\left[\frac{(P_2)}{(P_1)}\right]}$$

B. $$\text{efficiency}_{poly} = \frac{\frac{k-1}{k}}{\frac{n-1}{n}}$$

C. $$\text{head}_{poly} = \left(\frac{\text{ft-lb}_1}{\text{lb}_m}\right) = \frac{1545}{mw} \times T_1 \times \frac{n}{n-1} \times Z_{avg} \times \left[\left(\frac{p_2}{p_1}\right)^{\frac{(n-1)}{n}} - 1\right]$$

5. Compare to previous value. If decreasing trend exists greater than 10%. Inspect at first opportunity.

Figure 32.25 Compressor performance condition monitoring

Condition monitoring parameters and their alarm limits

Compressor liquid seal

Parameter	Limits
1. Gas side seal oil/gas ΔP	
■ bushing	< 12ft. (3.5m)
■ mechanical contact	< 20 psi (140 kpa)
2. Atmospheric bushing oil drain temperature	200°F (95°C)
3. Seal oil valve* position	> 25% position change
4. Gas side seal oil leakage	> 20 gpd per seal

*supply valve = + 25%
return valve = -25%

Note this assumes compressor reference gas pressure stays constant

Figure 32.26 Condition monitoring parameters and their alarm limits

Condition monitoring parameters and their alarm limits

Compressor liquid seal oil systems

Parameter	Limits
1. Oil viscosity	Off spec 50%
2. Oil flash point	Below 200°F (100°C)
3. Auxiliary oil pump operating yes/no	Operating
4. Bypass valve position (p.d pumps)	Change > 20%
5. Temperature control valve position	Closed, supply temperature > 130°F (55°C)
6. Filter Δp	> 25 psid (170 kpag)
7. Seal oil valve position	Change > 20% open (supply) > 20% closed (return)
8. Seal oil drainer condition	(Proper operation)
■ constant level (yes/no)	level should be observed
■ observed level (yes/no)	level should not be constant
■ time between drains	approximately 1 hour (depends on drainer volume)

Figure 32.27 Condition monitoring parameters and their alarm limits

Conversion to metric system

Unit Nomenclature

Symbol	Name	Quantity
a	annum (year)	time
bar	bar	pressure
°C	degree Celsius	temperature
°	degree	plane angle
d	day	time
g	gram	mass
h	hour	time
Hz	hertz	frequency
J	joule	work, energy
K	kelvin	temperature
kg	kilogram	mass
litre	litre	volume
m	metre	length
min.	minute	time
N	newton	force
Pa	pascal	pressure
rad	radian	plane angle
s	second	time
t	tonne	mass
V	volt	electric potential
W	watt	power

Conversion tables – Customary to metric (SI) units

Quantity and SI unit	Customary unit	Metric unit	Conversion factor Multiply customary unit by factor to get metric unit	
Space, time				
Length – m	mi	km	1.609 344	E+00
	m	m	1	
	yd	m	9.144	E–01
	ft	m	3.048	E–01
	in	mm	2.54	E+01
	cm	mm	1.0	E+01
	mm	mm	1	
	mil	μm	2.54	E+01
	micron (μ)	μm	1	
Area – m^2	mi^2	km^2	2.589 988	E+00
	yd^2	m^2	8.361 274	E–01
	ft^2	m^2	9.290 304	E–02
	in^2	mm^2	6.451 6	E+02
	cm^2	mm^2	1.0	E+02
	mm^2	mm^2	1	
Volume, capacity – m^3	m^3	m^3	1	
	yd^3	m^3	7.645 549	E–01
	bbl (42 US gal)	m^3	1.589 873	E–01
	ft^3	m^3	2.831 685	E–02
	UK gal	m^3	4.546 092	E–03
	US gal	m^3	3.785 412	E–03
	litre	dm^3	1	
	UK qt	dm^3	1.136 523	E+00
	US qt	dm^3	9.463 529	E–01
	UK pt	dm^3	5.682 609	E–01
	US pt	dm^3	4.731 765	E–01
	US fl oz	cm^3	2.957 353	E+01
	UK fl oz	cm^3	2.841 305	E+01
	in^3	cm^3	1.638 706	E+01
	ml	cm^3	1	

Quantity and SI unit	Customary unit	Metric unit	Conversion factor Multiply customary unit by factor to get metric unit	
Plane angle – rad	rad	rad	1	
	deg (°)	rad	1.745 329	E–02
	min (′)	rad	2.908 882	E–04
	sec (″)	rad	4.848 137	E–06

Conversion to the metric (SI) system

Quantity and SI unit	Customary unit	Metric unit	Conversion factor Multiply customary unit by factor to get metric unit	
Mass, amount of substance				
Mass – kg	UK ton	t	1.016 047	E+00
	US ton	t	9.071 847	E–01
	UK cwt	kg	5.080 234	E+01
	US cwt	kg	4.535 924	E+01
	kg	kg	1	
	lb	kg	4.535 924	E–01
	oz (troy)	g	3.110 348	E+01
	oz (av)	g	2.834 952	E+01
	g	g	1	
	grain	mg	6.479 891	E+01
	mg	mg	1	
	μg	μg	1	
Amount of substance – mol	lb mol	kmol	4.535 924	E–01
	g mol	kmol	1.0	E–03
	std m^3 (0°C, 1atm)	kmol	4.461 58	E–02
	std m^3 (15°C, 1atm)	kmol	4.229 32	E–02
	std ft^3 (60°C, 1 atm)	kmol	1.195 30	E–03

Quantity and SI unit	Customary unit	Metric unit	Conversion factor Multiply customary unit by factor to get metric unit	
Calorific value, heat, entropy, heat capacity				
Calorific value – J/kg	Btu/lb	MJ/kg	2.326 000	E–03
(Mass basis)		kJ/kg	2.326 000	E+00
	cal/g	kJ/kg	4.184	E+00
	cal/lb	J/kg	9.224 141	E+00
Calorific Value –J/mol	kcal/g mol	kJ/kmol	4.184	E+03
(Mole basis)	Btu/lb mol	MJ/kmol	2.326 000	E–03
		kJ/kmol	2.326 000	E+00
Calorific value – J/m3	therm/UK gal	MJ/m3	2.320 800	E+04
(Volume basis –		kJ/m3	2.320 800	E+07
solids and liquids)	Btu/US gal	MJ/m3	2.787 163	E–01
		kJ/m3	2.787 163	E+02
	Btu/UK gal	MJ/m3	2.320 800	E–01
		kJ/m3	2.320 800	E+02
	Btu/ft3	MJ/m3	3.725 895	E–02
		kJ/m3	3.725 895	E+01
	kcal/m3	MJ/m3	4.184	E–03
		kJ/m3	4.184	E+00
	cal/ml	MJ/m3	4.184	E+00
	ft•lb$_f$/US gal	kJ/m3	3.581 692	E–01
Calorific value – J/m3	cal/ml	kJ/m3	4.184	E+03
(Volume basis – gases)	kcal/m3	kJ/m3	4.184	E+00
	Btu/ft3	kJ/m3	3.725 895	E+01
Specific entropy – J/kg•K	Btu/lb••R	kJ/kg•K	4.186 8	E+00
	cal/g••K	kJ/kg•K	4.184	E+00
	kcal/kg••C	kJ/kg•K	4.184	E+00
Specific heat	kW·h/kg••C	kJ/kg••C	3.6	E+03
capacity – J/kg•K	Btu/lb••F	kJ/kg••C	4.186 8	E+00
(Mass basis)	kcal/kg••C	kJ/kg••C	4.184	E+00
Specific heat	Btu/lb mol••F	kJ/kmol••C	4.186 8	E+00
capacity – J/mol•K	cal/g mol••C	kJ/kmol••C	4.184	E+00
(Mole basis)				

Quantity and SI unit	Customary unit	Metric unit	Conversion factor Multiply customary unit by factor to get metric unit	
Temperature, pressure, vacuum				
Temperature – K	°R	K	5/9	
(Absolute)	°K	K	1	
Temperature – K	°F	°C	5/9 (°F – 32)	
(Traditional)	°C	°C	1	
Temperature – K	°F	°C	5/9	
(Difference)	°C	°C	1	
Temperature/Length – K/m (Geothermal gradient)	°F per 100 ft	mK/m	1.822 689	E+01
Pressure – Pa	atm	MPa	1.013 250	E–01
		kPa	1.013 250	E+02
	bar	MPa	1.0	E–01
		kPa	1.0	E+02
	at (kg_f/cm^2)	MPa	9.806 650	E–02
		kPa	9.806 650	E+01
	lb_f/in^2 (psi)	MPa	6.894 757	E–03
		kPa	6.894 757	E+00
	in Hg (60°F)	kPa	3.376 85	E+00
	in H_2O (39.2°F)	kPa	2.490 82	E–01
	in H_2O (60°F)	kPa	2.488 4	E–01
	mm Hg = torr (0°C)	kPa	1.333 224	E–01
	cm H_2O (4°C)	kPa	9.806 38	E–02
	lb_f/ft^2 (psf)	kPa	4.788 026	E–02
	μm Hg (0°C)	Pa	1.333 224	E–01
	μbar	Pa	1.0	E–01
	dyn/cm^2	Pa	1.0	E–01
Vacuum, draft – Pa	in Hg (60°F)	kPa	3.376 85	E+00
	in H_2O (39.2°F)	kPa	2.490 82	E–01
	in H_2O (60°F)	kPa	2.488 4	E–01
	mm Hg = torr (0°C)	kPa	1.33 224	E–01
	cm H_2O (4°C)	kPa	9.806 38	E–02
Liquid head – m	ft	m	3.048	E–01
	in	mm	2.54	E+01
Pressure drop/length – Pa/m	psi/ft	kPa/m	2.262 059	E+01
	psi/100 ft	kPa/m	2.262 059	E–01

Quantity and SI unit	Customary unit	Metric unit	Conversion factor Multiply customary unit by factor to get metric unit	
Density, specific volume, concentration, dosage				
Density (gases) – kg/m^3	lb/ft^3	kg/m^3	1.601 846	E+01
		g/m^3	1.601 846	E+04
Density (liquids) – kg/m^3	lb/US gal	kg/dm^3	1.198 264	E–01
	lb/UK gal	kg/dm^3	9.977 644	E–02
	lb/ft^3	kg/dm^3	1.601 846	E–02
	g/cm^3	kg/dm^3	1	
Density (solids) – kg/m^3	lb/ft^3	kg/dm^3	1.601 846	E–02
Specific volume – m^3/kg (gases)	ft^3/lb	m^3/kg	6.242 796	E–02
		m^3/g	6.242 796	E–05
Specific volume – m^3/kg (liquids)	ft^3/lb	dm^3/kg	6.242 796	E+01
	UK gal/lb	dm^3/kg	1.022 241	E+01
	US gal/lb	dm^3/kg	8.345 404	E+00
Specific volume – m^3/mol (mole basis)	litre/g mol	$m^3/kmol$	1	
	ft^3/lb mol	$m^3/kmol$	6.242 796	E–02
Concentration – kg/kg (mass/mass)	wt %	kg/kg	1.0	E–02
		g/kg	1.0	E–05
	wt ppm	mg/kg	1	
Concentration – kg/m^3 (mass/volume)	lb/bbl	kg/m^3	2.853 010	E+00
	g/US gal	kg/m^3	2.641 720	E–01
	g/UK gal	kg/m^3	2.199 692	E–01

Quantity and SI unit	Customary unit	Metric unit	Conversion factor Multiply customary unit by factor to get metric unit	
Facility throughput, capacity				
Throughput – kg/s (mass basis)	million lb/yr	t/a	4.535 924	E+02
	UK ton/yr	t/a	1.016 047	E+00
	US ton/yr	t/a	9.071 847	E–01
	UK ton/d	kg/h	1.016 047	E+00
	US ton/d	t/d	9.071 847	E–01
	UK ton/h	t/h	1.016 047	E+00
	US ton/h	t/h	9.071 847	E–01
	lb/h	t/d	4.535 924	E–01
Throughput – m^3/s (volume basis)	bbl/d	t/a	5.803 036	E+01
		m^3/h	6.624 471	E–03
	ft^3/d	m^3/h	1.179 869	E–03
	bbl/h	m^3/h	1.589 873	E–01
	ft^3/h	m^3/h	2.831 685	E–02
	UK gal/h	m^3/h	4.546 092	E–03
	US gal/h	m^3/h	3.785 412	E–03
	UK gal/min	m^3/h	2.727 655	E–01
	US gal/min	m^3/h	2.271 247	E–01
Throughput – mol/s (mole basis)	lb mol/h	kmol/h	4.535 924	E–01

Quantity and SI unit	Customary unit	Metric unit	Conversion factor Multiply customary unit by factor to get metric unit	
Flow rate				
Flow rate – kg/s (mass basis)	UK ton/min	kg/s	1.693 412	E+01
	US ton/min	kg/s	1.511 974	E+01
	UK ton/h	kg/s	2.822 353	E–01
	US ton/h	kg/s	2.519 958	E–01
	UK ton/d	kg/s	1.175 980	E–02
	US ton/d	kg/s	1.049 982	E–02
	million lb/yr	kg/s	5.249 912	E+00
	UK ton/yr	kg/s	3.221 864	E–05
	US ton/yr	kg/s	2.876 664	E–05
	lb/s	kg/s	4.535 924	E–01
	lb/min	kg/s	7.559 873	E–03
	lb/h	kg/s	1.259 979	E–04
Flow rate – m^3/s (volume basis)	bbl/d	dm^3/s	1.840 131	E–03
	ft^3/d	dm^3/s	3.277 413	E–04
	bbl/h	dm^3/s	4.416 314	E–02
	ft^3/h	dm^3/s	7.865 791	E–03
	UK gal/h	dm^3/s	1.262 803	E–03
	US gal/h	dm^3/s	1.051 503	E–03
	UK gal/min	dm^3/s	7.576 820	E–02
	US gal/min	dm^3/s	6.309 020	E–02
	ft^3/min	dm^3/s	4.719 474	E–01
	ft^3/s	dm^3/s	2.831 685	E+01
Flow rate – mol/s (mole basis)	lb mol/s	kmol/s	4.535 924	E–01
	lb mol/h	kmol/s	1.259 979	E–04

Quantity and SI unit	Customary unit	Metric unit	Conversion factor Multiply customary unit by factor to get metric unit	
Energy, work, power				
Energy, work – J	therm	MJ	1.055 056	E+02
		kJ	1.055 056	E+05
		kJ	3.6	E+03
	Btu	kJ	1.055 056	E+00
	kcal	kJ	4.184	E+00
	cal	kJ	4.184	E–03
	ft•lb_f	kJ	1.355 818	E–03
	lb•ft	kJ	1.355 818	E–03
	J	kJ	1.0	E–03
	lb•ft^2/s^2	kJ	4.214 011	E–05
	erg	J	1.0	E–07
Power – W	million Btu/h	MW	2.930 711	E–01
	ton of refrigeration	kW	3.516 853	E+00
	Btu/s	kW	1.055 056	E+00
	kW	kW	1	
	hydraulic horsepower–hhp	kW	7.460 43	E–01
	hp (electric)	kW	7.46	E–01
	Btu/min	kW	1.758 427	E–02
	ft•lb_f/s	kW	1.355 818	E–03
	kcal/h	W	1.162 222	E+00
	Btu/h	W	2.930 711	E–01
	ft•lb_f/min	W	2.259 697	E–02
Power/area – W/m^2	Btu/s•ft^2	kW/m^2	1.135 653	E+01
	Cal/h•cm^2	kW/m^2	1.162 222	E–02
	Btu/h•ft^2	kW/m^2	3.154 591	E–03
Cooling duty – W/W (machinery)	Btu/bhp•h	W/kW	3.930 148	E–01
Specific fuel – kg/J consumption (mass basis)	lb/hp•h	mg/J	1.689 659	E–01
Specific fuel – m^3/J consumption (volume basis)	m^3/kW•h	dm^3/MJ	2.777 778	E+02
	US gal/hp•h	dm^3/MJ	1.410 089	E+00

Quantity and SI unit	Customary unit	Metric unit	Conversion factor Multiply customary unit by factor to get metric unit	
Mechanics				
Velocity (linear) – m/s	mi/h	km/h	1.609 344	E+00
speed	m/s	m/s	1	
	ft/s	m/s	3.048	E–01
	ft/min	m/s	5.08	E–03
	ft/h	mm/s	8.466 667	E–02
	in/s	mm/s	2.54	E+01
	in/min	mm/s	4.233 333	E–01
Corrosion rate – mm/a	in/yr (ipy)	mm/a	2.54	E+01
Rotational frequency – rev/s	rev/s	rev/s	1	
	rev/min	rev/s	1.666 667	E–02
Acceleration – m/s^2	ft/s^2	m/s^2	3.048	E–01
(linear)	gal (cm/s^2)	m/s^2	1.0	E–02
Acceleration – rad/s^2 (rotational)	rad/s^2	rad/s^2	1	
Momentum – kg•m/s	lb•ft/s	kg•m/s	1.382 550	E–01
Force – N	UK ton_f	kN	9.964 016	E+00
	US ton_f	kN	8.896 443	E+00
	kg_f (kp)	N	9.806 650	E+00
	lb_f	N	4.448 222	E+00
	N	N	1	
Bending moment, – N•m	US ton_f•ft	kN•m	2.711 636	E+00
torque	kg_f•m	N•m	9.806 650	E+00
	lb_f•ft	N•m	1.355 818	E+00
	lb_f•in	N•m	1.129 848	E–01
Bending – N•m/m	lb_f•ft/in	N•m/m	5.337 866	E+01
moment/length	kg_f•m/m	N•m/m	9.806 650	E+00
	lb_f•in/in	N•m/m	4.448 222	E+00
Moment of – kg•m^2	lb_f•ft^2	kg•m^2	4.214 011	E–02
inertia	in^4	cm^4	4.162 314	E+01

Quantity and SI unit	Customary unit	Metric unit	Conversion factor Multiply customary unit by factor to get metric unit	
Mechanics continued				
Stress – Pa	US ton_f/in^2	MPa	1.378 951	E+01
	kg_f/mm^2	MPa	9.806 650	E+00
	US ton_f/ft^2	MPa	9.576 052	E-02
	lb_f/in^2 (psi)	MPa	6.894 757	E-03
	lb_f/ft^2 (psf)	kPa	4.788 026	E-02
	dyn/cm^2	Pa	1.0	E-01
Mass/length – kg/m	lb/ft	kg/m	1.488 164	E+00
Transport properties				
Thermal resistance	°C•m²•h/kcal	°C•m²/kW	8.604 208	E+02
– K•m²/W	°F•ft²•h/Btu	°C•m²/kW	1.761 102	E+02
Heat flux – W/m²	Btu/h•ft²	kW/m²	3.154 591	E-03
Thermal – W/m•K	cal/s•cm²•°C/cm	W/m•°C	4.184	E+02
conductivity	Btu/h•ft²•°F/ft	W/m•°C	1.730 735	E+00
	Kcal/h•m²•°C/m	W/m•°C	1.162 222	E+00
	Btu/h•ft²•°F/in	W/m•°C	1.442 279	E-01
	cal/h•cm²•°C/cm	W/m•°C	1.162 222	E-01
Heat transfer – W/m²•K	cal/s•cm²•°C	kW/m²•°C	4.184	E+01
coefficient	Btu/s•ft²•°F	kW/m²•°C	2.044 175	E+01
	cal/h•cm²•°C	kW/m²•°C	1.162 222	E-02
	Btu/h•ft²•°F	kW/m²•°C	5.678 263	E-03
	Btu/h•ft²•°R	kW/m²•K	5.678 263	E-03
	kcal/h•m²•°C	kW/m²•°C	1.162 222	E-03
Volumetric heat – W/m³•K	Btu/s•ft³•°F	kW/m³•°C	6.706 611	E+01
transfer coefficient	Btu/h•ft³•°F	kW/m³•°C	1.862 947	E-02
Surface tension – N/m	dyn/cm	mN/m	1	
Viscosity – Pa•s	lb_f•s/in²	Pa•s	6.894 757	E+03
(dynamic)	lb_f•s/ft²	Pa•s	4.788 026	E+01
	kg_f•s/m²	Pa•s	9.806 650	E+00
	lb/ft•s	Pa•s	1.488 164	E+00
	dyn•s/cm²	Pa•s	1.0	E+01
	cP	Pa•s	1.0	E-03
	lb/ft•h	Pa•s	4.133 789	E-04

Quantity and SI unit	Customary unit	Metric unit	Conversion factor Multiply customary unit by factor to get metric unit	
Transport properties continued				
Viscosity – m^2/s (kinematic)	ft^2/s	mm^2/s	9.290 304	E+04
	in^2/s	mm^2/s	6.451 6	E+02
	m^2/h	mm^2/s	2.777 778	E+02
	cm^2/s	mm^2/s	1.0	E+02
	ft^2/h	mm^2/s	2.580 64	E+01
	cSt	mm^2/s	1	

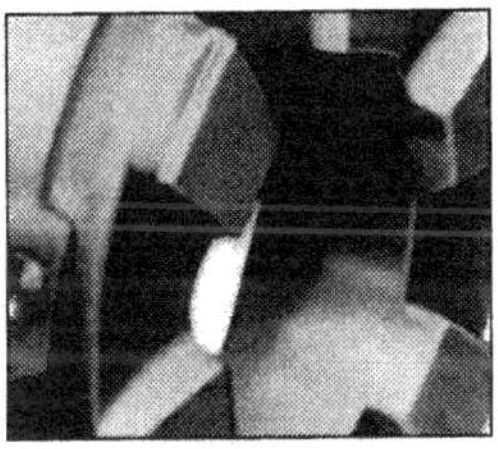

Index

FSC
www.fsc.org
MIX
Papier | Fördert gute Waldnutzung
FSC® C083411

Kolibri 360 GmbH
Ferdinand-Jühlke-Straße 7
99095 Erfurt, Deutschland
produktsicherheit@kolibri360.de